Adaption of Microbial Life to Environmental Extremes

Helga Stan-Lotter • Sergiu Fendrihan
Editors

Adaption of Microbial Life to Environmental Extremes

Novel Research Results and Application

Second Edition

Springer

Editors
Helga Stan-Lotter
FB Molekulare Biologie
Universität Salzburg
Salzburg
Austria

Sergiu Fendrihan
Bioresource Centre & Advanced Research Association
Bucharest
Romania

Research-Development Institute for Plant Protection
Bucharest
Romania

Vasile Goldiş
Western University of Arad
Arad
Romania

ISBN 978-3-319-83914-1 ISBN 978-3-319-48327-6 (eBook)
DOI 10.1007/978-3-319-48327-6

Softcover reprint of the hardcover 2nd edition 2017

Printed on acid-free paper

This Springer imprint is published by Springer Nature
The registered company is Springer International Publishing AG
The registered company address is: Gewerbestrasse 11, 6330 Cham, Switzerland

Preface to the 2nd Edition

Extremophilic life has seen a significant increase in interest of many scientists and even non-scientists. One reason is surely the unprecedented large number of exoplanets discovered during the last few years – nearly 3500 (May 2016) confirmed planets since 1988 and even more suspected candidate planets. These discoveries have intensified interest in the search for extraterrestrial life. Planets which orbit in a star's habitable zone, where it is possible for liquid water to exist on the surface, are of special interest. Planetary habitability has come into focus and with it the range of environmental factors which determine the suitability for life.

Biotechnology and particularly fields like nanotechnology are also benefitting from extremophilic microorganisms. The underpinning of many applications is provided by a careful taxonomic identification of extremophilic strains. The increase in novel isolates has been quite remarkable, considering the extensive work which is still required for a formal description of a taxon. For example, 47 novel species from desert environments and 96 novel haloarchaeal species have been published since the first edition of this book.

All authors are thanked for having added about 15–20 % new material into their chapters. Thanks go to the new author groups Mirete et al. and Lee et al. for their chapter contributions.

We are also very grateful to Silvia Herold and Kumar Athiappan from Springer-Verlag/Springer Global for the helpful suggestions and wonderful cooperation.

Salzburg, Austria
Bucuresti, Romania
September 2016

Helga Stan-Lotter
Sergiu Fendrihan

Preface to the 1st Edition

Life on the edge, life at the physicochemical limits, and weird and eccentric life – these are the descriptive terms assigned to extremophiles. The first European workshop entitled "Microbial adaptation to extreme environments" was held at the University of Nijmegen in the Netherlands in 1973 (Heinen et al. 1974). The first use of the term "extremophiles" is credited to Robert MacElroy, a NASA exobiologist, who participated in this workshop (MacElroy 1974).

Many more conferences on extremophiles followed, the journal *Extremophiles* was launched, and the CAREX initiative (Coordination Action for Research Activities on Life in Extreme Environments) was started (www.carex-eu.org), providing an extensive database and recently a new roadmap.

The often surprising ranges of physicochemical factors, within which life is possible, stimulated many scientists to explore astrobiological perspectives, with the reasoning that findings on the evolution and mechanisms of adaptation of life at extremes would help to understand the environments of other planets or moons. These astrobiological aspects are dealt with in several chapters of this volume (Billi; Mapelli et al.; Gomez and Parro; Moissl-Eichinger; Stan-Lotter).

The biodiversity of extreme environments appears of unexpected and enormous size; its magnitude is still largely unknown. Classical and molecular approaches are used for its investigation and described by Enache et al., Fendrihan and Negoiţă, Mapelli et al., Heulin et al., Hreggvidsson et al., and Pearce.

Several updated lists of taxonomic descriptions of extremophilic species and strains are provided: halophilic archaea and bacteria (Enache et al.), desiccation-resistant and thermotolerant bacteria (Heulin et al.), thermophilic bacteria (Hreggvidsson et al.), and psychrophilic bacteria, archaea, algae, and fungi (Fendrihan and Negoiţă; Pearce).

Adaptations to extreme environmental conditions, especially on the molecular level, are a topic of uninterrupted fascination since the first investigations were made and are described by Billi, Heulin et al., Fendrihan and Negoiţă, and Pearce.

Numerous applications of extremophiles for biotechnological purposes have already been implemented, although the whole potential is certainly not realized yet – treatments of this subject are included in the chapters by Billi, Enache et al., Fendrihan and Negoiţă, and Pearce.

Sadly, our dear colleague Teodor Negoiță passed away during the completion of this volume (on March 23, 2011, at the age of 64). Dr. Negoiță was the founder of the Romanian Institute of Polar Research. He had organized several international expeditions into the Arctic and in the year 2005 managed to establish the Romanian Law-Racoviță Research Station in the Antarctic.

Our thanks go to the initiators and organizers of the CAREX project – Cynan Ellis-Evans, Nicolas Walter, and Petra Rettberg – and for the opportunity of participating in great meetings at wonderful places, Sant Feliu de Guixols and Sasbachwalden, which facilitated the interactions between people who would otherwise not have met and which was also the partial foundation for the contributions in this book.

We are also very grateful to Amrei Strehl and Eva-Maria Oberhauser from Springer-Verlag GmbH for the helpful suggestions and excellent cooperation.

Salzburg, Bucharest, May 2011

Helga Stan-Lotter
Sergiu Fendrihan

References

Heinen, W, Skinner FA, Tempest WD, van den Ende G, Vogels GD, van den Drift C, Dundas I, Harder W, Weerkamp A, Gounot AM, van Beeumen J, MacElroy RD, Knoope P, Schwartz AW, Schlegel HG (1974) Proceedings of the first European workshop on microbial adaptation to extreme environments. Biosystems 6: 57–80.

MacElroy RD (1974) Some comments on the evolution of extremophiles. Biosystems 6: 74–75.

Contents

Physicochemical Boundaries of Life

1

Helga Stan-Lotter

Contents

1.1 Introduction

All life we know no matter how freaky in other respects, is still based on organic molecules dissolved in water, and we all use the same basic cellular machinery. Extremophiles haven't fundamentally changed the way we think about strategies to look for life, but they have bolstered the optimism with which we search. Right now anywhere with liquid water is considered a possible habitat, and this guides our quest.

David Grinspoon. Lonely planets. The natural philosophy of alien life. 2003.

H. Stan-Lotter
University of Salzburg, Department of Molecular Biology, 5020 Salzburg, Austria
e-mail: helga.stan-lotter@sbg.ac.at

H. Stan-Lotter, S. Fendrihan (eds.), *Adaption of Microbial Life to Environmental Extremes*, DOI 10.1007/978-3-319-48327-6_1

When attempting to define extremophiles – meaning organisms living in an extreme environment – it is obvious that a strongly anthropocentric component emerges. Extreme environments are considered "hostile to higher forms of life" and "uninhabitable by other organisms" (Richard Johnson, NASA director). Thomas D. Brock, a microbiologist, who pioneered in the 1970s the study of life in thermal springs in Yellowstone National Park, USA, defined the characteristics of extreme habitats from the taxonomists's point of view: "environments with a restricted species diversity and the absence of some taxonomic groups" (Brock 1969).

Why there is a limited diversity and why some groups are missing will be comprehensible when the physicochemical factors, which are characteristic for extreme environments (see Table 1.1), are examined. These parameters define niches, which allow occupancy by only certain groups of organisms; none can be expected to survive the whole range of conditions. An early ecologist, Victor Shelford, presented already in 1913 the observation that organisms will usually be limited by abiotic factors in his "law of tolerance" (cited in Krebs 2008). Each particular factor that an organism responds to in an ecological system has what he called limiting effects. The factors function within a range, that is, a maximum and a minimum value for the factors exists, which Shelford designated "limits of tolerance" (see Table 1.1). For an organism to succeed in a given environment, each of a complex set of conditions must remain within the tolerance range of that organism, and if any condition exceeds the minimum or maximum tolerance of that organism, it will fail to thrive.

The recent years have seen an unprecedented expansion of knowledge about the physicochemical limits of life. Several books and reviews have been published on extremophilic microorganisms, occasionally including higher organisms (e.g., Seckbach 2000; Rothschild and Mancinelli 2001; Oren and Rainey 2006; Gerday and Glansdorff 2007; Seckbach and Walsh 2009; Rampelotto 2016; Seckbach et al. 2013) and viruses from extreme environments (Le Romancer et al. 2007). Dedicated scientific journals are available, e. g., *Extremophiles*, *Archaea*, and *Astrobiology*. More specific books and articles, pertaining to certain groups of extremophiles, are mentioned in the chapters of this volume.

The purpose of this chapter is providing an overview of the extreme environmental factors which influence life on Earth, a short survey of the various types of extremophiles – including viruses – as well as an update of current records for the limits of growth on the one hand and survival of extreme conditions on the other hand.

1.2 Brief History of Life on Earth

The Earth is about 4.6 billion years old; the most ancient rocks were dated to be 3.5 to 3.86 billion years old (Westall 2005). They are located in Southern Africa (Swaziland and Barberton Greenstone Belt), Western Australia (Warrawoona series and Pilbara formation), and Greenland (Isua rocks). Besides volcanic and carbonaceous rocks, the sedimentary rocks are of particular interest, since their formation required liquid water, probably in the form of oceans, and therefore, the conditions for life did exist.

Table 1.1 Classes and examples of extremophilic prokaryotes

Physicochemical factor	Descriptive term	Genus/species	Lineage	Habitat	Minimum	Optimum	Maximum	Reference
Temperature								
High	Hyper-thermophile	*Pyrolobus fumarii*	Archaea	Hydrothermal	90 °C	106 °C	113 °C	Blöchl et al. (1997)
High	"	Strain 121	Archaea	Black smoker	85 °C	?	121 °C	Kashefi and Lovley (2003)
High	"	*Methanopyrus kandleri* strain 116	Archaea	Black smoker fluid	90 °C	105 °C	122 °C	Takai et al. (2008)
Low	Psychrophile	Permafrost bacteria	Bacteria	Siberian Permafrost	−20 °C	?	5 °C	Rivkina et al. (2000)
pH low	Acidophile	*Picrophilus oshimae*	Archaea	Acidic hot spring	pH −0.06	pH 0.7	pH 4	Schleper et al. (1995)
	"	*Ferroplasma acidarmanus*	Archaea	Acid mine drainage	pH 0	pH 1.2	pH 2.5	Edwards et al. (2000)
pH high	Alkaliphile	*Alkaliphilus transvaalensis*	Bacteria	Deep gold mine	pH 8.5	pH 10	pH 12.5	Takai et al. (2001)
Hydrostatic pressure	Piezophile	*Moritella yayanosii*	Bacteria	Ocean sediment	50 MPa	70 MPa	110 MPa	Nogi and Kato (1999)
	Piezophile	*Pyrococcus* CH1	Archaea	Black Smoker	20 MPa	52 MPa	120 MPa	Zeng et al. (2009)
Salt (NaCl)	Halophile	*Halobacterium salinarum*	Archaea	Saltern	15 % NaCl	25 % NaCl	32 % NaCl	Grant et al. (2001)
Water activity	Xerophile	*Xeromyces bisporus*	Fungus	Moldy fruit	a_w 0.61	a_w 0.82	a_w 0.92	Hocking and Pitt (1999)
	"	Halophiles	Archaea, Bacteria	Brines	a_w 0.611	a_w 0.755	$a_w < 0.9$	Stevenson et al. (2015)

The evidence for early microbial life is derived from fossilized remains of cells in ancient rocks and on an abundance of the lighter carbon isotope (^{12}C), due to its generally assumed preference over the heavier isotope (^{13}C) by microorganisms (see Westall 2005, for a discussion). Microfossils, which were found in old sedimentary rocks, are similar in shape, size, and arrangements as modern bacteria; the earliest forms were probably cocci (spheres) and short straight or curved rods (Westall 2005).

The environments on the early Earth were different than today; especially temperatures were likely much higher than now. It is thought that surface temperatures during the first 200 million years exceeded 100 °C. The accumulation of water occurred only later, when the Earth was cooling down. How much time this process took is not known; if life originated at the end of the first half million years – as the microfossils suggest – Earth was still fairly hot, and therefore the first microorganisms were likely thermophilic or at least thermotolerant.

The atmosphere of the early Earth did not contain oxygen; the main gases were carbon dioxide, water vapor, and hydrogen sulfide, with probably smaller amounts of nitrogen, methane, and carbon monoxide (Westall 2005). Only much later, about 2 billion years ago, oxygen was produced by phototrophic bacteria, and its concentration in the atmosphere was slowly rising to reach the present value of 20.95 % (Nealson and Conrad 1999). Early life forms were thus anaerobes; even up to now, there are still many niches where anaerobic microorganisms are thriving. The lack of oxygen meant that there was no protective layer of ozone in the upper atmosphere, and therefore, a high flux of ultraviolet radiation reached the surface of the Earth (Westall 2005). Larger unicellular organisms did not exist before about 1 billion years ago, as was deduced from microfossils. Multicellular organisms probably did not appear before 600–700 million years ago.

The early Earth presented extreme conditions compared to the general situation today. But life was present on Earth throughout most of its history, and that life was solely microbial for very long periods of geological time, developing a high diversity and adaptations to various environmental extremes.

1.3 Prokaryotes, Eukaryotes, the Tree of Life, and Viruses

Living organisms consist of cells, and much of what is known about cell structures is due to information from microscopy. Microscopic examination revealed early the presence of two fundamental types of cells – eukaryotic and prokaryotic cells – to denote the presence or absence of a cell nucleus, a feature which was easily identifiable. Figure 1.1 contains a general scheme of a eukaryotic and a prokaryotic cell. The nucleus encloses the genetic material of the eukaryotic cell, which is present in the form of chromosomes; prokaryotic cells contain their genetic material as an aggregated mass of DNA, which is called nucleoid. Other differences between the two types of cells are the sizes (eukaryotic cells are on average 10–20 times larger than the typical prokaryotic cell) and the presence of organelles in the eukaryotes. Organelles are small membrane-enclosed structures within cells, which were traced

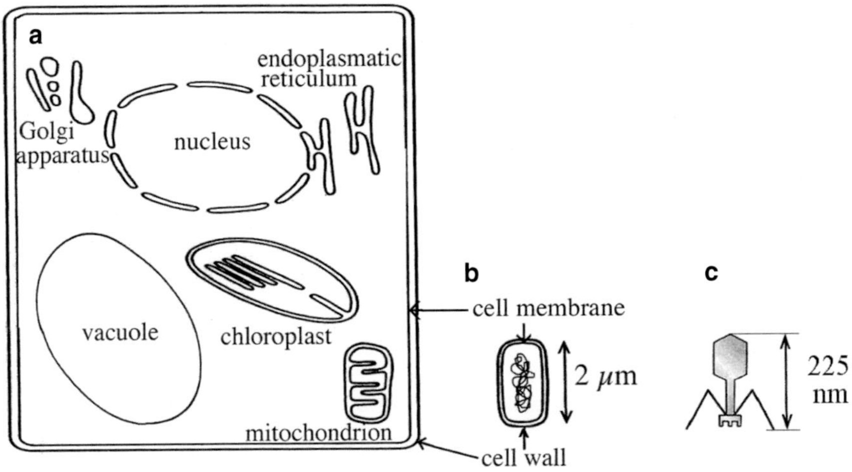

Fig. 1.1 Schemes of a eukaryotic cell from a plant (**a**), a prokaryotic cell (**b**), and a bacterial virus (**c**). The presence of organelles, nucleus, and inner compartments (vacuole, Golgi apparatus, endoplasmatic reticulum) in the eukaryotic cell is indicated. Mitochondria and chloroplasts (**a**) are of similar sizes as prokaryotic cells (**b**) and possess inner membranes. The DNA of prokaryotes is a coiled structure inside the cell (**b**). Viruses are much smaller than prokaryotes (note the difference of scale) and of simple composition

back to bacteria-like precursors; they were incorporated into early eukaryotic cells and became the mitochondria and chloroplasts of modern cells.

Prokaryotic cells are simpler; besides the lack of a nucleus and lack of organelles, their morphology, protein synthesis, cell proliferation, and genetic apparatus are less diverse and more streamlined than comparative features of eukaryotic cells. On the other hand, prokaryotes display a stunning diversity of metabolic functions; they can exist in a wide range of inhospitable environments, where eukaryotes would not be able to survive. Their influence on all processes of the biosphere, including global climate, is probably still severely underestimated; their biomass alone is thought to be equal to or even supersede the total plant mass on Earth (Whitman et al. 1998; Pedersen 2000; Parkes et al. 2014).

Viruses are separate entities – they need a host for replication; without one they are inert particles, and therefore the question if they are alive or not has been debated for years. Newer insights have prompted suggestions (see also below) to include them into the living world (Le Romancer et al. 2007; Forterre 2016). Viruses are generally much smaller than even most prokaryotic cells (Fig. 1.1c) and contain often only a nucleic acid (DNA or RNA) and some proteins.

Relationships between organisms and knowledge about their evolution can be obtained by comparing their nucleic acid sequences. Figure 1.2 shows a phylogenetic tree, which is based on the sequences of a molecule which is present in similar form in all organisms, the ribonucleic acid (RNA) of the small subunit of ribosomes. The greater the differences in the sequences between the molecules from

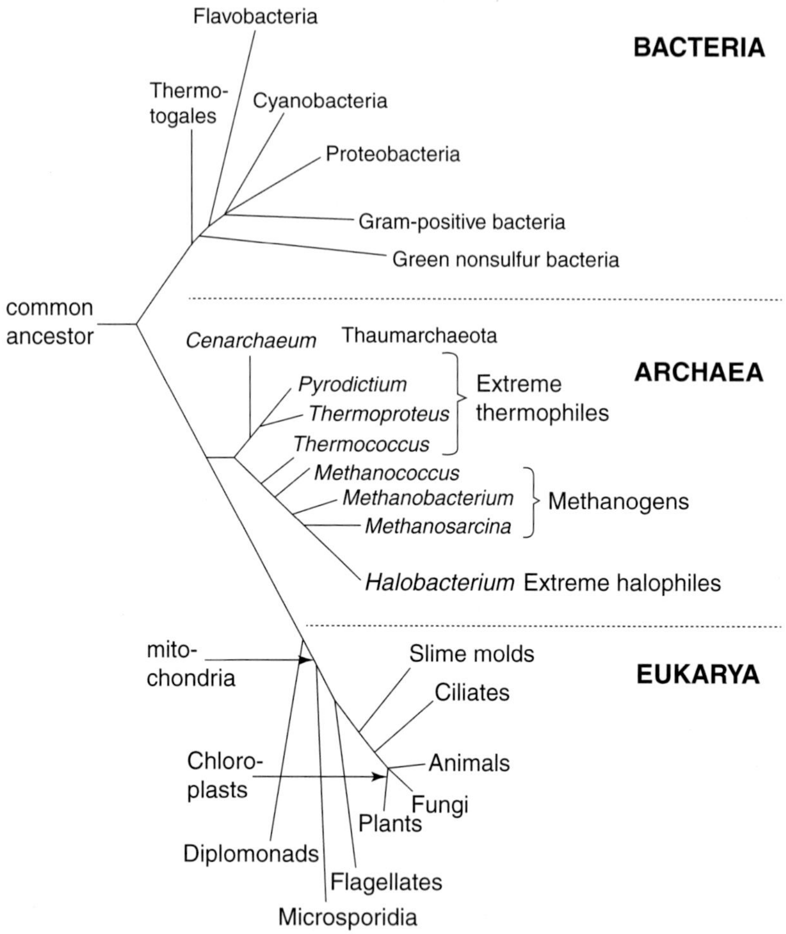

Fig. 1.2 Phylogenetic tree of all organisms, based on sequences of small ribosomal RNA genes. Prokaryotes form two lineages (Archaea and Bacteria); eukaryotes (Eukarya) form one lineage. The length of the branches represents the approximate evolutionary distances between groups, based on the number and positions of different bases in their small rRNAs

two or more organisms, the greater is their evolutionary distance. This is reflected in the length of the branches of the tree. Three main lineages of organisms have emerged from these analyses – the Eukarya (or eukaryotes), which include all animals, plants, fungi, and many unicellular microorganisms, and two lineages of prokaryotes, the Bacteria (or eubacteria) and the Archaea (or archaebacteria). The length of the archaeal branch is shorter than the bacterial and eukaryal branches (Fig. 1.2), which is generally interpreted as a sign of closeness to the common ancestor of all life. The Archaea were classified historically into three main groups, which were termed extreme halophiles, extreme thermophiles, and methanogens, hinting at the preference of extreme habitats by many Archaea. Although Archaea existing in moderate environments, such as the Thaumarchaeota (Fig. 1.2), are being increasingly identified (DeLong 1998; Brochier-Armanet et al. 2008), many record holders with respect to living conditions are found only within the extremophilic Archaea (see below).

Nucleic acid sequences showed also clearly the similarity between bacteria and the organelles – mitochondria and chloroplasts – in Eukarya (Fig. 1.2). The uptake of the forerunners of mitochondria is assumed to have occurred about 1.9 billion years ago; the uptake of photosynthetic endosymbionts, which became modern chloroplasts, occurred somewhat later in Earth's history.

Viruses are not included in this general tree of life, since they do not have the genes for small subunit rRNAs on which the tree is based. In fact, there is no common informational molecule which could be the basis for a phylogenetic tree of viruses (Forterre and Prangishvili 2009a). It is becoming increasingly clear that modern viruses are not fragments of genetic materials that escaped from mother cells but descendants of an ancient "virosphere" that possibly even preceded the origin of modern cells (Le Romancer et al. 2007; Forterre and Prangishvili 2009b), and the concept of a "virocell," where viral information is actively expressed and reproduced, has been suggested (Forterre 2016). Like all other organisms, extremophiles serve as hosts for viral replication, and a great diversity of extremophilic viruses is now known (Le Romancer et al. 2007; Parikka et al. 2016). It is thus reasonable to study their characteristics and to view life on early Earth considering the existence of a possible virosphere.

1.4 Extreme Environments and Their Inhabitants

1.4.1 Temperature Ranges of Microorganisms

Microorganisms can be grouped into broad, although not very precise, categories, according to their temperature ranges for growth (Burgess et al. 2007; Rothschild and Mancinelli 2001):

- Psychrophiles (cold loving) can grow at 0 °C and some even as low as −10 °C or maybe −16 °C; their upper limit is often about 20–25 °C.
- Mesophiles grow in the moderate temperature range, from about 20 °C (or lower) to 45 °C.
- Thermophiles are heat loving, with an optimum growth temperature of 50 °C or more, a maximum of up to 70 °C or more, and a minimum of about 20 °C.
- Hyperthermophiles have an optimum above 75–106 °C and thus can grow at the highest temperatures tolerated by any organism (Stetter 2002). Some will not grow below 80 °C.

Temperatures lower than 55 °C are widespread on Earth and are associated with sun-heated habitats, but temperatures higher than 55–60 °C are much rarer and are almost exclusively associated with geothermal habitats (Brock 1986). Therefore, a "thermophile boundary" of 55–60 °C has sometimes been suggested for prokaryotes, beyond which thermophiles are growing (Brock 1986).

Hot springs and geothermal vents are found in several parts of the world, such as in Yellowstone National Park (Wyoming, USA), Iceland, Southern Italy, New Zealand, Japan, and on the floors of the oceans, where spreading zones occur and superheated fluids are emitted (see also Hreggvidsson et al. 2017, this volume).

Life beyond the boiling point of water was unthinkable well into the 1970s, until Stetter (1982) transcended this boundary by reporting on a hyperthermophile which grew at 105 °C. Subsequently, numerous microorganisms have been described, which flourish at temperatures up to 113 °C and possibly even at 122 °C (Table 1.1). The upper boundary of thermophilic life is still a matter of debate (Cowan 2004). Several isolated enzymes from hyperthermophiles ("thermozymes") were shown to be active and stable up to 130 °C (Daniel 1996; Lévêque et al. 2000), but microbial growth at that temperature has not been identified so far.

With respect to low temperature limits of life, psychrophiles (see Pearce 2017; Fendrihan and Negoiță 2017, both in this volume) were reported to perform metabolic activities (uptake of ^{14}C-labeled acetate) at −20 °C under defined conditions (Rivkina et al. 2000; Gilichinsky 2002; D'Amico et al. 2006). The low temperature limit for microbial metabolism is thus considered to be −20 °C (see Table 1.1). A recent investigation by Clarke et al. (2013) based on vitrification experiments showed that the high internal viscosity of cells causes slowing of diffusion of oxygen and metabolites such that cellular metabolism ceases, but vitrified cells do survive at temperatures between −10 °C and even −26 °C, depending on the species. A low temperature limit for microbial reproduction has not been established (Clarke et al. 2013). Mykytczuk et al. (2013) reported growth and division of the Arctic permafrost bacterium *Planococcus halocryophilus* at −15 °C, together with a thorough genomic and transcriptomic analysis of its specific adaptations.

1.4.2 Low Water Activity: Halophiles, Osmophiles, and Xerophiles

High concentrations of salt and sugar have traditionally been used to prevent spoilage of food by microbial growth. The availability of water is greatly reduced by such additions, yet several types of halophilic bacteria, osmophilic yeasts, and xerophilic fungi are known to grow under these circumstances (Gilmour 1990; Grant 2004a). Inhibition of many fungi and yeasts occurs at water activities (a_w) between 0.8 and 0.75, but the mold *Xeromyces bisporus* is exceptional in being able to grow at a lower water activity (a_w 0.61) than any other organism described (Grant 2004a; Table 1.1). Water activity as a life-limiting factor has come into focus recently, since a_w of 0.61 from the record holder *Xeromyces bisporus* was due to high sugar concentrations, which is not characteristic for most prokaryotic environments on Earth and maybe beyond. Data from the deep hypersaline anoxic lakes of the Mediterranean Ridge and culture-based as well as culture-independent studies suggested multiplication and metabolic activity of halophilic Archaea and Bacteria down to an a_w of 0.611 (Yakimov et al. 2015; Stevenson et al. 2015) (see also Mapelli et al. 2017, this volume).

Other environments of low water activity are hypersaline sites, e.g., the Dead Sea in Israel; the Great Salt Lake in Utah, USA; and natural and man-made salterns, where water availability is limited by a high concentration of salts (usually NaCl). These hypersaline waters are populated by halophilic microorganisms, mainly haloarchaea, which require substantial amounts of salt for growth (at least 12–15 % NaCl) and are able to grow up to saturation (Table 1.1; see also Enache et al. 2017, this volume). Numerous strains were isolated from such sites (Grant 2004a); for a regularly updated listing of halophilic genera and species – as well as other prokaryotes – see the website "List of Prokaryotic Names with Standing in Nomenclature" maintained by Aidan C. Parte, Curator, at http://www.bacterio.net.

When hypersaline waters are evaporating, halophilic microorganisms become entrapped within fluid inclusions in halite crystals (Norton and Grant 1988; Fendrihan et al. 2009). Haloarchaea remain viable within the fluid inclusions for considerable periods of time; laboratory cultures kept in this way have been recovered after more than 10 years (Grant 2004a). Ancient halite deposits, which are widely distributed on Earth (see Grant 2004a), range in age up to the Precambrian, while the most massive salt sediments – in the order of an estimated 1.3 million cubic kilometers – stem from the late Permian and early Triassic period (ca. 245–290 million years old; Zharkov 1981). There have been several reports of viable halophilic prokaryotes, being recovered from such geologically old halite deposits (e.g., Grant et al. 1998; McGenity et al. 2000; Stan-Lotter et al. 1999, 2002; Gruber et al. 2004; Schubert et al. 2010; Gramain et al. 2011), suggesting that the microorganisms might be of the same age as the sediments and were originally entrapped when the ancient brines dried down. These extreme halophiles can be considered desiccation-resistant, oligotrophic, and long-term survivors.

1.4.3 Extremes of pH and Low Nutrients

Many familiar biological processes are taking place at pH values around neutral, but organisms from all three domains contain representatives, which thrive at pH values on both ends of the scale.

Low pH Acidophiles are extremophiles living in conditions with a pH of 3.0 or below. Many species of acidobacteria have been found in poor or polluted acidic soils and in acid mine drainages. Numerous acidophilic archaeal species are known, which are found in geothermal sulfurous sites, e.g., acidic springs (Hreggvidsson et al. 2017, this volume) or the Rio Tinto River (Mirete et al. 2017, this volume). Ferroplasma acidarmanus has been described growing at pH 0 in acid mine drainage (Table 1.1). The genome of one isolated strain of the species was sequenced and compared with sequence data from an environmental population of the same species, revealing genomic heterogeneity within the population (Allen et al. 2007). Acidophiles have evolved efficient mechanisms that pump protons out of the intracellular space to keep the cytoplasm at or near neutral pH (see Mirete et al. 2017, this volume). There are some exceptions: the recently described acidophilic archaean

Picrophilus torridus grows optimally at pH of 0.7 and apparently maintains an intracellular pH of 4.6 (Ciaramella et al. 2004).

High pH Alkaliphiles are extremophiles thriving in conditions with a pH of 9.0 or above. Very stable alkaline environments on Earth are soda lakes and soda deserts, which occur on all continents (Grant 2004b). In particular, numerous alkaline lakes are situated in the East African Rift Valley, which, due to their high content of NaCl and Na_2CO_3, have pH values of 10.5–11.5 (Grant 2004b). Haloalkaliphilic bacteria-like cyanobacteria, proteobacteria and halomonads, were isolated from these waters as well as Archaea of the genera *Natronobacterium*, *Natronomonas*, and *Natronococcus* (Grant 2004b). Another highly alkaline ecosystem is the Lost City hydrothermal field, which is located on the seafloor mountain Atlantis massif, near the Mid-Atlantic Ridge (Kelley et al. 2005). Reactions between seawater and upper mantle peridotite produce methane- and hydrogen-rich fluids whose pH values range from 9 to 11, with temperatures ranging from 40° to 90 °C. A variety of microorganisms live in, on, and around the vents. Archaeal populations include *Methanosarcinales* which form thick biofilms inside the vents, subsisting on methane and hydrogen. Bacteria related to *Firmicutes* live also inside the vents; on the outside, methane- and sulfur-oxidizing proteobacteria are present (Kelley et al. 2005).

Low Nutrients Microorganisms have evolved certain physiological and metabolic strategies for growing in nutrient-poor environments. In the open ocean, particularly in the deepwater column, in deep sediments, and in deep crustal environments, carbon and energy sources are extremely scarce, but active microorganisms are present. Heterotrophic prokaryotes that can reproduce at very low levels of organic carbon concentrations are known as oligotrophs (their counterparts, which would reproduce at high organic carbon concentrations, are called copiotrophs). There is no generally accepted definition of these categories, but typically the value for distinction is 1–10 mg of C per liter for oligotrophic environments (Giovannoni and Rappé 2000). Recently, *Pelagibacter ubique*, a representative of one of the most cosmopolitan microorganisms in oligotrophic oceans, was found to grow only in the *in situ* micromolar concentrations of organic carbon. This bacterium has one of the smallest genomes known from free-living organism. *P. ubique* and related marine oligotrophs are setting the lower limits of concentration of organic compounds that can support the growth of heterotrophs (Giovannoni et al. 2005).

1.4.4 High and Low Pressure

"Barophilic" and "piezophilic" are sometimes used interchangably, but barophilic should refer to an organism that lives and thrives under high barometric (atmospheric) pressure, and organisms, which live in the deep oceans, are called piezophilic following an agreement in 1995 (Simonato et al. 2006). The hydrostatic pressure in the oceans increases by 1 atm for every 10 m of depth. The maximum depth is in the Mariana Trench; with its 10 900 m depth, the pressure is about 1100 atm (ca. 110 MPa). Even from this site, viable microorganisms have been isolated. The extremely piezophilic bacterium *Moritella yayanosii* is unable to grow at pressures below 500 atm; its optimum growth occurs at 700 to 800 atm, and it can grow up to 1035 atm, which is the pressure of its natural habitat (Table 1.1). Other isolates from the ocean floors and sediments of about 4000–6000 m depth grow optimally at pressures of 400–600 atm but have usually retained the capacity for growth at normal pressure (1 atm). In laboratory experiments, model organisms like *Saccharomyces cerevisiae* and *Escherichia coli* were subjected for short times to pressures of 200 and 275 MPa, respectively, and their survival and responses were tested (reviewed in Simonato et al. 2006).

Low pressure is an emerging extreme environmental factor which will acquire more significance as astrobiological missions are planned. Schuerger and Nicholson (2016) investigated 23 species of bacterial hypobarophiles, defined as growing at low pressures (approximately 1–2 kPa). The hypobaric assays were conducted in polycarbonate desiccators connected to an external pump and controller (Schuerger et al. 2013). The strains were capable of growth at 0.7 kPa, which is present at the Martian surface, and belonged mostly to the genus *Serratia*. Other strains were members of the genera *Carnobacterium*, *Exiguobacterium*, *Leuconostoc*, *Paenibacillus*, and *Trichococcus*. So far only Bacteria, no Archaea or fungi, were found to be hypobarophilic. The data are also significant for judging bacterial metabolic activity and growth at the equivalent pressure of 34 km in the middle stratosphere of Earth.

1.4.5 Microgravity and Hypergravity

Microgravity can be considered an extreme environmental factor which has significant effects on microbial physiological processes, such as cell growth, gene expression, and certain biotechnological products such as antibiotics. The effect of microgravity on microorganisms has been a particular active area of research because it is relevant to human health during space flight (for a recent review, see Taylor 2015). Microgravity can be simulated with a variety of devices (Horneck et al. 2010), or it can be achieved in the real world, for example, by conducting experiments on the International Space Station. The majority of studies indicated that microgravity stimulates the growth of some microorganisms, enhanced or suppressed production of antibiotics, and increased virulence of pathogens in a murine infection model (Wilson et al. 2007).

Experiments in hypergravity were performed exclusively in simulated environments. Deguchi et al. (2011) studied the effect of hypergravity by exposing microbial cells to acceleration produced by centrifugal rotation. *Escherichia coli*, *Paracoccus denitrificans*, *Shewanella amazonensis*, *Lactobacillus delbrueckii*, and *Saccharomyces cerevisiae* were cultivated while being rotated in an ultracentrifuge at high speeds corresponding to 403,627 g (i.e., 403,627 times the gravity experienced on Earth). *P. denitrificans* and *E. coli* displayed not only survival but also robust cellular growth by binary fission under conditions of hyperacceleration. Such hypergravity is usually found only in cosmic environments, such as on very massive stars or in the shock waves of supernovas. Deguchi et al. (2011) discussed also the implications of these data for panspermia, the putative transport of bacteria between planets, which would involve the asteroidal impact on a donating planet followed by ejection of bacteria-bearing rocks. Mastrapa et al. (2001) had previously found 40–100 % survival of *Bacillus* spores and *Deinococcus* cells in ultracentrifugal and ballistic experiments where the microorganisms were subjected briefly to similarly high accelerations.

1.4.6 Viruses

A review on viruses in extreme environments was published by Le Romancer et al. (2007), who described viruses from the following sites: hypersaline (Dead Sea, solar salterns), alkaline and hypersaline (soda lakes; Mono Lake), deserts (Sahara), polar regions (lakes in the Taylor Valley of Antarctica; Arctic and Antarctic sea ice), high temperature (terrestrial hot springs in Yellowstone National Park, Iceland and Japan; deep sea hydrothermal vents), and a deep subsurface sediment (drilling hole near the West Canadian coast).

It is perhaps noteworthy that so far only double-stranded DNA viruses were isolated from extreme environments, but no RNA viruses, and it has been suggested that this very stable form of genome may be necessary to face the harsh constraints of extreme habitats (Le Romancer et al. 2007).

So far, no viral fossils have been detected, but interestingly, silicification of viruses (bacteriophage T4) under conditions of silica-depositing hot spring environments has been achieved recently in the laboratory (Laidler and Stedman 2010). Viral morphology and elemental biosignatures were preserved, thus raising the exciting possibility of finding viruses in the geological record (Laidler and Stedman 2010). In addition, silicification of viruses, including the extremophilic SSV-K from *Sulfolobus*, appeared to be a protective mechanism for subsequent possibly global distribution, making virus particles resistant to desiccation while preserving their activity (Laidler et al. 2013).

1.4.7 Extremophilic Multicellular Organisms

Halophilic representatives occur in all three domains of life – besides prokaryotes, there are halophilic or halotolerant fungi, algae, and animals (Oren 2007). Similarly, thermophilic members are found in all three branches, although their maximum temperatures of growth vary from 38 °C for vertebrates (fish) to 50 °C (insects; mosses) to >100 °C (prokaryotes), depending on the taxonomic group as outlined by Brock (1986).

There is no comprehensive review yet on extremophilic higher organisms, but there are books in the COLE series (Cellular Origin and Life in Extreme Habitats), edited by Joseph Seckbach, which contain much information on this subject (e.g., Seckbach and Chapman 2010; Seckbach and Grube 2010). Some further sources and a few examples of special interest are given here:

Weber et al. (2007) addressed in their review the metabolism of extremophilic algae and plants, with some coverage of metazoa and fungi. A website from 1998 "Eukaryotes in extreme environments" (http://www.nhm.ac.uk/research-curation/research/projects/euk-extreme/; accessed May 2016) provides some older literature and still timely discussion on the subject.

Alvinella pompejana, the Pompeii worm, was detected by Desbruyeres and Laubier (1986) near hydrothermal vents. It is living at 80 °C, possesses chemolithotrophic thermophilic epi-bacteria as symbionts (Jeanthon and Prieur 1990), and is probably the most thermotolerant animal known.

Tardigrades (meaning "slow walkers"), known as water bears, are microscopic, water-dwelling, segmented animals with eight legs. The adults may reach a body length of 1.5 mm. They are found on lichens and mosses, on dunes, beaches, soil, and in marine or freshwater sediments. The survival capacities of tardigrades are stunning – several species are able to survive temperatures of −180 or −196 °C, respectively (Jönsson 2007 and references therein), about 1000 times more radiation than other animals (Horikawa et al. 2006), almost a decade without water (Guidetti and Jönsson 2002), and storage at −20 °C for 30 years (Tsujimoto et al. 2016). In 2007, tardigrades were taken into low Earth orbit on the FOTON-M3 mission and were exposed to the vacuum of space and solar radiation for 10 days. Back on Earth, it was discovered that many of them survived and laid eggs that hatched normally (Jönsson et al. 2008). The review by Jönsson (2007) deals, besides tardigrades, also with the radiation and desiccation tolerance of a few more animals, such as larvae from *Polypedilum vanderplanki* (an insect), eggs from the brine shrimps *Artemia salina* and *Artemia franciscana*, larvae and adults of nematodes, and rotifers (genera *Mniobia* and *Macrotrachela*).

1.5 Microbial Survival of Extreme Conditions

Table 1.2 shows several examples of microorganisms, which survived exposure to extreme physicochemical factors. The time of exposure varied greatly, for example, *Deinococcus radiodurans* will, depending on the source of gamma radiation, survive a dose of about 20,000 Gy, obtained by a pulse lasting a few hours (Ito et al. 1983). These levels of radiation are not found naturally on present-day Earth. Several more species of *Deinococcus* are known, which possess high resistance against radiation and desiccation, and several genomes have been sequenced (see Heulin et al. 2017, this volume). Interestingly, irradiation at a temperature of −79 °C – simulating Martian surface conditions – increased the resistance of *Deinococcus radiodurans* markedly, compared to irradiation at room temperature or on ice (Dartnell et al. 2010).

The survival of the bacterium *Streptococcus mitis*, which has traveled to the Moon during the Apollo missions and apparently survived there for 2.5 years in a camera, has, unfortunately, not been rigorously documented – in some descriptions,

Table 1.2 Microbial survival of extreme conditions

Microorganism	Physicochemical parameter	Time of exposure	Other information	Reference
Deinococcus radiodurans	Ionizing radiation	5–20 h	ca. 20,000 Gy	Ito et al. (1983)
Streptococcus mitis	Surface of the Moon	2.5 years	In a camera	Website[a]
Numerous microorganisms	Minus 20 °C	ca. 10^6 years	Permafrost	Gilichinsky (2002)
Numerous microorganisms[b]	Minus 193 °C	> 10 years	Liquid N_2	Gherna (1994)
Numerous microorganisms[b]	$a_w < 0.75$	> 10 years	Vacuum	Gherna (1994)
Halococcus salifodinae	NaCl >30 %	>10^6 years (?)	In salt crystals	Denner et al. (1994); McGenity et al. (2000)
Endospores (*Bacillus*, *Clostridium*)	Heat; chemicals	>3000 y	Sediments	Madigan et al. (2009)
Endospores (*Bacillus*)	Outer space	6 years	Surface of space probe	Horneck et al. (1994)

[a]http://science.nasa.gov/science-news/science-at-nasa/1998/ast01sep98_1/ (accessed May 2016)
[b]With protective substances (e.g., 25–40 % glycerol)

survival of 20–500 cells is mentioned, however, without reference to the method of determination. Nevertheless, the fact of its apparent resistance to space conditions over several years is of profound importance.

Careful documentation of survival was carried out with endospores from *Bacillus subtilis*, which were fully exposed to space conditions in the LDEF (Long Duration Exposure Facility) mission for 6 years (from 1984 to 1990), from which a significant fraction was recovered in a viable state (Horneck et al. 1994). Samples of the same mission, which were shielded against UV light with metal foils, exhibited about 70 % survivors, following the 6 years in space (Horneck et al. 1994; reviewed by Horneck et al. 2010).

Permafrost is an environment, where many viable microorganism have been found; some stem from sediments which are several million years old (Gilichinsky 2002). Their metabolism is, due to the frozen state, extremely slow or, perhaps in some cases, nonexistent. Some microorganisms in permafrost soils are apparently forming resting stages with thick walls, but these are not endospores (Soina et al. 2004). Freezing will prevent microbial growth, but it does not necessarily kill microbes. Culture collections are using freezing in the presence of protective substances for long-term preservation of microbial strains and other biological materials (Gherna 1994). Prerequisites are the addition of water-miscible liquids such as glycerol or dimethyl sulfoxide in concentrations of 20 % or more; they will penetrate the cells and protect them, mainly by hindering crystal formation, upon freezing. Temperatures around that of liquid nitrogen (−196 °C) are survived by strains for up to 30 years (Gherna 1994).

Endospores are a survival form of certain bacteria (mostly from the genera *Bacillus* and *Clostridium*), which are formed when they find themselves in an unfavorable environment, such as low nutrients or high concentrations of salt. Endospores are perhaps the most resilient life forms on Earth (Madigan et al. 2009). They are resistant to extreme temperatures (an autoclave operating at 121 °C will kill endospores of most species, but exceptional endospores survive up to 150 °C), to most disinfectants, radiation, and drying. They can survive for thousands of years in this dormant state, since they have been found in Egyptian mummies and 8000-year-old lake sediments (Madigan et al. 2009). They may survive even longer, since it was reported that *Bacillus* species were revived from bees encased in 25–30 million-year-old amber (Cano and Borucki 1995). However, since phylogenetic analysis showed that the isolates from ancient amber were very similar to extant microbes, some caution is advised, and an independent confirmation of these results would be desirable.

Still older microbial isolates were reported from salt sediments and from other deep subterranean sources (bore cores, mines), which are of geological ages up to 290 million years (McGenity et al. 2000). Haloarchaea are believed to survive in fluid inclusions of halite (Gramain et al. 2011; Lowenstein et al. 2011; Fendrihan et al. 2012; Jaakkola et al. 2016). If any metabolic maintenance activities of these microorganisms, including potential repair of damaged DNA, are occurring, awaits further research.

The survival of endospores in adverse conditions, including the space environment (Nicholson et al. 2000), raises the possibility that bacterial endospores could travel to Mars on the surface of spacecraft and survive on or in Martian soil. This could seriously compromise future efforts to establish whether there is, or has been, life on Mars, as it would be difficult for researchers to know whether any endospores found originated from Earth or Mars. Similarly, survival of nonspore-forming microorganisms (*Streptococcus mitis*, permafrost bacteria, various subterranean bacteria, haloarchaea) under Martian or other space conditions should be explored in detail.

There are regulations already agreed upon by all space-faring nations, notably by COSPAR (the International Committee of Space Research), with the goal of preventing contamination by spacecrafts (see Moissl-Eichinger 2017, this volume); the rules are to be updated and revised, taking into account the potential of extreme survival of microorganisms.

1.6 Life from Meteorites?

Meteorites are messengers from the early solar system and can provide a wealth of important informations (Hofmann 2010). Several thousands of tons of meteorites are estimated to fall on Earth each year (see Lee et al. 2017, this volume). There have been reports of biogenic materials within meteorites. Intensely debated is the presence of fossilized nanobacteria in the ALH84001 meteorite from Mars and in the Tatahouine meteorite (Lee et al. 2017, this volume). The claims for the presence of living bacteria are generally refuted, since evidence for growth of Earth microbes, which invaded the meteorite after/during weathering, was found in many cases.

Extreme properties would be required of microorganisms which would travel through space in a meteorite, e.g., survival of the impact shock of a meteorite, resistance to or survival of prolonged desiccation, space radiation, hyperacceleration, hypergravity, and hypervelocity. Some of these features are known to be attributed to prokaryotes, as described in several chapters of this book.

1.7 Concluding Remarks

Organisms in all three domains of life have adapted to many terrestrial extremes. High-temperature, low-pH, and high-salinity environments represent probably very ancient sites, as may frozen environments. Extreme environments are not rare at all: most of the ocean is cold and deep, and a vast portion of the subsurface of Earth is hot. Extreme environments are characterized by one or several physicochemical parameters. Species diversity in extreme environments may be lower than in moderate environments; however, improved methods have led to the identification of numerous novel phylogenetic groups, including viruses, and these data can be expected to increase substantially in the future.

Extremophilic microorganisms survive exposure to space conditions, desiccation, high and low temperatures, and low-nutrient environments, perhaps for millions of years. They may therefore also survive space travel and be capable to live in extraterrestrial environments, which suggests that the old concept of panspermia is an acceptable idea for the distribution of life in the universe.

Acknowledgments Thanks go to former and present coworkers and students for their contributions. Support of my laboratory work by the Austrian Science Foundation (FWF), grants P16260 and P18256, and the Austrian Research Promotion Agency (FFG), grant ASAP 815141, is gratefully acknowledged.

References

Allen EE, Tyson GW, Whitaker RJ, Detter JC, Richardson PM, Banfield JF (2007) Genome dynamics in a natural archaeal population. Proc Natl Acad Sci U S A 104:1883–1888

Blöchl E, Rachel R, Burggraf S, Hafenbradl D, Jannasch HW, Stetter KO (1997) *Pyrolobus fumarii*, gen. and sp. nov., represents a novel group of archaea, extending the upper temperature limit for life to 113 °C. Extremophiles 1:14–21

Brochier-Armanet C, Boussau B, Gribaldo S, Forterre P (2008) Mesophilic Crenarchaeota: proposal for a third archaeal phylum, the Thaumarchaeota. Nat Rev Microbiol 6:245–252

Brock TD (1969) Microbial growth under extreme conditions. Symp Soc Gen Microbiol 19:15–42

Brock TD (1986) Introduction: An overview of the thermophiles. In: Brock TD (ed) Thermophiles. General, molecular and applied microbiology. Wiley, New York, pp. 1–16

Burgess EA, Wagner ID, Wiegel J (2007) Thermal environments and biodiversity. In: Gerday C, Glansdorff N (eds) Physiology and biochemistry of extremophiles. ASM Press, Washington, DC, pp. 13–29

Cano RJ, Borucki MK (1995) Revival and identification of bacterial spores in 25- to 40-million-year-old Dominican amber. Science 268:1060–1064

Ciaramella M, Napoli A, Rossi M (2004) Another extreme genome: how to live at pH 0. Trends Microbiol 13:49–51

Clarke A, Morris GJ, Fonseca F, Murray BJ, Acton E, Price HC (2013) A low temperature limit for life on Earth. PLoS One 8(6):e66207. doi:10.1371/journal.pone.0066207

Cowan DA (2004) The upper temperature for life – where do we draw the line? Trends Microbiol 12:58–60

D'Amico S, Collins T, Marx JC, Feller G, Gerday C (2006) Psychrophilic microorganisms: challenges for life. EMBO Rep 7:385–389

Daniel RM (1996) The upper limits of enzyme thermal stability. Enzyme Microb Technol 19:74–79

Dartnell LR, Hunter SJ, Lovell KV, Coates AJ, Ward JM (2010) Low-temperature ionizing radiation resistance of *Deinococcus radiodurans* and Antarctic Dry Valley bacteria. Astrobiology 10:717–732

Deguchi S, Shimoshige H, Tsudome M, Mukai S, Corkery RW, Ito S, Horikoshi K (2011) Microbial growth at hyperaccelerations up to 403,627 × *g*. Proc Natl Acad Sci U S A 108:7997–8002. doi:10.1073/pnas.1018027108

DeLong EF (1998) Everything in moderation: archaea as 'non-extremophiles'. Curr Opin Genet Dev 8:649–654

Denner EBM, McGenity TJ, Busse H-J, Wanner G, Grant WD, Stan-Lotter H (1994) *Halococcus salifodinae* sp.nov., an archaeal isolate from an Austrian salt mine. Int J Syst Bacteriol 44:774–780

Desbruyeres D, Laubier L (1986) Les Alvinellidae, une famille nouvelle d'annelides polychetes infeodees aux sources hydrothermales sous-marines: systematique, biologie et ecologie. Can J Zool 64:2227–2245

Edwards KJ, Bond PL, Gihring TM, Banfield JF (2000) An archaeal iron-oxidizing extreme acidophile important in acid mine drainage. Science 287:1796–1799

Enache M, Teodosiu G, Itoh T, Kamekura M, Stan-Lotter H (2017) Halophilic microorganisms from man-made and natural hypersaline environments: physiology, ecology and biotechnological potential. In: Stan-Lotter H, Fendrihan S (eds) Adaption of microbial life to environmental extremes. Springer International Publishing, Cham

Fendrihan S, Negoiţă TG (2017) Psychrophilic microorganisms as important source for biotechnological processes. In: Stan-Lotter H, Fendrihan S (eds) Adaption of microbial life to environmental extremes. Springer International Publishing, Cham

Fendrihan S, Berces A, Lammer H, Musso M, Ronto G, Polacsek TK, Holzinger A, Kolb C, Stan-Lotter H (2009) Investigating the effects of simulated Martian ultraviolet radiation on *Halococcus dombrowskii* and other extremely halophilic archaebacteria. Astrobiology 9: 104–112

Fendrihan S, Dornmayr-Pfaffenhuemer M, Gerbl FW, Holzinger A, Grösbacher M, Briza P, Erler A, Gruber C, Plätzer K, Stan-Lotter H (2012) Spherical particles of halophilic archaea correlate with exposure to low water activity – implications for microbial survival in fluid inclusions of ancient halite. Geobiology 10:424–433

Forterre P (2016) To be or not to be alive: How recent discoveries challenge the traditional definitions of viruses and life. Stud Hist Philos Biol Biomed Sci. 2016. pii: S1369–8486(16)30008–5. doi: 10.1016/j.shpsc.2016.02.013.

Forterre P, Prangishvili D (2009a) The origin of viruses. Res Microbiol 160:466–472

Forterre P, Prangishvili D (2009b) The great billion-year war between ribosome- and capsid-encoding organisms (cells and viruses) as the major source of evolutionary novelties. Natural Genetic Engineering and Natural Genome Editing. Ann NY Acad Sci 1178: 65–77

Gerday C, Glansdorff N (eds) (2007) Physiology and biochemistry of extremophiles. ASM Press, Washington, DC

Gherna RL (1994) Culture preservation. In: Gerhardt P, Murray RGE, Wood WA, Krieg NR (eds) Manual of methods for general microbiology. American Society for Microbiology, Washington, DC, pp. 278–292

Gilichinsky DA (2002) Permafrost. In: Bitton G (ed) Encyclopedia of environmental microbiology. Wiley, New York, pp. 2367–2385

Gilmour D (1990) Halotolerant and halophilic microorganisms. In: Edwards C (ed) Microbiology of extreme environments. Open University Press, Milton Keynes, pp. 147–177

Giovannoni S, Rappé M (2000) Evolution, diversity and molecular ecology of marine prokaryotes. In: Kirchman DL (ed) Microbial ecology of the oceans. Wiley, New York, pp. 47–84

Giovannoni SJ, Tripp HJ, Givan S, Podar M, Vergin KL, Baptista D, Bibbs L, Eads J, Richardson TH, Noordewier M, Rappé MS, Short JM, Carrington JC, Mathur EJ (2005) Genome streamlining in a cosmopolitan oceanic bacterium. Science 309:1242–1245

Gramain A, Chong Díaz GC, Demergasso C, Lowenstein TK, McGenity TJ (2011) Archaeal diversity along a subterranean salt core from the Salar Grande (Chile). Environ Microbiol 13: 2105–2121

Grant WD (2004a) Life at low water activity. Phil Trans R Soc Lond B 359:1249–1267

Grant WD (2004b) Introductory chapter: half a lifetime in soda lakes. In: Ventosa A (ed) Halophilic microorganisms. Springer, Berlin/Heidelberg, pp. 17–32

Grant WD, Gemmell RT, McGenity TJ (1998) Halobacteria: the evidence for longevity. Extremophiles 2:279–287

Grant WD, Kamekura M, McGenity TJ, Ventosa A (2001) Order I. Halobacteriales Grant and Larsen 1989b, 495VP (effective publication: Grant and Larsen 1989a, 2216). In: Boone DR, Castenholz RW, Garrity GM (eds) Bergey's manual of systematic bacteriology, vol 1, 2nd edn. Springer, Berlin/Heidelberg/New York, pp. 294–299

Grinspoon D (2003) Lonely planets. The natural philosophy of alien life. HarperCollins Publishers, New York, p. 139

Gruber C, Legat A, Pfaffenhuemer M, Radax C, Weidler G, Busse H-J, Stan-Lotter H (2004) *Halobacterium noricense* sp. nov., an archaeal isolate from a bore core of an alpine Permian salt deposit, classification of *Halobacterium* sp. NRC-1 as a strain of *H. salinarum* and emended description of *H. salinarum*. Extremophiles 8:431–439

Guidetti R, Jönsson KI (2002) Long-term anhydrobiotic survival in semi-terrestrial micrometazoans. J Zool 257:181–187

Heulin T, De Luca G, Barakat M, Gommeaux M, de Groot A, Blanchard L, Ortet P, Achouak W (2017) Bacterial adaptation to hot and dry deserts. In: Stan-Lotter H, Fendrihan S (eds) Adaption of microbial life to environmental extremes. Springer International Publishing, Cham

Hocking AD, Pitt JI (1999) *Xeromyces bisporus* Frazer. In: Robinson RK, Batt CA, Patel PD (eds) Encyclopaedia of food microbiology, vol 3. Academic, London, pp. 2329–2333

Hofmann BA (2010) Meteorites: messengers from the early solar system. Chimia (Aarau) 64:736–740

Horikawa DD, Sakashita T, Katagiri C, Watanabe M, Kikawada T, Nakahara Y, Hamada N, Wada S, Funayama T, Higashi S, Kobayashi Y, Okuda T, Kuwabara M (2006) Radiation tolerance in the tardigrade *Milnersium tardigradum*. Int J Radiat Biol 82:843–848

Horneck G, Bücker H, Reitz G (1994) Long-term survival of bacterial spores in space. Adv Space Res 14:41–45

Horneck G, Klaus DM, Mancinelli RL (2010) Space microbiology. Microbiol Mol Biol Rev 74:121–156

Hreggvidsson GO, Petursdottir SK, Stefansson SK, Björnsdottir SH, Fridjonsson OH (2017) Microbial speciation in the geothermal ecosystem. In: Stan-Lotter H, Fendrihan S (eds) Adaption of microbial life to environmental extremes. Springer International Publishing, Cham

Ito H, Watanabe H, Takeshia M, Iizuka H (1983) Isolation and identification of radiation-resistant cocci belonging to the genus *Deinococcus* from sewage sludges and animal feeds. Agric Biol Chem 47:1239–1247

Jaakkola ST, Ravantti JJ, Oksanen HM, Bamford DH (2016) Buried alive: microbes from ancient halite. Trends Microbiol 24:148–160. doi:10.1016/j.tim.2015.12.002

Jeanthon C, Prieur D (1990) Susceptibility to heavy metals and characterization of heterotrophic bacteria isolated from two hydrothermal vent polychaete annelids, *Alvinella pompejana* and *Alvinella caudata*. Appl Environ Microbiol 56:3308–3314

Jönsson KI (2007) Tardigrades as a potential model organism in space research. Astrobiology 7:757–766

Jönsson KI, Rabbow E, Schill RO, Harms-Ringdahl M, Rettberg P (2008) Tardigrades survive exposure to space in low Earth orbit. Curr Biol 18:R729–R731

Kashefi K, Lovley DR (2003) Extending the upper temperature limit for life. Science 301:934

Kelley DS, Karson JA, Früh-Green GL, Yoerger DR, Shank TM, Butterfield DA, Hayes JM, Schrenk MO, Olson EJ, Proskurowski G, Jakuba M, Bradley A, Larson B, Ludwig K, Glickson D, Buckman K, Bradley AS, Brazelton WJ, Roe K, Elend MJ, Delacour A, Bernasconi SM, Lilley MD, Baross JA, Summons RE, Sylva SP (2005) A serpentinite-hosted ecosystem: the Lost City hydrothermal field. Science 307:1428–1434

Krebs CJ (2008) The ecological world view. CSIRO Publishing, Collinwood, pp. 19–39

Laidler JR, Stedman KM (2010) Virus silicification under simulated hot spring conditions. Astrobiology 10:569–576

Laidler JR, Shugart JA, Cady SL, Bahjat KS, Stedman KM (2013) Reversible inactivation and desiccation tolerance of silicified viruses. J Virol 87:13927–13929. doi:10.1128/JVI.02825-13

Le Romancer M, Gaillard M, Geslin C, Prieur D (2007) Viruses in extreme environments. Rev Environ Sci Biotechnol 6:17–31

Lee N, Fritz J, Fries MD, Gil JF, Beck A, Pellinen-Wannberg A, Schmitz B, Steele A, Hoffmann BA (2017) The extreme biology of meteorites: Their role in understanding the origin and distribution of life on earth and in the universe. In: Stan-Lotter H, Fendrihan S (eds) Adaption of microbial life to environmental extremes. Springer International Publishing, Cham

Lévêque E, Janecek S, Haye B, Belarbi A (2000) Thermophilic archaeal amylolytic enzymes. Enzyme Microb Technol 26:3–14

Lowenstein TK, Brian A, Schubert BA, Timofeeff MN (2011) Microbial communities in fluid inclusions and long-term survival in halite. GSA Today 21:4–9. doi:10.1130/GSATG81A.1

Madigan MT, Martinko JM, Dunlap PV, Clark DP (2009) Chapter one: microorganisms and microbiology. In: Brock biology of microorganisms, 12th edn. Pearson Benjamin Cummings, San Francisco, pp. 1–24

Mapelli F, Barozzi A, Michoud G, Merlino G, Crotti E, Borin S, Daffonchio D (2017) An updated view of the microbial diversity in deep hypersaline anoxic basins. In: Stan-Lotter H, Fendrihan S (eds) Adaption of microbial life to environmental extremes. Springer International Publishing, Cham

Mastrapa RME, Glanzberg H, Head JN, Melosh HJ, Nicholson WL (2001) Survival of bacteria exposed to extreme acceleration: implications for panspermia. Earth Planet Sci Lett 189:1–8

McGenity TJ, Gemmell RT, Grant WD, Stan-Lotter H (2000) Origins of halophilic microorganisms in ancient salt deposits. Environ Microbiol 2:243–250

Mirete S, Morgante V, González-Pastor JE (2017) Acidophiles: diversity and mechanisms of adaptation to acidic environments. In: Stan-Lotter H, Fendrihan S (eds) Adaption of microbial life to environmental extremes. Springer International Publishing, Cham

Moissl-Eichinger C (2017) Extremophiles in spacecraft assembly cleanrooms. In: Stan-Lotter H, Fendrihan S (eds) Adaption of microbial life to environmental extremes. Springer International Publishing, Cham

Mykytczuk NC, Foote SJ, Omelon CR, Southam G, Greer CW, Whyte LG (2013) Bacterial growth at -15 °C; molecular insights from the permafrost bacterium *Planococcus halocryophilus* Or1. ISME J 7:1211–1226. doi:10.1038/ismej.2013.8

Nealson KH, Conrad PG (1999) Life: past, present and future. Phil Trans R Soc Lond B 354:1923–1939

Nicholson WL, Mukenata N, Horneck G, Melosh HJ, Setlow P (2000) Resistance of *Bacillus* endospores to extreme terrestrial and extraterrestrial environments. Microbiol Mol Biol Rev 64:548–563

Nogi Y, Kato C (1999) Taxonomic studies of extremely barophilic bacteria isolated from the Mariana Trench and description of *Moritella yayanosii* sp. nov., a new barophilic bacterial isolate. Extremophiles 3:71–77

Norton CF, Grant WD (1988) Survival of halobacteria within fluid inclusions in salt crystals. J Gen Microbiol 134:1365–1373

Oren A (2007) Biodiversity in highly saline environments. In: Gerday C, Glansdorff N (eds) Physiology and biochemistry of extremophiles. ASM Press, Washington, DC, pp. 223–231

Oren A, Rainey F (eds) (2006) Methods in microbiology, Extremophiles, vol 35. Elsevier, Oxford

Parikka KJ, Le Romancer M, Wauters N, Jacquet S (2016) Deciphering the virus-to-prokaryote ratio (VPR): insights into virus-host relationships in a variety of ecosystems. Biol Rev Camb Philos Soc. doi:10.1111/brv.12271

Parkes RJ, Cragg B, Roussel E, Webster G, Weightman A, Sass H (2014) A review of prokaryotic populations and processes in sub-seafloor sediments, including biosphere: geosphere interactions. Mar Geol 352:409–425

Pearce D (2017) Extremophiles in Antarctica; life at low temperatures. In: Stan-Lotter H, Fendrihan S (eds) Adaption of microbial life to environmental extremes. Springer International Publishing, Cham

Pedersen K (2000) MiniReview. Exploration of deep intraterrestrial microbial life: current perspectives. FEMS Microbiol Lett 185:9–16

Rampelotto PH (ed) (2016) Biotechnology of extremophiles. Advances and challenges. Springer International Publishing, Cham

Rivkina EM, Friedmann EI, McKay CP, Gilichinsky DA (2000) Metabolic activity of permafrost bacteria below the freezing point. Appl Environ Microbiol 66:3230–3233

Rothschild LJ, Mancinelli RL (2001) Life in extreme environments. Nature 409:1092–1101

Schleper C, Puehler G, Holz I, Gambacorta A, Janekovic D, Santarius U, Klenk HP, Zillig W (1995) *Picrophilus* gen. nov., fam. nov.: a novel aerobic, heterotrophic, thermoacidophilic genus and family comprising archaea capable of growth around pH 0. J Bacteriol 177:7050–7059

Schubert BA, Lowenstein TK, Timofeeff MN, Parker MA (2010) Halophilic archaea cultured from ancient halite, Death Valley, California. Environ Microbiol 12:440–454

Schuerger AC, Nicholson WL (2016) Twenty-three species of hypobarophilic bacteria recovered from diverse ecosystems exhibit growth under simulated Martian conditions at 0.7 kPa. Astrobiology 16:335–347

Schuerger AC, Ulrich R, Berry BJ, Nicholson WL (2013) Growth of *Serratia liquefaciens* under 7 mbar, 0 °C, and CO_2-enriched anoxic atmospheres. Astrobiology 13:115–131

Seckbach J (ed) (2000) Journey to diverse microbial worlds. Adaptation to exotic environments, Cellular origins and life in extreme habitats, vol 2. Kluwer, Dordrecht

Seckbach J, Chapman DJ (eds) (2010) Red algae in the genomic age, Cellular origin and life in extreme habitats and astrobiology, vol 13. Springer, Heidelberg

Seckbach J, Grube M (eds) (2010) Symbioses and stress, Cellular origin and life in extreme habitats and astrobiology, vol 17. Springer, Heidelberg

Seckbach J, Walsh M (eds) (2009) From fossils to astrobiology, Cellular origin and life in extreme habitats and astrobiology, vol 12. Springer, Heidelberg

Seckbach J, Oren A, Stan-Lotter H (eds) (2013) Polyextremophiles. Life under multiple form of stress, Cellular origins, life in extreme habitats and astrobiology. Springer, Dordrecht/Heidelberg/New York/London

Simonato F, Campanaro S, Lauro FM, Vezzi A, D'Angelo M, Vitulo N, Valle G, Bartlett DH (2006) Review. Piezophilic adaptation: a genomic point of view. J Biotechnol 126:11–25

Soina VS, Mulyukin AL, Demkina EV, Vorobyova EA, El-Registan GI (2004) The structure of resting bacterial populations in soil and subsoil permafrost. Astrobiology 4:345–358

Stan-Lotter H, McGenity TJ, Legat A, Denner EBM, Glaser K, Stetter KO, Wanner G (1999) Very similar strains of *Halococcus salifodinae* are found in geographically separated Permo-Triassic salt deposits. Microbiology 145:3565–3574

Stan-Lotter H, Pfaffenhuemer M, Legat A, Busse H-J, Radax C, Gruber C (2002) *Halococcus dombrowskii* sp. nov., an Archaeal isolate from a Permian alpine salt deposit. Int J Syst Evol Microbiol 52:1807–1814

Stetter KO (1982) Ultrathin mycelia-forming organisms from submarine volcanic areas having an optimum growth temperature of 105 °C. Nature 300:258–260

Stetter KO (2002) Hyperthermophilic microorganisms. In: Horneck G, Baumstark-Khan C (eds) Astrobiology. The quest for the conditions of life. Springer, Berlin/New York, pp. 169–184

Stevenson A, Burkhardt J, Cockell CS, Cray JA, Dijksterhuis J, Fox-Powell M, Kee TP, Kminek G, McGenity TJ, Timmis KN, Timson DJ, Voytek MA, Westall F, Yakimov MM, Hallsworth JE (2015) Multiplication of microbes below 0.690 water activity: implications for terrestrial and extraterrestrial life. Environ Microbiol 17:257–277

Takai K, Moser DP, Onstott TC, Spoelstra N, Pfiffner SM, Dohnalkova A, Fredrickson JK (2001) *Alkaliphilus transvaalensis* gen. nov., sp. nov., an extremely alkaliphilic bacterium isolated from a deep South African gold mine. Int J Syst Evol Microbiol 51:1245–1256

Takai K, Nakamura K, Toki T, Tsunogai U, Miyazaki M, Miyazaki J, Hirayama H, Nakagawa S, Nunoura T, Horikoshi K (2008) Cell proliferation at 122 °C and isotopically heavy CH_4 production by a hyperthermophilic methanogen under high-pressure cultivation. Proc Natl Acad Sci U S A 105:10949–10954

Taylor PW (2015) Impact of space flight on bacterial virulence and antibiotic susceptibility. Infect Drug Resist 8:249–262. doi:10.2147/IDR.S67275

Tsujimoto M, Imura S, Kanda H (2016) Recovery and reproduction of an Antarctic tardigrade retrieved from a moss sample frozen for over 30 years. Cryobiology 72:78–81

Weber APM, Horst RJ, Barbier GG, Oesterhelt C (2007) Metabolism and metabolomics of eukaryotes living under extreme conditions. Int Rev Cytol 256:1–34

Westall F (2005) Early life on Earth and analogies to Mars. In: Tokano T (ed) Water on Mars and life, Advances in Astrobiology and Biogeophysics. Springer, Berlin/Heidelberg, pp. 45–64

Whitman WB, Coleman DC, Wiebe WJ (1998) Prokaryotes: the unseen majority. Proc Natl Acad Sci U S A 95:6578–6583

Wilson JW, Ott CM, Höner zu Bentrup K, Ramamurthy R, Quick L, Porwollik S, Cheng P, McClelland M, Tsaprailis G, Radabaugh T, Hunt A, Fernandez D, Richter E, Shah M, Kilcoyne M, Joshi L, Nelman-Gonzalez M, Hing S, Parra M, Dumars P, Norwood K, Bober R, Devich J, Ruggles A, Goulart C, Rupert M, Stodieck L, Stafford P, Catella L, Schurr MJ, Buchanan K, Morici L, McCracken J, Allen P, Baker-Coleman C, Hammond T, Vogel J, Nelson R, Pierson DL, Stefanyshyn-Piper HM, Nickerson CA (2007) Space flight alters bacterial gene expression and virulence and reveals a role for global regulator Hfq. Proc Natl Acad Sci U S A 104: 16299–16304

Yakimov MM, La Cono V, La Spada G, Bortoluzzi G, Messina E, Smedile F, Arcadi E, Borghini M, Ferrer M, Schmitt-Kopplin P, Hertkorn N, Cray JA, Hallsworth JE, Golyshin PN, Giuliano L (2015) Microbial community of the deep-sea brine Lake Kryos seawater–brine interface is active below the chaotropicity limit of life as revealed by recovery of mRNA. Environ Microbiol 17:364–382

Zeng X, Birrien JL, Fouquet Y, Cherkashov G, Jebbar M, Querellou J, Oger P, Cambon-Bonavita MA, Xiao X, Prieur D (2009) *Pyrococcus* CH1, an obligate piezophilic hyperthermophile: extending the upper pressure-temperature limits for life. ISME J 3:873–876

Zharkov MA (1981) History of paleozoic salt accumulation. Springer, Berlin

An Updated View of the Microbial Diversity in Deep Hypersaline Anoxic Basins

2

Francesca Mapelli, Alan Barozzi, Grégoire Michoud, Giuseppe Merlino, Elena Crotti, Sara Borin, and Daniele Daffonchio

Contents

F. Mapelli • E. Crotti • S. Borin
Department of Food, Environmental and Nutritional Sciences (DeFENS), University of Milan, 20133 Milan, Italy

A. Barozzi • G. Michoud • G. Merlino
King Abdullah University of Science and Technology (KAUST), Biological and Environmental Sciences and Engineering Division (BESE), Thuwal 23955-6900, Kingdom of Saudi Arabia

D. Daffonchio (✉)
Department of Food, Environmental and Nutritional Sciences (DeFENS), University of Milan, 20133 Milan, Italy

King Abdullah University of Science and Technology (KAUST), Biological and Environmental Sciences and Engineering Division (BESE), Thuwal 23955-6900, Kingdom of Saudi Arabia
e-mail: daniele.daffonchio@kaust.edu.sa; daniele.daffonchio@unimi.i

H. Stan-Lotter, S. Fendrihan (eds.), *Adaption of Microbial Life to Environmental Extremes*, DOI 10.1007/978-3-319-48327-6_2

2.1 Introduction

Deep hypersaline anoxic basins (DHABs) represent one of the most fascinating and extreme habitats on Earth. The first DHABs were discovered on the seafloor of different oceanographic areas starting from the end of the 1970s of the last century, and currently the discovery of new DHABs is still reported by multidisciplinary researchers' teams (La Cono et al. 2011; Yakimov et al. 2013).

Microbiologists demonstrated the occurrence of life in DHABs discovered around the world. The first studies describing the microbial populations inhabiting DHABs focused on Bacteria and Archaea, and most of our knowledge about life in DHABs still concerns prokaryotes (Antunes et al. 2008; Borin et al. 2009; Daffonchio et al. 2006; Dickins and Van Vleet 1992). More recently increasing efforts were devoted to the characterization of microeukaryotic communities (Bernhard et al. 2014; Edgcomb et al. 2011; Stoeck et al. 2014). DHABs are ecosystems where microorganisms face multiple extreme conditions spanning from anoxia to high pressure, salinity, and the absence of light (van der Wielen et al. 2005). Recently, active microbial communities were described in a newly discovered DHAB in the Mediterranean Sea named Kryos, demonstrating that life can flourish under environmental conditions previously considered unsuitable and shedding a new light on the possibility that life, as we know it, occurs elsewhere in the solar system (Yakimov et al. 2015).

An interesting aspect of DHABs is the occurrence of a stable interface that separates the overlying seawater and the brine body, due to the different densities of these two layers. Such interface generally hosts, in few meters, an oxy-picno-chemocline, representing a natural laboratory that allows microbial ecologists to study the ability of microbial communities to adapt along a gradient of changing environmental conditions (Borin et al. 2009; Daffonchio et al. 2006; Ngugi et al. 2015). The enhanced power of "omics" approaches, together with the increasing number of studies exploiting them in the last years (Bougouffa et al. 2013; Ferrer et al. 2012), generated a large amount of information about DHAB microbial communities, in turn providing better insight on the metabolic and ecologically relevant processes performed in such extreme ecosystems.

A crucial aspect to unravel microbial processes in DHABs, as well as in other deep-sea ecosystems, is the capacity to analyze samples that have to be maintained at conditions as much as possible similar to those occurring at the original site. Usually, deep seawater, interface, and brines are sampled from DHABs using Niskin bottles located on a rosette sampler equipped with conductivity-temperature-depth (CTD) and pressure sensors (Fig. 2.1). As soon as the rosette sampler is on board of the oceanographic ship, scientists collect the samples to immediately process them, which in case of DNA-/RNA-based analyses means to perform a filtration. However, this widely used approach does not permit to maintain the samples under the in situ conditions (especially concerning pressure). To overcome this inconvenience, a team of scientists recently designed a Microbial Sampler-Submersible Incubation Device (MS-SID) that allows directly in situ (i) to filter water samples and (ii) to add a chemical preservative for the storage of samples before nucleic acid

Fig. 2.1 The picture shows the equipment generally used for sampling of brine and interface from DHABs during oceanographic cruises. Niskin bottles of 10 L capacity are located on a rosette sampler equipped with conductivity-temperature-depth (CTD) and pressure sensors

extraction (Edgcomb et al. 2014). The researchers compared the data obtained by metatranscriptomics on cells collected from seawater at about 2000 meters depth using both the traditional Niskin method and the MS-SID sampler. The data showed the influence of the collection method on certain microbial taxa and gene expression patterns, highlighting the importance of sampling issues while studying deep-sea ecosystems, including DHABs.

The occurrence in DHABs of physicochemical parameters at the limits for life's establishment implies the existence of a strong selective pressure on the exposed microbes. As a consequence, the microbial communities inhabiting DHABs are very interesting also for biotechnologists. Indeed, their putative applications range from the ability to produce an array of products of interest for the industry (e.g., compatible solutes, extremozymes) to the possible use of specific strains as biocatalysts operating under harsh conditions that conventional strains are unable to cope with.

In this chapter, we present an overview of the microbial diversity associated to DHABs located in different geographical areas around the world (Fig. 2.2), with a focus on those placed in the Eastern Mediterranean Sea and the Red Sea. Extended information on the geochemical setting of the investigated DHABs are also reported and linked to the available biodiversity studies.

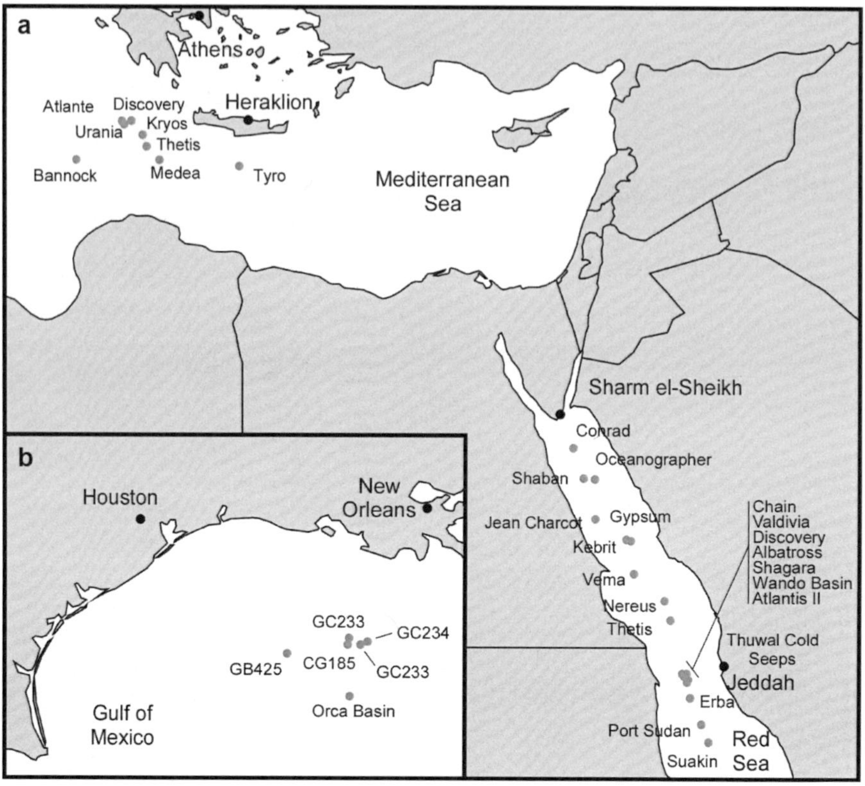

Fig. 2.2 Localization of DHABs in the (**a**) Eastern Mediterranean Sea and Red Sea, (**b**) Gulf of Mexico. The position of Thuwal Cold Seep is also shown in panel **a**

2.2 DHABs of the Red Sea

2.2.1 Localization and Geochemical Features of DHABs

With about 25 brine bodies situated between latitudes of 19° to 27° N, the Red Sea is the part of the world that presents the highest number of DHABs (Fig. 2.2a). They are distributed in the range of depths between 1200 and 2900 m and may present hydrothermal flows (Guan et al. 2015). The high number of DHABs in the Red Sea is due to an active rift process that started 25 million years ago between the Arabian and African tectonic plates (Schardt 2015). During this time, evaporitic deposits accumulated during the Late Miocene underwent redissolution along the central Red Sea. In some points, this dissolution increased the deep-water density resulting in distinct stratification depending on the reached salinity. Furthermore, hydrothermal flows entrapped in this dense layer started to generate highly mineralized brines (Gurvich 2006). The geochemical composition of these flows is directly related to the general physicochemical composition of each basin (Gurvich 2006). One of the most studied is the Atlantis II Deep, located at a latitude of 21°N and characterized by a temperature of 68 °C and very anoxic conditions in the brine body (Ngugi et al. 2015) (Fig. 2.2a). This DHAB is also characterized by a high concentration of metalliferous sediments, enriched especially in copper, zinc, iron, and heavy metals (Anschutz et al. 2000). On the contrary, the Kebrit basin, situated approximately 400 km North at a latitude of 24°N, has a maximum temperature of 23.4 °C and is characterized by a relatively high concentration of oxygen (2.3 μM) compared to most of the other DHABs studied in the Red Sea and high concentrations of hydrogen sulfide (Hartmann et al. 1998; Mapelli et al. 2012; Ngugi et al. 2015). The differences in accumulation of metals and nutrients, especially in the seawater-brine interfaces of the different DHABs (Larock et al. 1979; Eder et al. 2002; Guan et al. 2015), support the presence of different ecological niches exploited by highly diverse microorganisms with specialized metabolisms in relation to the specific geochemical composition of the basin.

2.2.2 Microbial Diversity of Red Sea DHABs

Nearly all the DHABs of the Red Sea were discovered 50 years ago with the notable exception of the Thuwal Cold Seep which was discovered in 2010. This is located not in the central Red Sea but much closer to the coast, near the city of Thuwal (Kingdom of Saudi Arabia) (Batang et al. 2012; Fig. 2.2a, Sect. 2.3).

Despite a large effort in their geochemical characterization, a relatively low number of studies focused on the microbial diversity of DHABs, using two different approaches. Firstly, researchers concentrated on the cultivation of microorganisms usually found in the brine-seawater interface, an environment less harsh than the brine where the majority of phylogenetic and metabolic biodiversity occurs (Fiala et al. 1990; Huber and Stetter 2001; Antunes et al. 2011). Secondly, with the advance of "omics" techniques, especially the advent of 16S rRNA high-throughput sequencing, studies started to evaluate the global diversity of DHAB microbiota from different perspectives, e.g., dissecting a basin along depth (e.g., brine body, brine-seawater interface, deep seawater) or comparing geochemically different DHABs (Siam et al. 2012; Guan et al. 2015; Ngugi et al. 2015). The majority of these studies concentrated on few brines pools, including Atlantis II Deep, Discovery, Kebrit, Nereus, and Erba.

Atlantis II Deep possesses different transition layers (interfaces) separating four convective layers where the concentration of salt and minerals is constant. As a consequence, the microbial diversity is very diverse in these convective/transition zones and the normalized ratio of Bacteria/Archaea varies from 0.28 in the brine-seawater interface to around 81, 1.67, and 6.4 in the first, second, and the deepest convective layer, respectively (Guan et al. 2015). Interestingly, alpha diversity (measure of the species richness within a community) was higher in the lower layer, where the temperature (68 °C) and salinity (24.8 %) are more extreme than in any other part of the different convective/transition layers (Bougouffa et al. 2013). This could be explained by the extreme conditions that could prevent the dominance of specific taxa, thus increasing the proportion of the "rare biosphere" (Bougouffa et al. 2013). Another explanation could be that the main organic matter is trapped in the other convective/transition zones hence increasing the proportion of dormancy forms like spores that cannot grow in such conditions (Qian et al. 2011).

As it was also reported for the Mediterranean brine pools (see Sect. 2.3.3), the majority of the microbial communities in the Red Sea DHABs are dominated by *Bacteria* rather than *Archaea*, with the exception of the Kebrit and Atlantis II Deep brine-seawater interface (Bougouffa et al. 2013; Guan et al. 2015). In the interface of these two basins, the concentration of dissolved oxygen which is two to three times higher than in other interfaces could allow the proliferation of ammonia-oxidizing Archaea (AOA; Ngugi et al. 2015). 16S rRNA gene pyrosequencing performed on the samples collected from the interfaces in the different basins of the Red Sea showed that Archaea are mainly dominated by the phylum *Thaumarchaeota* that comprises from 64 % (Erba Deep) to 99 % (Atlantis II and Discovery Deeps) of the total archaeal community (Ngugi et al. 2015). Among the other Archaea present in the brine bodies, members of the candidate division Mediterranean Sea Brine Lakes 1 (MSBL1) are particularly interesting since this uncultured archaeal lineage is present in nearly all the DHABs that were studied in the Eastern Mediterranean Sea, Red Sea, and Gulf of Mexico (Antunes et al. 2011; Yakimov et al. 2013), originally considered responsible of methane production in the brines. Although members of the MSBL1 are found in environments where the concentration of methane is high, a recent study indicated that they are not methanogens but probably sugar-fermenting organisms capable of autotrophic growth (Mwirichia et al. 2016). Indeed, the fact that MSBL1 seems to be metabolically versatile and able to grow either heterotrophically or autotrophically by CO_2 fixation could explain why they are particularly abundant in DHABs and could give insights in new culture methodology (Mwirichia et al. 2016).

The bacterial communities in the brines of the five characterized Red Sea DHABs are mainly represented by the phyla *Proteobacteria*, *Bacteroidetes*, *Deferribacteres* and *Chloroflexi*. Interestingly, the class *Deltaproteobacteria* is common in Kebrit, Erba, and Nereus basins, while it is less abundant in the hot brines of Atlantis II and Discovery. These two pools are mostly inhabited by *Nitrospirae*-like bacteria in the interface and by *Gammaproteobacteria* in the lower convective layers (Bougouffa et al. 2013; Guan et al. 2015).

Different species of microorganisms were isolated from the Red Sea DHABs, such as *Flexistipes sinusarabici*, a Gram-negative bacterium isolated from Atlantis II Deep. *F. sinusarabici* is a moderate halophile and thermophile (growth at 3–18 % NaCl and 30–53 °C), belonging to the phylum *Deferribacteres* and showed a fermentative metabolism (Fiala et al. 1990). Subsequent isolation experiments from the Kebrit Deep resulted in the isolation of further two anaerobic fermenting and halophile bacteria, affiliated to the genus *Halanaerobium* (Eder et al. 2001). Other isolates were obtained from the Shaban Deep, like *Salinisphaera shabanensis* and *Marinobacter salsuginis* that shared common metabolisms such as the capacity to assimilate aliphatic hydrocarbons (Antunes et al. 2008). *S. shabanensis* belongs to the order *Salinisphaerales*, a deep branching lineage of *Gammaproteobacteria*.

This bacterium is capable of growing on both a wide range of salinity and temperature (1–28 % NaCl and 5–42 °C) with a large substrate spectrum such as n-alkanes (dodecane) or in the presence of thiosulfate as electron donor (Antunes et al. 2003). Similarly, *M. salsuginis*, also belonging to the *Gammaproteobacteria* class, is a heterotrophic, facultative anaerobic bacterium capable of fermentation and nitrate reduction (Antunes et al. 2008). Cultivation-based studies in DHABs showed nevertheless little agreement with the results of cultivation-independent approaches. Indeed, many isolated strains were not found during 16S rRNA molecular surveys, and conversely most taxonomic groups sequenced from DHAB metagenome cannot be cultivated, confirming the well-known cultivation paradox (Rappé and Giovannoni 2003).

2.2.3 Thuwal Cold Seeps

How the deep-sea brines were generated by specific tectonic activities is well understood. However, some other extreme deep-sea environments can be found under different geodynamic situations. One interesting example is a cold brine seep system on the Saudi Arabian continental margin of the central Red Sea, discovered in 2010 and named Thuwal Seep (Batang et al. 2012) (Fig. 2.2a).

2.2.3.1 Localization, Origin, and Geochemical Features

The Thuwal Seep system lies close to Saudi Arabian shore (about 20 km). It consists of two cold seeps, Seep I (22°17.3′ N, 38°53.8′ E) and Seep II (22°16.9′ N, 38°53.9′ E), located about 830 m apart along the base of a wall rock. The two seeps discharge cold hypersaline fluids and originate, on an underlying slight depression (within 840–850 m depths), a small brine pool. The hypersaline waters released by the two seeps, likely originate from evaporitic deposits of submarine geological formations that flowed from the faulting system at the base of the rocky scarp (Cao et al. 2015). The brine pool is characterized by a surface of about 2.2 km^2 area, with a temperature of 21.7 °C, comparable to the overlying deep water. With these features, the Thuwal Seep is the coldest and least saline brine system identified so far in the Red Sea. The recorded depth is about 1.0 m in most places of the basin (Cao et al. 2015). The conditions in the brine were reported to be almost anoxic with concentration of oxygen of 0.5 % compared to 25 % of those typical of seawater (Cao et al. 2015), while intermediate levels were measured in the seeping water from Seep II (17.7 %; Lee et al. 2014). The salinity of the brine was reported to be 96 psu in 2011 and 77 psu in 2013, while that of the seeping water at Seep II was reported to be 125 psu in 2011 and 100 psu in 2013 (Yang et al. 2015; Cao et al. 2015). The chemical composition of the brine water showed the enrichment in the brine of metals and metalloids like manganese, iron and arsenic, and hydrogen sulfide at more than tenfold concentrations relative to deep seawater (Batang et al. 2012; Wang et al. 2014). Compared to the water column and the brine body, the seeping water was characterized by higher temperature (28.7 °C) and high level of manganese (0.79 mg/l), in contrast with the previously reported lower levels of Mn^{2+} in the brine (0.14 ppm; Batang et al. 2012).

This could suggest the active presence in the seeping water flowing in the brine body of manganese-reducing microorganisms able to thrive in the anoxic condition of the brine using Mn^{2+} as electron acceptor. Other interesting features of the seeping water were the total absence of iron and high concentrations of sulfate (Lee et al. 2014). Similarly to manganese, the high level of sulfate detected could correlate with the high concentration of hydrogen sulfide measured in the brine water (> 200 μM; Wang et al. 2014), suggesting strong sulfate-reducing activity.

2.2.3.2 Microbial Diversity Associated to the Thuwal Seeps

Despite the apparently prohibitive conditions, the Thuwal Seeps constitutes a hot spot for life compared to the typical oligotrophic conditions of the Red Sea. A remarkable feature of the basin is the extensive presence of white microbial mats on the margins of the brine pool as well as along the small streams of fluids coming from the seeps (Batang et al. 2012), an evident proof of microbial life thriving in this extreme environment.

Three studies were focused on the description of the dynamics through time and space of the microbial assemblages of this ecosystem. Yang et al. (2015) compared the microbial composition of the normal seawater, the brine pool, and the seeping water in 2011 and 2013, using 16S rRNA pyrosequencing and correlating these data with the chemical characterization of the water. Significant differences in both the structure of the microbial communities and the environmental milieu were detected between 2011 and 2013. Lower abundance of (i) *Halobacteriales*, (ii) heterotrophic Archaea typical of saline-rich environments, and (iii) *Desulfobacterales*, which comprises known anaerobic sulfate-reducing bacterium, was reported in the brine pool samples collected in 2013 compared to those collected in 2011, a data consistent with the observed decrease in the salinity and sulfate of the brine water between the 2 years. The recorded weakening of the seeping water (reduced salinity and sulfate concentration) suggested that the variation of the composition of the fluids coming from the seeps was the main driving force influencing the changes of the microbial community.

Cao et al. (2015) described the microbial community changes along the environmental gradient of the seep system, collecting samples from the seeping water at Seep II, the microbial mat along the seeping stream, and the sediments at the bottom of the brine pool in comparison with deep-sea sediments and seawater collected close to the pool margins. As expected, substantial differences of proportions of microorganisms were detected among the different habitats down to the genus level. For instance, uncultured *Thaumarchaeota* dominated the deep seawater and the seeping water (62.5 % and 19.27 %, respectively), while they were less abundant in the marine sediments and on the microbial mat (0.34 % and 1.7 %, respectively). Ammonia-oxidizing Archaea (AOA) varied significantly along the environmental gradient. Interestingly, low abundance of AOA was detected in the brine, even though a high concentration of ammonia was measured, which was consistent with the absence of oxygen in the brine body (Cao et al. 2015).

The microbial eukaryotic community was investigated in this DHAB by Wang et al. (2014) with a similar sampling setup adopted by Cao and coauthors (Cao et al. 2015).

Evidences of the presence of potential novel species of annelids and nematodes possibly thriving on the biomass of the microbial mat were found, as well as of novel *Euglenozoa* and *Alveolata* species in the brine-seawater interface. Fungi, as previously described, mainly populated the oxic marine sediments. These organisms were also detected in brine pool sediments, closely related to species identified in similar environments in other locations, suggesting a wider distribution in aquatic environments than previously imagined.

2.3 DHABs of the Mediterranean Sea

Till few years ago, few DHABs were known in the Eastern Mediterranean Sea. This area has been widely explored leading to the discovery of three new DHABs since 2011. The studies performed on the three novel basins (i.e., Thetis, Medee, and Kryos) contributed to extend our understanding of such unique ecosystems and deepened the knowledge on the ability of microbial communities to adapt to extreme conditions previously considered inappropriate for life establishment.

2.3.1 Localization and Origin of DHABs

DHABs in the Mediterranean Sea are located at depths higher than 3000 meters in the Eastern basin, along the Mediterranean Ridge that is subjected to tectonic collision between the African, Eurasian, and Anatolian plates (Fig. 2.2a). The origin of Mediterranean DHABs was determined by the dissolution, followed to tectonic events, of salt evaporites. The originated brines filled in local depressions and anoxia developed in these basins (Wallmann et al. 1997). Moreover, compositional changes of brines occurred during time, due to both biochemical and biological processes.

Tyro and Bannock were the first DHABs discovered in the Eastern Mediterranean Sea (Cita et al. 1985; Jongsma et al. 1983), followed by Urania, Discovery, and L'Atalante (Medriff Consortium 1995) and, more recently, by Medee, Thethis, and Kryos (La Cono et al. 2011; Yakimov et al. 2013, 2015).

2.3.2 Geochemical Features of DHABs

As typically observed in all DHABs around the world, brines are well separated from the overlaying seawater. Such separation is due to the different densities of the two water masses that are interfaced by a chemo-oxy-halocline. The key aspect of DHABs chemistry is represented by the high salinity values of the brines. Typically, along the interface that is only few meters thick, salinity varies from seawater values to those of the hypersaline brines, 7–10 times higher. Moreover, seawater-brine interfaces host a gradient of oxygen, which falls below the detection limit in the brines, and of several reduced and oxidized chemical species. While salinity variation is linear with depth, the shifts of ion (e.g., nitrate, ammonium, sulfate) concentration deviate from linearity, suggesting a role of microbial populations in their production/consumption (Borin et al. 2009, 2013; Daffonchio et al. 2006).

In the Eastern Mediterranean, most of the DHABs' brines have thalassohaline composition, although Discovery and Kryos show athalassohaline brines enriched of magnesium. Discovery basin is filled by a near-saturated $MgCl_2$ brine, which concentration (around 5 M) is the highest reported in marine environments (van der Wielen et al. 2005). In this basin, the high magnesium concentration determines a very low water activity value and extremely high chaotropicity, considered unsuitable for life (Hallsworth et al. 2007). The recent characterization of brine in the Kryos basin showed that it is similar to that of Discovery, especially in terms of magnesium concentration (4.38 M), although its sodium and sulfate concentrations are higher and apparently involved in compensating the effect of the high $MgCl_2$ values, making this environmental niche habitable by extremophilic microorganisms (Yakimov et al. 2015). Comparative studies showed that brines of different DHABs can be distinguished according to their chemical composition, highlighting that each DHAB hosts brines with specific chemical features involved in the selection of adapted microbial communities (van der Wielen et al. 2005). However, similarities can be observed between different brines, as in the case of the recently characterized Thetis brine that is similar to those of Bannock and L'Atalante basins (La Cono et al. 2011). The physical separation of the body brine from the rest of the deep-water column, for thousands of years, makes DHAB ecosystems particularly intriguing for microbial diversity and adaptation studies, as discussed in the following Sect. 2.3.3.

2.3.3 Microbial Communities of the Newly Discovered DHABs in the Eastern Mediterranean Sea

Most of the microbial surveys performed in Eastern Mediterranean DHABs focused on seawater-brine interface (Borin et al. 2009; Daffonchio et al. 2006; Yakimov et al. 2007). Such water transition zones accumulate nutrients and microbial cells due to the different density from the deep oxic water column, representing a hot spot in terms of microbial abundance and diversity. However, active microbial communities have been reported also in brines (La Cono et al. 2011; Yakimov et al. 2015) and in surface sediments below the halocline of Urania, L'Atalante, and Discovery basins (Kormas et al. 2015; Sorokin et al. 2016). Clone libraries of 16S rRNA gene from the upper and lower interface and from the brine of the Thetis basin described the structure of bacterial and archaeal communities. In the interface samples, *Proteobacteria* were the more abundant taxonomic group within *Bacteria*, mainly represented by *Epsilonproteobacteria* (La Cono et al. 2011), although several other phyla were detected along Thetis interface by metagenomic study (Ferrer et al. 2012). *Deltaproteobacteria* and the division KB (Kebrit Deep)-1 were prevalent in the Thetis brine; *Deltaproteobacteria* adapted to high salinity values of Thetis brine were similar to those previously retrieved from the brines of other DHABs (van der Wielen et al. 2005), and a possible role in sulfate reduction has been hypothesized for them, according to the brine geochemistry (La Cono et al. 2011). Bacteria and Archaea showed similar abundance in Thetis brine (Ferrer et al. 2012). Also Archaea composition shifted across the halocline, showing differential distribution of the populations between upper and lower interface and brine. The latter was mainly composed by the candidate divisions MSBL (Mediterranean Sea Brine Lake)-1 and HC (Halophilic Cluster)-1. MSBL-1 and HC-1 were detected also in the lower interface (salinity value: 25–28 %), but at different relative abundance, while they were absent in the upper interface, at lower salinity values of 4.8–11 %. The archaeal community in the upper interface was represented by *Halobacteriaceae* and *Thaumarchaeota* (La Cono et al. 2011). As previously reported for Urania basin (Borin et al. 2009), sulfur cycling was the main metabolic process shaping the structure of prokaryotic communities also in Thetis basin (La Cono et al. 2011), and its importance was further corroborated by metagenomic-based survey (Ferrer et al. 2012). Besides sulfate reduction, also methanogenesis was a key process in Thetis ecosystem (La Cono et al. 2011), again showing analogies with the Urania DHAB. Genetic signatures of anaerobic methane oxidation were also reported (La Cono et al. 2011), as previously retrieved in the deepest layers of the Bannock basin (Daffonchio et al. 2006). Moreover, anaerobic ammonium oxidation (anammox), a metabolic process discovered only in 1999 (Strous et al. 1999), was indicated as possibly relevant for the productivity of Urania interface (Borin et al. 2009). Indeed, anammox activity was measured at Bannock and L'Atalante basins, where putative new branches of anammox populations were retrieved by fluorescence in situ hybridization (FISH) and PCR assays using specific primer sets for anammox *16S rRNA* and hydrazine synthase genes (Borin et al. 2013). Despite the different salinity, thalassohaline vs athalassohaline, the structure of archaeal community along the interface and brine of Kryos basin was similar to what observed in Thetis. The saltiest layers of the chemocline, approaching Kryos brine, were dominated by MBSL-1 and HC-1 candidate divisions, while the upper interface displayed a higher abundance of Marine group I *Thaumarchaeota* (Yakimov et al. 2013). However, Bacteria dominated Kryos interface (Yakimov et al. 2013). The highest diversity within bacterial communities was measured at interface layers down to those corresponding to the "chaotropicity window for life" (CHW), while the deeper layer corresponding to the water activity (A_w) limit for life (AWW) had the lowest bacterial diversity, mainly represented by *Deltaproteobacteria* and members of KB-1 division. Both the upper interface and the CHW layers were dominated by *Epsilonproteobacteria* and sulfate-reducing bacteria belonging to *Deltaproteobacteria* different from those colonizing the deeper layers of Kryos interface (Yakimov et al. 2013). Chemoautotrophic *Epsilonproteobacteria* are generally important members of bacterial communities in Eastern Mediterranean DHABs and include taxa considered highly adapted to dynamic conditions, putatively contributing to the sulfur cycle in the basin. However, molecular studies did not detect them neither in Bannock nor in Medee DHABs, both showing a wider interface compared to the rest of DHABs (Daffonchio et al. 2006; Yakimov et al. 2013). Bacteria represent a percentage of the prokaryotic assemblages varying between 62.5 and 90 %, respectively, in the less salty layer and in the brine of Medee (Yakimov et al. 2013). Along the interface of this basin, *Gammaproteobacteria* outcompeted *Epsilonproteobacteria*, and also SAR11 division was highly abundant. Methyl-coenzyme M reductase (*mcrA*), functional marker gene for methanogenesis, was detected in Medee brine, although methane production rate was very low compared to the other thalassohaline Eastern Mediterranean DHABs. According to the geochemical characterization and molecular investigation of interface and brine samples, Yakimov et al. (2013) proposed that methane production in Medee basin could derive from the reductive degradation of the compatible solute glycine betaine performed by the KB-1 division, generating acetate and trimethylamine, the latter consumed by putative methanogens belonging to the MBSL-1 division. A consortium composed by three microbial partners, able to perform the above-described process, has been recently enriched from Thetis brine, proving the link between anaerobic glycine betaine degradation and methanogenesis in DHABs (La Cono et al. 2015).

Besides prokaryotes, the viruses have been also studied in the DHABs. Corinaldesi et al. (2014) showed that virus-induced mortality of prokaryotes was lower in oxic sediments compared to anoxic ones collected in the Medee and L'Atalante basins and, accordingly, the concentration of extracellular DNA as well as DNase activity was higher in DHABs sediments than outside the basin. Extracellular DNA can represent a source of nitrogen and phosphorous for prokaryotic communities in DHABs sediments (Corinaldesi et al. 2014), and its role as reservoir of genes acquirable by horizontal gene transfer events has been proposed in DHAB brines (Borin et al. 2008). Interestingly, according to the taxonomic affiliation, the pool of *16S rRNA* genes detected in extracellular DNA was different from that retrieved from microbial cells in terms of phylogenetic diversity, even though 85 % of the extracellular DNA was derived by virus-induced mortality of prokaryotes (Corinaldesi et al. 2014). On the other side, the fate of prokaryotic populations in DHABs can be regulated by eukaryotic grazers. Several studies were performed on the Eastern Mediterranean DHABs, unraveling the distribution of eukaryotic populations in DHAB brines, interfaces, and the underlying sediments (Bernhard et al. 2014; Filker et al. 2013). Overall, the prevalent protists retrieved in DHAB brines were alveolates, mainly dinoflagellates and ciliates, although molecular investigations also showed the presence of several other groups, namely, chlorophytes, jakobids, cryptophytes, haptophytes, radiolaria, and euglenozoans (Stoeck et al. 2014). Dinoflagellates and ciliates were the most abundant eukaryotes in the Discovery basin, followed by fungi. RNA-based studies showed that fungi are the most diverse eukaryotes in Thetis, where also ciliates and stramenopiles showed high taxonomic diversity (Edgcomb and Bernhard 2013). Protist diversity in seawater-brine interface and brine was investigated in the Discovery, Urania, Thetis, Tyro, and Medee DHABs by Terminal Restriction Length Polymorphism (T-RFLP). Protist communities in the brines and interfaces clustered differently, and those of the Thetis and Discovery DHABs were separated from the others (Filker et al. 2013). According to the analysis performed by the authors, the composition of protist communities was not significantly correlated to distance, while it was to the environmental setting occurring in the analyzed samples (Filker et al. 2013). Specifically, the concentration of sulfate, magnesium, and oxygen, besides salinity resulted involved in shaping the protist assemblages. In a different study, the pyrosequencing of the hypervariable V4 fragment of the small subunit ribosomal *RNA* gene was then applied to samples collected at the interfaces and brines of Tyro, Thetis, Urania, and Medee (Stock et al. 2013), confirming the sharp separation of DHAB brines and interfaces. Moreover, the authors showed that, while the protist communities present in brines of the four basins are distinct, in the interfaces they are quite similar, except in the Tyro basin (Stock et al. 2013). Pyrosequencing allowed the detection of about 100 ciliate genera and showed that (i) several taxa were present exclusively in one or more brines, (ii) no taxa were common between all samples, and, as also observed for prokaryotes, (iii) amplicons affiliated to different taxonomic groups were differentially distributed in the studied brines and interfaces (Stock et al. 2013).

2.4 Microbial Diversity of Hypersaline Brines in the Gulf of Mexico

Since the first description of a DHAB in 1977 (Shokes et al. 1977), other similar extreme ecosystems have been discovered in the deep sea of different oceanic regions all over the world. However, DHABs situated in enclosed sea basins, such as the Mediterranean and the Red Sea (Van der Wielen et al. 2005; Daffonchio et al. 2006; Borin et al. 2009; Antunes et al. 2011; Ngugi et al. 2015; Guan et al. 2015), have historically been studied more extensively, compared to the ones identified and described in other locations.

Another geographic area characterized by a high density of hypersaline bodies (such as brine pools, mud volcanos, and cold seeps) is the continental slope of the Gulf of Mexico (GoM) (Fig. 2.2b). In this area, the origin of many hypersaline basins was due to geological processes known as "salt tectonics." For a description of the events and mechanisms leading to this extensive seepage area and of the megafauna associated to different hypersaline bodies, refer to Cordes et al. (2009) and references therein.

Despite such high concentration of extreme deep-sea ecosystems identified so far on the seafloor of the continental slope of the GoM, still little information are available about structure, dynamics, and functionality of the microbial communities inhabiting such environments. The main studies published so far, and previously already discussed (Mapelli et al. 2012), were focused on the microbial and molecular ecology of resident manganese and iron reducers and on the metabolic processes related to methane, specifically anaerobic oxidation of methane (AOM) and methanogenesis. The studies focused on few hypersaline habitats, such as the Orca Basin (Shokes et al. 1977; Van Cappellen et al. 1998) and two brine bodies associated to either quiescent or active mud volcano systems, namely, NR1/GC233 and GB425 (MacDonald et al. 2000; Joye et al. 2005, 2009).

Recently, two new studies (Lloyd et al. 2006; Yan et al. 2006) focused on few sites of the continental slope of the GoM. These studies investigated the role and functionality of anaerobic methane oxidizers (ANMEs) and of aerobic methanotrophs in the cycling of methane in this hypersaline hydrocarbon- and methane-seepage-rich area. Lloyd et al. (2006) studied the involvement in the anaerobic oxidation of methane of the resident microbial community in the top sediments of the seafloor near a mud volcano in Green Canyon lease block 205 (GC205, 27.71672778° N, 90.5334627° W). The significance of this study consists in the attempt to shed light for the first time on the interaction between archaeal ANME and sulfate-reducing bacteria (SRB) populations within the peculiar geochemical gradient characteristic of different seeping sites along the continental slope of the GoM. It is known that in this region, sediments cover salt diapirs, and fractures along such sediment layers allow the upward migration of hypersaline fluids rich in petroleum and methane. The dynamics of the two groups investigated was evaluated through the analysis of functional genes involved in AOM and sulfate reduction: respectively, the genes *mcrA*, coding for the α-subunit of the enzyme methyl coenzyme M reductase, and *dsrAB*, coding for the α- and β- subunits of the dissimilatory (bi)sulfite reductase. The authors described for the first time a seafloor sediment microbial community where the archaeal ANME population was strongly dominated by representatives belonging to the group ANME-1b, whereas previous studies typically described the co-occurrence of representatives of groups ANME-1a, ANME-1b, and ANME-2. The authors hypothesized that the presence of only the ANME-1b subpopulation could be due to the high salinity that characterizes site GC205, roughly four times higher than that of seawater and sediments nearby (2200 mM of Cl^- against 550 mM of Cl^-). Indeed, previous studies on ANME populations in the Black Sea by Nauhaus et al. (2005) showed that at normal seawater salinity (470 mM of Cl^-), the ANME population was co-dominated by ANME-1 and ANME-2, while at increasing values of salinity (up to 945 mM of Cl^-), the two groups showed a sharp decline. Moreover, *16S rRNA* gene analysis showed that the whole community was strongly affected by high salinity, since all the sequences retrieved belonged to the uncultured groups KB-1, HS-I, and HS-II, usually identified only in hypersaline environments like DHABs.

A comparison of the methanotrophs in different locations of the continental slope of the GoM was carried out by Yan et al. (2006). Four sites were investigated: two gas hydrates (GC185 and GC234), one hydrocarbons seep at a brine pool (GC233), and one normal seawater location (NBP). Four clone libraries were constructed amplifying the functional gene *pmoA*, encoding the particulate methane monooxygenase marker for methanotrophs. The phylogenetic analysis carried out on the clone libraries showed the detection of a majority of novel methanotrophs in the GoM (84.8 % of total OTUs) unrelated to those of other deep-sea environments, suggesting that the methanotrophs community of this environment may significantly differ from the ones of other environments previously studied. A strong spatial variation was observed in particular between the normal seawater site NBP and the other three hydrocarbons and hydrate sites (GC185, GC234, and GC233), with six of the OTUs unique to the hydrates and hydrocarbon seep. Moreover, the methanotrophic communities detected at GC185, GC234, and GC233 were characterized by dominant OTUs which differed among sites. The authors speculated that such variation in the chemical compositions detected in all these locations could strongly contribute to the spatial variations observed in functional bacterial groups. For instance, sulfate was completely depleted below 2 cm in the sediments at G234, whereas it was more abundant (18–25 mM) in sediments below 5 cm at GC233 and NBP. Conversely, sulfide concentrations were higher, with values of 3 to 5 mM, at GC234; intermediate, between 1.4–2.1 mM, at GC233; and below 0.01 mM at NBP.

2.5 Concluding Remarks

The advance of knowledge about microbial diversity in DHABs, occurred in the last decade, was mainly enabled by the advent of high-throughput methods through DNA- and RNA-based studies. Although Archaea and Bacteria are still the most investigated microbes in DHABs around the world, recent studies have enlarged our perspective focusing on eukaryotes and viruses inhabiting these fascinating ecosystems. A number of studies enlarged the overall knowledge about the metabolic processes occurring at DHABs, sometimes contributing to understand processes of pivotal ecological impact in salty anoxic environments. Furthermore, the studies performed by multidisciplinary research teams on the microbial communities inhabiting DHABs could be a huge source of information potentially useful for boosting "blue biotechnologies," although still few extremophiles have been brought into culture until now and efforts are required in this direction. Indeed, besides ecological studies, some authors already focused on the biotechnological potential of microbes isolated from DHABs (De Vitis et al. 2015; Ferrer et al. 2005). Finally, the recent discovery and characterization of novel DHABs, including the Eastern Mediterranean Sea Kryos, changed our knowledge about environmental limits for life establishment, shedding a new light on the opportunities provided by seafloor exploration, including speculations about life in the early Earth and in extraterrestrial bodies.

References

Anschutz P, Blanc G, Monnin C, Boulègue J (2000) Geochemical dynamics of the Atlantis II Deep (Red Sea): II. Composition of metalliferous sediment pore waters. Geochem Cosmochim Acta 64:3995–4006

Antunes A, Eder W, Fareleira P et al (2003) *Salinisphaera shabanensis* gen. nov., sp. nov., a novel, moderately halophilic bacterium from the brine-seawater interface of the Shaban Deep, Red Sea. Extremophiles 7:29–34

Antunes A, Francça L, Rainey FA et al (2008) *Marinobacter salsuginis* sp. nov., isolated from the brine-seawater interface of the Shaban Deep, Red Sea. Int J Syst Evol Microbiol 57:1035–1040

Antunes A, Ngugi DK, Stingl U (2011) Microbiology of the Red Sea (and other) deep-sea anoxic brine lakes. Environ Microbiol Rep 3:416–433

Batang ZB, Papathanassiou E, Al-Suwailem A et al (2012) First discovery of a cold seep on the continental margin of the central Red Sea. J Mar Syst 94:247–253

Bernhard JM, Kormas K, Pachiadaki MG et al (2014) Benthic protists and fungi of Mediterranean deep hypersaline anoxic basin redoxcline sediments. Front Microbiol 5:605. doi:10.3389/fmicb.2014.00605

Borin S, Crotti E, Mapelli F et al (2008) DNA is preserved and maintains transforming potential after contact with brines of the deep anoxic hypersaline lakes of the Eastern Mediterranean Sea. Saline Syst 4:10

Borin S, Brusetti L, Mapelli F et al (2009) Sulfur cycling and methanogenesis primarily drive microbial colonization of the highly sulfidic Urania deep hypersaline basin. Proc Natl Acad Sci U S A 106:9151–9156

Borin S, Mapelli F, Rolli E et al (2013) Anammox bacterial populations in deep marine hypersaline gradient systems. Extremophiles 17:289–299

Bougouffa S, Yang JK, Lee OO et al (2013) Distinctive microbial community structure in highly stratified deep-sea brine water columns. Appl Environ Microbiol 79:3425–3437

Cao H, Zhang WP, Wang Y, PY Q (2015) Microbial community changes along the active seepage site of one cold seep in the Red Sea. Front Microbiol 6:1–11

Cita MB, Aghib FS, Cambi A et al (1985) Precipitazione attuale di gesso in un bacino anossico profondo; prime osservazioni geologiche, idrologiche, paleontologiche sul Bacino Bannock (Mediterraneo orientale). Giornal Geol 47:143–163

Cordes EE, Bergquist DC, Fisher CR (2009) Macro-ecology of Gulf of Mexico cold seeps. Ann Rev Mar Sci 1:143–168

Corinaldesi C, Tangherlini M, Luna GM, Dell'Anno A (2014) Extracellular DNA can preserve the genetic signatures of present and past viral infection events in deep hypersaline anoxic basins. Proc Roy Soc B Biol Sci 281:20133299

Daffonchio D, Borin S, Brusa T et al (2006) Stratified prokaryote network in the oxic–anoxic transition of a deep sea halocline. Nature 440:203–207

De Vitis V, Guidi B, Contente ML et al (2015) Marine microorganisms as source of stereoselective esterases and ketoreductases: kinetic resolution of a prostaglandin intermediate. Marine Biotechnol 17:144–152

Dickins HD, Van Vleet ES (1992) Archaebacterial activity in the Orca basin determined by the isolation of characteristic isopranyl ether-linked lipids. Deep-Sea Res 39:521–536

Eder W, Jahnke LL, Schmidt M, Huber R (2001) Microbial diversity of the brine-seawater interface of the Kebrit Deep, Red Sea, studied via 16S rRNA gene sequences and cultivation methods. Appl Environ Microbiol 67:3077–3085

Eder W, Schmidt M, Koch M et al (2002) Prokaryotic phylogenetic diversity and corresponding geochemical data of the brine-seawater interface of the Shaban Deep, Red Sea. Environ Microbiol 4:758–763

Edgcomb VP, Bernhard JM (2013) Heterotrophic protists in hypersaline microbial mats and deep hypersaline basin water columns. Life 3:346–362

Edgcomb VP, Orsi W, Breiner HW et al (2011) Novel kinetoplastids associated with hypersaline anoxic lakes in the Eastern Mediterranean deep-sea. Deep-Sea Res I 58:1040–1048

Edgcomb VP, Taylor C, Pachiadaki MG et al (2014) Comparison of Niskin vs. in situ approaches for analysis of gene expression in deep Mediterranean Sea water samples. Deep-Sea Res. II. http://dx.doi.org/10.1016/j.dsr2.2014.10.020i

Ferrer M, Golyshina OV, Chernikova TN et al (2005) Microbial enzymes mined from the Urania deep-sea hypersaline anoxic basin. Chem Biol 12:895–904

Ferrer M, Werner J, Chernikova TN et al (2012) Unveiling microbial life in the new deep-sea hypersaline Lake Thetis. Part II: a metagenomic study. Environ Microbiol 14:268–281

Fiala G, Woese CR, Langworthy TA, Stetter KO (1990) *Flexistipes sinusarabici*, a novel genus and species of eubacteria occurring in the Atlantis II Deep brines of the Red Sea. Arch Microbiol 154:120–126

Filker S, Stock A, Breiner HW et al (2013) Environmental selection of protistan plankton communities in hypersaline anoxic deep-sea basins, Eastern Mediterranean Sea. MicrobiologyOpen 2:54–63

Guan Y, Hikmawan T, Antunes A et al (2015) Diversity of methanogens and sulphate-reducing bacteria in the interfaces of five deep-sea anoxic brines of the Red Sea. Res Microbiol 166:688–699

Gurvich E (2006) Metalliferous Sediments of the World Ocean: Fundamental Theory of Deep-Sea Hydrothermal Sedimentation. Springer, Berlin, Heidelberg, 416 p

Hallsworth JE, Yakimov MM, Golyshin PN et al (2007) Limits of life in $MgCl_2$-containing environments: chaotropicity defines the window. Environ Microbiol 9:801–813

Hartmann M, Scholten J, Stoffers P, Wehner F (1998) Hydrographic structure of brine-filled deeps in the Red Sea new results from the Shaban, Kebrit, Atlantis II, and Discovery Deep. Mar Geol 144:311–330

Huber H, Stetter K (2001) Class I. Deferribacteres class. nov. Bergey's manual of systematics of Archaea and Bacteria. doi:10.1002/9781118960608.cbm00022

Jongsma D, Fortuin AR, Huson W et al (1983) Discovery of an anoxic basin within the Strabo trench, eastern Mediterranean. Nature 305:795–797

Joye SB, MacDonald IR, Montoya JP, Piccini M (2005) Geophysical and geochemical signatures of Gulf of Mexico seafloor brines. Biogeosciences 2:295–309

Joye SB, Samarkin VA, Orcutt BN et al (2009) Metabolic variability in seafloor brines revealed by carbon and sulphur dynamics. Nat Geosci 2:349–354

Kormas KA, Pachiadaki MP, Karayanni H et al (2015) Inter comparison of the potentially active prokaryotic communities in the halocline sediments of Mediterranean deep sea hypersaline basins. Extremophiles 19:949–960

La Cono V, Smedile F, Bortoluzzi G et al (2011) Unveiling microbial life in new deep-sea hypersaline Lake Thetis. Part I: prokaryotes and environmental settings. Environ Microbiol 13:2250–2268

La Cono V, Arcadi E, La Spada G (2015) A three-component microbial consortium from deep-sea salt-saturated anoxic Lake Thetis links anaerobic glycine betaine degradation with methanogenesis. Microorganisms 3:500–517

Larock PA, Lauer RD, Schwarz JR et al (1979) Microbial biomass and activity distribution in an anoxic, hypersaline basin. Appl Environ Microbiol 37:466–470

Lee OO, Wang Y, Tian RM et al (2014) In situ environment rather than substrate type dictates microbial community structure of biofilms in a cold seep system. Sci Rep 4:3587

Lloyd KG, Lapham L, Teske A (2006) An anaerobic methane-oxidizing community of ANME-1b archaea in hypersaline Gulf of Mexico sediments. Appl Environ Microbiol 72:7218–7230

MacDonald IR, Buthman D, Sager WW, Peccini MB, Guinasso NL (2000) Pulsed oil discharged from a mud volcano. Geology 28:907–910

Mapelli F, Borin S, Daffonchio D (2012) Microbial diversity in deep hypersaline anoxic basins. Adapt Microb Life Environ Extrem Nov Res Results Appl:21–36. doi:10.1007/978-3-211-99691-1_2

MEDRIFF Consortium (1995) Three brine lakes discovered in the seafloor of the eastern Mediterranean. Eos Trans AGU 76:313–318

Mwirichia R, Alam I, Rashid M et al (2016) Metabolic traits of an uncultured archaeal lineage -MSBL1- from brine pools of the Red Sea. Sci Rep 6:19181

Nauhaus K, Treude T, Boetius A, Kruger M (2005) Environmental regulation of the anaerobic oxidation of methane: a comparison of ANME-1 and ANME-II communities. Environ Microbiol 7:98–106

Ngugi DK, Blom J, Alam I et al (2015) Comparative genomics reveals adaptations of a halotolerant thaumarchaeon in the interfaces of brine pools in the Red Sea. ISME J 9:396–411

Qian PY, Wang Y, Lee OO et al (2011) Vertical stratification of microbial communities in the Red Sea revealed by 16S rDNA pyrosequencing. ISME J 5:507–518

Rappé MS, Giovannoni SJ (2003) The uncultured microbial majority. Annu Rev Microbiol 57:369–394

Schardt C (2015) Hydrothermal fluid migration and brine pool formation in the Red Sea: the Atlantis II Deep. Miner Depos 51:89–111

Shokes RF, Trabant PK, Presley BJ, Reid DF (1977) Anoxic, hypersaline basin in the Northern Gulf of Mexico. Science 196:1443–1446

Siam R, Mustafa GA, Sharaf H et al (2012) Unique prokaryotic consortia in geochemically distinct sediments from red sea Atlantis II and discovery deep brine pools. PLoS One 7:e42872

Sorokin DY, Kublanov IV, Gavrilov SN et al (2016) Elemental sulfur and acetate can support life of a novel strictly anaerobic haloarchaeon. ISME J 10:240–252

Stock A, Edgcomb V, Orsi W et al (2013) Evidence for isolated evolution of deep-sea ciliate communities through geological separation and environmental selection. BMC Microbiol 13:150

Stoeck T, Filker S, Edgcomb V et al (2014) Living at the limits: evidence for microbial eukaryotes thriving under pressure in deep anoxic, hypersaline habitats. Adv Ecol Article ID 532687, doi:10.1155/2014/532687

Strous M, Fuerst JA, Kramer EHM et al (1999) Missing lithotroph identified as new planctomycete. Nature 400:446–449

Van Cappellen P, Viollier E, Roychoudhury A et al (1998) Biogeochemical cycles of manganese and iron at the oxic–anoxic transition of a stratified marine basin (Orca Basin, Gulf of Mexico). Environ Sci Technol 32:2931–2939

Van der Wielen PWJJ, Bolhuis H, Borin S et al (2005) The enigma of prokaryotic life in deep hypersaline anoxic basins. Science 307:121–123

Wallmann K, Suess E, Westbrook GH et al (1997) Salty brines on the Mediterranean sea floor. Nature 387:31–32

Wang Y, Zhang W-P, Cao H (2014) Diversity and distribution of eukaryotic microbes in and around a brine pool adjacent to the Thuwal cold seeps in the Red Sea. Front Microbiol 6:1–12

Yakimov MM, La Cono V, Denaro R et al (2007) Primary producing prokaryotic communities of brine, interface and seawater above the halocline of deep anoxic lake L'Atalante, Eastern Mediterranean Sea. ISME J 1:1–13

Yakimov MM, La Cono V, Slepak VZ et al (2013) Microbial life in the Lake Medee, the largest deep-sea salt-saturated formation. Sci Rep 3:3554

Yakimov MM, La Cono V, La Spada G et al (2015) Microbial community of the deep-sea brine Lake Kryos seawater–brine interface is active below the chaotropicity limit of life as revealed by recovery of mRNA. Environ Microbiol 17:364–382

Yan T, Ye Q, Zhou J, Zhang CL (2006) Diversity of functional genes for methanotrophs in sediments associated with gas hydrates and hydrocarbon seeps in the Gulf of Mexico. FEMS Microbiol Ecol 57:251–259

Yang B, Zhang WP, Tian RM, Wang Y, Qian PY (2015) Changing composition of microbial communities indicates seepage fluid difference of the Thuwal Seeps in the Red Sea. Antonie Van Leeuwenhoek 108:461–471

Divergence of Species in the Geothermal Environment

3

Gudmundur Oli Hreggvidsson, Solveig K. Petursdottir, Sigmar K. Stefansson, Snaedis H. Björnsdottir, and Olafur H. Fridjonsson

Contents

G.O. Hreggvidsson (✉)
Matís, Vínlandsleið 12, 113 Reykjavík, Iceland

University of Iceland, Department of Life and Environmental Sciences, Sturlugata 7, 101 Reykjavík, Iceland
e-mail: gudmundo@matis.is

S.K. Petursdottir • O.H. Fridjonsson
Matís, Vínlandsleið 12, 113 Reykjavík, Iceland

S.K. Stefansson • S.H. Björnsdottir
University of Iceland, Department of Life and Environmental Sciences, Sturlugata 7, 101 Reykjavík, Iceland

H. Stan-Lotter, S. Fendrihan (eds.), *Adaption of Microbial Life to Environmental Extremes*, DOI 10.1007/978-3-319-48327-6_3

3.1 Introduction

Geothermal areas are unique in many aspects as microbial habitats. They are rare on a global scale and geographically confined. They can be regarded as islands, ecologically separated by large distances and physicochemical dispersal barriers. In a sense the global geothermal ecosystem can be considered to be a world of widely dispersed, often very different 'archipelagos' with no mainland. These and other features make geothermal sites an attractive and perhaps ideal model system for studies of microbial divergence and speciation. Microbial speciation may even be more easily observable in geothermal habitats than in other ecosystems.

Geothermal ecosystems are the province of prokaryotes – of unicellular organism. In addition to the high temperature, conditions may range from one extreme to another in pH and other environmental variables. Chemical composition of the water varies enormously, even between adjacent sites, and steep environmental gradients are common. The geothermal gases provide abundant sources of chemical energy, in many instances leading to the formation of unique microbial mat communities. The geothermal biosphere is a very dynamic environment where many different forces are at play shaping intricate relationships in a rich tapestry of life.

It is commonly accepted that bacteria are clonal but it is also recognized that the population is the subject or unit of evolution. This presupposes a cohesive force keeping and reinforcing population boundaries. What is the nature of such a force and how universal is it? Are there more than one? One view maintains that prokaryotic evolution can be understood primarily in terms of clonal divergences and intraspecies competition and that periodic selection acts as the cohesive force by repeatedly purging a population – the species – of genetic diversity (Levin 1981; Cohan 2002). The proponents of this view advocate that a microbial species is an ecologically distinct population (Ward 1998). Another view embraces lateral gene transfer as a causal force, responsible for both patterns of similarities and differences between populations and species of bacteria (Gogarten et al. 2002). According to this view, lateral gene transfer acts as a cohesive force by genetic exchange and homologous recombination between related strains but also as a major force of divergence by mediating rarer acquisitions of foreign genes and a subsequent phenotypic or ecological divergence. It follows that a species may consist of groups of ecologically heterogeneous bacteria and that the overall species diversity is constrained by genetic exchange and homologous recombination (Gogarten et al. 2002; Lawrence 2002). This latter view is, essentially, the more practical view and conforms better to accepted but arbitrary species boundaries as they are demarcated by DNA hybridization values and 16S rRNA similarity.

The idea of bacterial species may remain elusive, but evolutionary studies of bacteria may not need a rigid concept of species. Perhaps it can be *dismissed* with altogether (Lawrence 2002), or the operational unit of evolution may be defined differently as 'ecologically distinct populations' that are 'species-like fundamental units of microbial communities' (Ward et al. 2008). In addition, it may be important to recognize that 'different microbial species may have evolved in different ways' (Cohan 2006; Ward et al. 2008).

The metabolic and physiological similarity between many thermophilic species is an issue that bears on the concept of species. The taxonomy of thermophiles has almost from the beginning been based on molecular methods. 16S rRNA sequencing and DNA:DNA hybridization have been fundamental for delineating taxa and establishing species status. Interestingly, this revealed the apparent phenotypic conformity within many widespread genera, such as *Thermus*, *Hydrogenobacter* and *Thermococcus*, but also between members of higher taxa, such as within the orders of *Sulfolobales* and *Aquificales* (Hreggvidsson and Kristjansson 2003; Kristjansson et al. 2000).

Since so many thermophilic genera appear to be collections of genospecies with few observed phenotypically distinguishing characteristics, it indicates that adaptive traits of thermophilic genospecies within a genus could be very few or involving subtle physiological or metabolic differences difficult to identify. This also raises related questions, e.g. what dictates the distribution of the different thermophilic species? Do geographical, topological or other physical dispersal barriers play the major role or is the distribution determined largely by physicochemical factors?

Culture-independent molecular methods are invaluable for analysing community structures; estimating diversity, evenness and abundance of taxa in geothermal areas; as well as studying their distribution and phenomena, such as dispersal mechanisms and barriers, endemism, migration and colonization. These methods have been helpful in revealing the geographic distribution of thermophilic bacteria and the relationships between environmental physicochemical conditions and species composition (Valverde et al. 2012; Miller et al. 2009; Skirnisdottir et al. 2000). They will continue to be important for answering questions about specific ecological adaptations, the roles of microorganisms in mineralization and cycling of nutrients as well as microbial interactions in hot springs.

Microbial evolutionary studies should involve an evaluation of the genetic structure of bacterial taxa, of the occurrence and the extent of gene flow and of the relevance or importance of different cell apparatus in directing the adaptive evolution of a species. Genomic studies play an ever-increasing role in this context. They have introduced important evolutionary concepts into microbiology, such as the pan-genome, the core genome and the dispensable genome (Tettelin et al. 2005). The results indicate that there is a certain fluidity of genetic information across species boundaries, and they have put to test and renewed theories about the concept of microbial species.

Studies on more dynamic aspects of thermophilic microbial populations in an ecological context have lagged behind studies with more taxonomic or systematic emphasis, often in the pursuit of the novel exotic species. Now, we have many, perhaps most, of the major pieces but too little of the context. Until now there has been scant information on species boundaries, genetic diversity and the distribution of thermophilic microorganisms. However, descriptions of entire microbial communities, their interactions with the physical environment and roles of constituent species are now becoming abundant as well as reflections on how geography and different evolutionary pressures may influence the population structure and divergence of thermophiles (Takacs-Vesbach et al. 2013; Inskeep et al. 2010; Hreggvidsson et al. 2006; Reysenbach et al. 2005; Ward and Cohan 2005; Whitaker et al. 2003; Petursdottir et al. 2000; Ward et al. 1998; Kristjansson and Hreggvidsson 1995).

In this chapter we describe various features of the geothermal biosphere and environmental factors that may influence speciation in these habitats. This includes physicochemical, geographic and topological aspects and other factors and forces that may influence the distribution and dispersal of thermophilic organisms. Lastly, we will give an example of the biogeography and ecology of *Thermus* to illustrate some of the questions and issues that have received attention in this particular field of study.

3.2 Geothermal Areas

Geothermal areas on earth are mainly connected with tectonic and volcanic activity caused by plate movements on the semi-fluid asthenosphere, where plates are moving apart, colliding or transforming (Fig. 3.1). The plate boundaries are commonly associated with geological events such as earthquakes and the creation of topographic features such as mountains, volcanoes, mid-ocean ridges and oceanic trenches. The hottest known geothermal regions and the majority of the world's active volcanoes occur along plate boundaries (Fig. 3.2). The best known and biologically most studied geothermal areas are in North America (Yellowstone National Park), New Zealand, Japan, Italy, the Kamchatka Peninsula and Iceland.

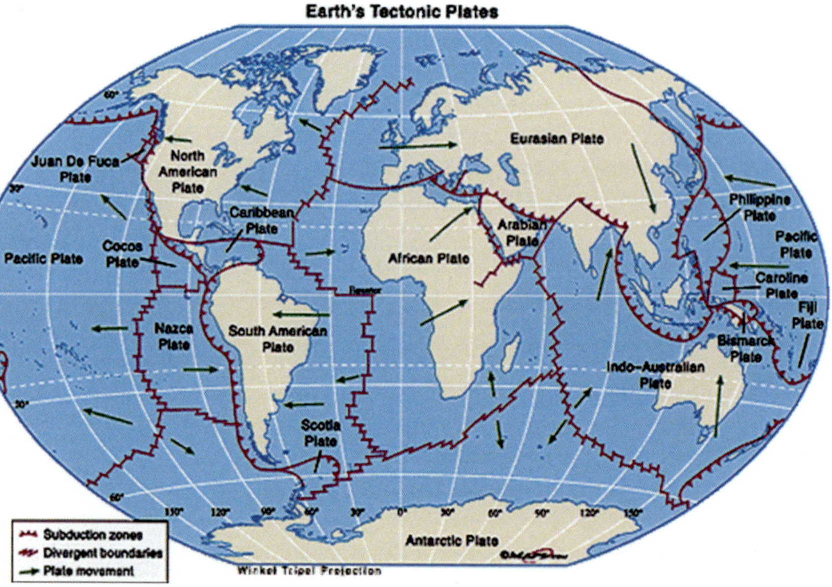

Fig. 3.1 Boundaries of tectonic plates on earth (credit: http://media.maps.com/magellan/Images/tectonic.gif)

Geothermal areas in different parts of the world vary greatly in geology and chemistry but belong mainly to two categories: firstly, the solfataric type characterized by acidic soils, sulfur, mud pots and fumaroles in high-temperature areas and secondly the neutral-alkaline type, characterized by freshwater hot springs and geysers in low-temperature areas (Kristjansson and Hreggvidsson 1995). The chemistry may vary significantly in many other aspects. For example, high arsenic concentration in geothermal waters is typically associated with acidic volcanic systems in continental settings, particularly argillaceous sediments where it is known to be preferentially partitioned. In contrast, geothermal springs in volcanic regions associated with magmas of predominantly basaltic composition, such as in Iceland, have much lower levels of arsenic compounds (Keller et al. 2014; Arnórsson 2003). Geothermal systems known to contain high arsenic concentrations include the Yellowstone Park, USA, and the El Tatio geothermal field, Chile, where arsenic concentrations have been reported to be as high as 150 mg/l and 50 mg/l, respectively (Romero et al. 2003; Langner et al. 2001). Arsenite, As(III), can be the sole or primary arsenic species in hot anaerobic source waters where it is rapidly converted to arsenate, As(V), due to microbial and chemical oxidations. The sulfur geobiochemical cycle is very prominent in the geothermal environment. H_2S is a major component of geothermal gases in high-temperature volcanic areas, and dissolved sulfate and sulfide are common in geothermal water. Other intermediate species have also been identified including polysulfides, thiosulfate, polythionates and sulfite (Kaasalainen and Stefansson 2011). These

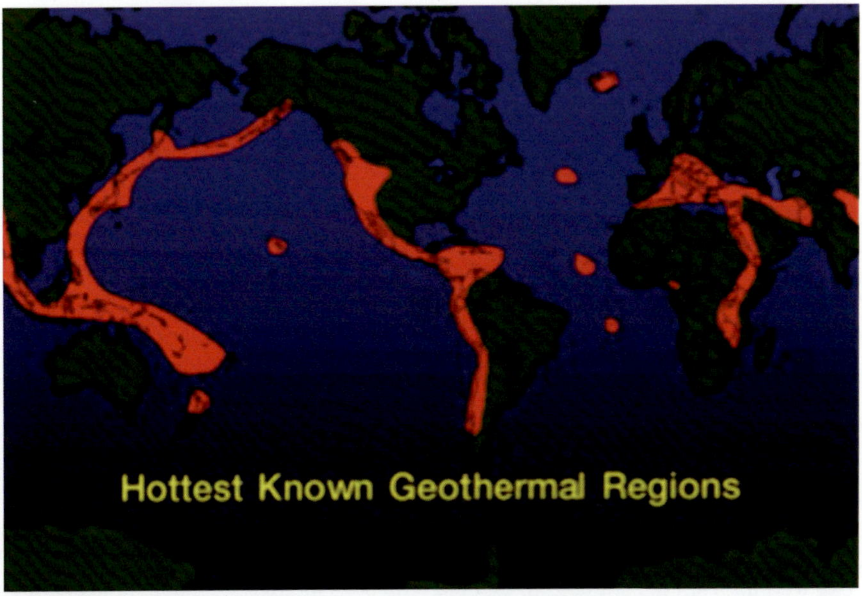

Fig. 3.2 The geothermal activities occur mainly along the plate boundaries (Slide 15 © 2000 Geothermal Education Office; http://geothermal.marin.org/geopresentation/)

compounds influence the microbial species composition in various ways, such as in directing the pH of a particular geothermal habitat, and depending on their oxidation state, they can serve as electron donors or acceptors in energy-yielding electron transport chains.

3.2.1 High- and Low-Temperature Fields

The two types of geothermal areas are the result of geological differences of the heat source. The high-temperature fields are associated with active central volcanoes and have magma chambers as the heat source. They are defined by temperatures above 200 °C at a depth of 1000 m and characterized by emissions of steam and volcanic gases on the surface. The gas is primarily N_2 and CO_2, but H_2S and H_2 can be up to 10 % each of the total gas fraction. Traces of ammonia, methane and carbon monoxide are also found. The pH of the subsurface steam is near neutrality because of the weak acids, CO_2 (pKa = 6.3) and H_2S (pKa = 7.2). Closer to the surface, the sulfide of the steam is oxidized chemically and biologically to sulfur ($H_2S + ½\, O_2 \leftrightarrow S + H_2O$). The oxidation can go further resulting in the formation of sulfuric acid ($H_2S + 2O_2 \leftrightarrow H_2SO_4$), thus lowering the pH and causing corrosion of the surrounding rocks and formation of the typical acidic mud of solfatara fields. As the temperature is high, there is little water coming to the surface, and the hot springs are mostly mud pools or fumaroles. These areas are generally unstable and openings emerge and disappear quite rapidly. As there is no outflow, the water is static and becomes saturated by gases from the geothermal steam (Palmason 2005; Kristjansson and Hreggvidsson 1995; Bödvarsson 1961).

Low-temperature areas are located at the flank of the volcanically active zones. These are defined by temperature lower than 150 °C at 1000 m depth. They are heated by high thermal gradients in the upper crust. Groundwater percolating into these hot areas is heated and returns up to the surface containing dissolved minerals such as silica and some dissolved gases. The concentration of sulfide in the water is low. The subsurface pH is near neutrality and maintains at or above neutral pH at the surface. There is little or no H_2S to be oxidized, and as the CO_2 escapes and silica precipitates, this results in increased pH. The thermal manifestations at the surface are warm or hot springs, under or above 50 °C, respectively, and individual hot springs are very constant in temperature and water flow. These areas are relatively stable as they are located outside the active volcanic zone. However, they may disappear temporarily or new created during periods of seismic activity (Palmason 2005; Kristjansson and Hreggvidsson 1995; Bödvarsson 1961).

Geothermal fields are also found on the seafloor, adding salinity, pressure and even sharper temperature gradients to the other factors. The sulfide is oxidized to sulfur and sulfuric acid, but it cannot affect the pH to the same extent as in the terrestrial fields due to the huge water mass. Hot geothermal water originating inland can well up from the sea bottom especially in coastal areas. A remarkable example of such submarine freshwater hot springs is found in the fjord Eyjafjördur on the north coast of Iceland. Up to 60 m high, cones of smectite rise from the sea bottom formed by the mixing of the hot SiO_2-rich geothermal fluid with the cold Mg-rich seawater (Marteinsson et al. 2001).

3.2.2 Origins of Hot Spring Water

The origin of the hot spring water is groundwater, rainwater or melting snow. Some of the rainwater that seeps deep into the earth will eventually surface in various hot springs. The water of hot springs can be ancient, flowing from the high grounds towards the sea for centuries.

In high-temperature fields, the geothermal steam originating from deep hot strata heats up the surface water from recently fallen rain and/or water seeping from glaciers or snow. The water volume of these hot springs can change rapidly with changing groundwater level.

In low-temperature areas, the groundwater is heated up by high thermal gradients in the upper crust. In both cases the water is enriched in minerals and dissolved chemicals from below. The final pH and temperature are variable according to the nature of the heat source.

All geothermal water contains silica (SiO_2). The level of dissolved silica in the water depends partly on temperature. As the water approaches the relatively cold surface, the silica precipitates and forms the typical silica sinters common in geothermal areas. On the other hand, calcium carbonate ($CaCO_3$) precipitates as the temperature gets higher and forms travertine. Sometimes precipitates of magnesium silicate are formed, but these are relatively rare (Palmason 2005).

3.2.3 Surface Characteristics of Geothermal Areas

The surface manifestations of high-temperature geothermal areas can be very colourful (Fig. 3.3 a–c). The different colours originate from the dissolved minerals and chemicals in the water and steam from deep below reacting with the oxygen dissolved in the aquifer closer to the surface. This is reflected in different colours at the surface of the hot spring areas. Elemental sulfur (S°) and sulfur compounds are characteristic for geothermal areas but these are more abundant in the acidic environments. The evident yellow colour is sulfur, the red is hematite (Fe_2O_3) and the dark grey colour of the clay is a compound of crystallized iron and sulfur (FeS).

White clay can be high in kaolinite, while dark clay is often high in smectite. Heavy metals are generally in high concentrations in the acidic environments. Silica sinters are common in geothermal alkaline areas, and travertines of calcium carbonate are generally found at pH around or slightly below neutral. Salts and minerals can be up to 2000 mg l^{-1} in water-rich alkaline hot springs. Nitrogen and phosphorus compounds are usually in high concentrations in alkaline hot springs (Palmason 2005).

The *water-poor acidic hot springs*, i.e. the boiling mud pools, steaming fumaroles and regions of hot humid and greatly transformed soil, are common in the high-temperature fields (Fig. 3.4). Geothermal steam from deep hot strata heats

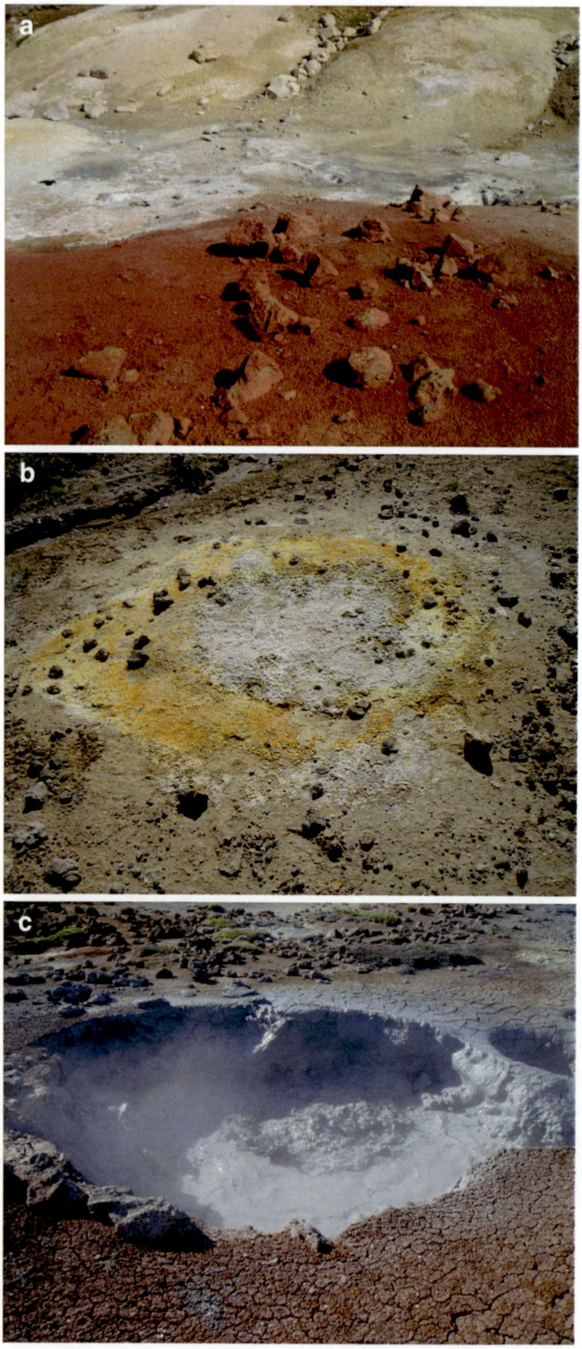

Fig. 3.3 (**a–c**) High-temperature geothermal areas are colourful: (**a**) a mud spot with fumaroles and red precipitates; (**b**) a fumarole with yellow sulfur precipitates; (**c**) a mud pool. The pictures were taken at Þeistareykir geothermal area in NE Iceland (Photos by Hreggvidsson)

Fig. 3.4 A colourful solfatara field in Þeistareykir geothermal area in NE Iceland (Photo by Hreggvidsson)

up the surface water, which is mainly recently fallen rain and/or water seeping from glaciers or snow. The water volume of these hot springs can change rapidly with weather. As there is no outflow, the water is static and becomes saturated by gases from the geothermal steam. The high-sulfide content of the geothermal steam (5–15 % H_2S) is oxidized chemically and biologically, first to sulfur and then to sulfuric acid. This lowers the pH, causing corrosion of the surrounding rocks (Fig. 3.5b) and the formation of the typical acidic mud of solfatara fields with pH around 2 (Kristjansson and Hreggvidsson 1995). This is the domain of strictly anaerobic archaea, and they mainly utilize reduced sulfur compounds and hydrogen as energy sources. Also, ubiquitous aerobic hyperthermophilic *Sulfolobus* species proliferate in mud pits and in the hot turbid acidic waters of the characteristic stagnant pools. At lower temperatures, heterotrophic archaea of *Thermoplasma* and *Picrophilus* can be found at pH around 3 and 0–1, respectively. Chemolithoautotrophic bacteria that belong to the genera *Thiomonas* and *Thiobacillus* and heterotrophic *Geobacillus* and *Deinococcus* species are found at the lowest temperatures.

Mud pools form in high-temperature fields where water is in relatively short supply. The available water rises to the surface and forms mud with the soil particles. The thickness of the mud depends on the water content (Fig. 3.5 a and b). These

Fig. 3.5 (**a**) A mud pool in the geothermal area in Vonarskard in Iceland (Photo by H. Jóhannesson); (**b**) mud pools in Þeistareykir geothermal area (Photo by Petursdottir)

pools are often inhabited by archaea of *Stygiolobus* as well as by *Hydrogenobaculum*, which are acidophilic members of *Aquificae.*

Fumaroles are small openings often located in hills high above the groundwater levels (Fig. 3.6). These emit steam and volcanic gases such as CO_2, SO_2 and H_2S.

Boiling pits are hot springs forming shallow depressions in the earth, often with gravel in the bottom and with clear boiling water. They can be found in both high- and low-temperature areas. In the high-temperature areas, the bubbles are caused by volcanic gases steaming through the water (Fig. 3.7), while in the low-temperature areas, the water is actually boiling.

Sulfide-rich hot springs with pH values of 5.5–6.5 are relatively rare. These are hot mineral springs with very high concentrations of sulfide in the source water. As the water flow is high and these hot springs have outlets, the acid does not accumulate as in the mud pools, so the pH is maintained below neutral. The temperature at the source is often around 80–85 °C but gradually lowers in the affluent where thick white sulfide utilizing microbial mats are often dominated by *Sulfurihydrogenibium* species. The growth seems to be limited to temperatures between 50 and 70 °C (Fig. 3.8 a and b) (Reysenbach et al. 2005; Skirnisdottir et al. 2000; Kristjansson and Hreggvidsson 1995).

Fig. 3.6 A fumarole in the geothermal area in Hveravellir in Iceland (Photo by H. Jóhannesson)

Fig. 3.7 Boiling pit in Vonarskard geothermal area in Iceland (Photo by H. Jóhannesson)

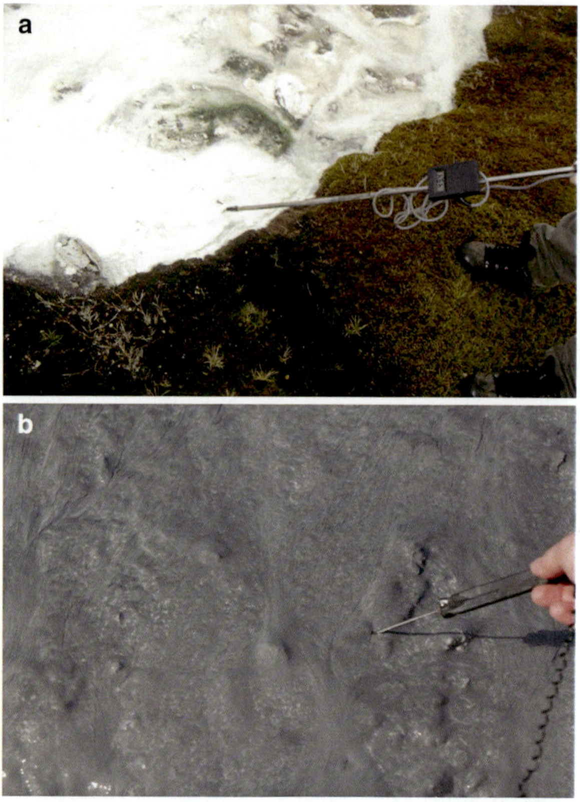

Fig. 3.8 (**a**) A thick, white microbial mat in a sulfide-rich hot spring in Vonarskard in Iceland (Photo by Pétursdóttir (2008)); (**b**) a microbial mat in a sulfide-rich hot spring in Seltun in Krisuvik in Iceland. The colour of the mat is white underneath the grey clay particles (Photo by Hreggvidsson)

Fig. 3.9 *Left*, a water-rich alkaline hot spring in Fludir in Iceland (Photo by H. Jóhannesson); *right*, thick microbial mats of cyanobacteria, chloroflexi and 'sulfur bacteria', well visible in an alkaline hot spring effluent in Iceland (Photo by M. Schmid)

Alkaline hot springs (>*50 °C*) *and warm springs* (<*50 °C*) are common in low-temperature fields, often located at the borders of the high-temperature fields, where the groundwater level is high. These are generally water-rich and have outlets (Fig. 3.9, left).

The chemical compounds directing the pH of the alkaline hot springs are HCO_3^-/CO_3^{2-} and SiO_2. The pH becomes rather high or in the range of pH ≥ 7–10. The

Fig. 3.10 *Left*, steam vents in Jarðbaðsholar in Myvatnssveit in Iceland. *Right*, a closer look at a steam vent's opening (Photos by Petursdottir)

sulfide concentration of the alkaline hot springs can be up to 1 mM. The abundant water in these springs runs off in small streams with marked temperature gradients. Large fluctuations in temperature can occur in the springs depending on the groundwater level that in turn depends on precipitation. The water level and water flow from the alkaline hot springs are therefore often affected seasonally.

Colourful microbial streamers and mats are common in circumneutral and alkaline hot springs, where different microbes dominate, and are often associated with characteristic colours: green (cyanobacteria), orange red (*Chloroflexus/Roseiflexus*), pink (*Thermocrinis ruber*, USA), white, grey, bluish and black (e.g. *Sulfurihydrogenibium* and *Thermocrinis albus*, Iceland). Sharp boundaries can sometimes be observed where one dominant organism replaces another. The colour change can be striking and reflecting differences in growth temperature ranges of the species (Fig. 3.9, right). It may also reflect radical changes in chemistries of the hot spring system. Certain chemicals exert strong selective pressures on mat communities as energy sources or by their toxicities. These chemicals include hydrogen sulfide and arsenic compounds (see below).

Ward and co-workers studied photosynthetic microbial mats common in alkaline hot springs. Using culture-independent molecular methods, they showed that the common *Synechococcus* morphology in the temperature range of 65–73 °C masks a considerable genetic diversity, also, that diversity decreased with increasing temperature and that genetically distinct clades lived in alkaline hot springs in different geographical regions of the world. They maintain that this distribution pattern cannot be explained by different chemical conditions, suggesting that geographical isolation is involved in diversification of hot spring cyanobacteria (Papke et al. 2003).

Steam vents are yet another type of geothermal surface manifestations in geothermal areas. These are relatively rare but sometimes found in low-temperature lava fields. Steam emerges from the hot groundwater below, up above the water table and through the porous lava, blowing slowly out from the openings (Fig. 3.10). The temperature is in the range of 55–85 °C and the pH between pH 7 and 8 in the soil around the openings. The soil is relatively untransformed, and the steam does usually not contain geothermal gases. Members of *Thermus*, *Chloroflexus*, *Actinobacteria* and *Acidobacteria* are commonly detected in these habitats.

3.3 Ecology of Thermophiles

Enormous diversity of chemical and physical properties influences and determines what kind of life can exist in a geothermal environment. It is an extreme habitat characterized by a high temperature, high or low pH and relatively high ionic strength. The species diversity in hot springs is generally low as estimated by culture-independent methods. Extreme conditions of temperature and pH create environmental pressures resulting in fewer species capable of coping with the relatively harsh environment. Regular temperature fluctuations appear to be common in hot springs, and up to 10 °C differences with approximately 6 h periodicity have been observed between the lowest and the highest temperature within a 24–60h period in many hot springs in Iceland (our unpublished results). Periodic disturbances of this kind may help to maintain a stable community structure, consisting of functionally near-equivalent groups, adapted to and active in different 'temperature windows' and tolerant of others. Other diverse environmental factors in hot springs are water content and ionic strength, the content of gases, minerals, dissolved oxygen, chemical compounds and light. Environmental gradients are common, especially in the effluents of the alkaline type of hot springs where water is abundant. These add still other variables for life to cope with or exploit in this environment (Hreggvidsson and Kristjansson 2003; Kristjansson et al. 2000).

3.3.1 Temperature

Temperature has been shown to be a key and perhaps the most important environmental factor affecting microbial diversity in geothermal areas (Wang et al. 2013; Miller et al. 2009). Traditionally, organisms that have maximum temperature for growth, T_{max}, above 50 °C are defined as thermophiles (Wiegel and Ljungdahl 1986), but Brock (1986) defined the thermophilic boundary at 55–60 °C based on two arguments: firstly, to restrict the thermobiome to temperatures rare in nature and to habitats associated with geothermal activity and, secondly, because eukaryotes do not grow above these temperatures. For the present discussion, the authors define moderate thermophiles as having $T_{max} > 50$ °C and <65 °C, thermophiles as having $T_{max} \geq 65$ °C and hyperthermophiles as having optimum temperature for growth, $T_{opt} \geq 80$ °C (Hreggvidsson and Kristjansson 2003; Stetter 1996; Brock 1986). The upper temperature limit of the thermobiome is above 100 °C in deep-sea hydrothermal vents where pressure comes into play (Kashefi and Lovley 2003).

In terrestrial habitats, the temperature of mud pools, fumaroles and solfataras is generally higher than that of alkaline hot springs and steam vents. The temperature of the surrounding environment is considerably lower than the temperature of a hot spring. Therefore, gradients form in all directions from the main heat source. These temperature gradients are usually very steep, in the order of 1–10 °C per mm and 10–50 °C per m in hot flowing water. This means that only a few millimetres away from the site, where the temperature is too high, microbes thrive and can build up massive mats (Kristjansson and Hreggvidsson 1995). Temperature fluctuations lead to periods of fast growth and build-up of a profuse biomass at permissive temperatures; then, periods of too high temperatures lead to death and decomposition of the same biomass providing an abundant source of nutrients for more thermophilic heterotrophs.

3.3.2 pH

The pH level of hot springs is determined by the origin and amount of available water as well as on the concentration of the volcanic gases such as H_2S. In the high-temperature areas, the H_2S is oxidized to H_2SO_4, and as water is usually scarce, the acid builds up in the pools or solfataric humid soil, which often stabilizes at pH values from 2 to 2.5. In sulfide-rich hot springs with abundant water and temperatures around 70 °C, the acid is washed away in the effluent resulting in pH values of around 6. In alkaline hot springs, the H_2S content of the source water is much lower than in the acidic geothermal areas. The chemical compounds directing the pH of the alkaline hot springs are HCO_3^-/CO_3^{2-} and SiO_2. The pH becomes rather high or in the range of pH $\geq$ 7–10. Thermoacidophily and thermoalkalophily are concepts reflecting the adaptation of microbes to extremes in pH at high temperatures (Hreggvidsson and Kristjansson 2003; Kristjansson et al. 2000; Kristjansson and Hreggvidsson 1995).

3.3.3 Energy Sources as Selective Pressures

The classification of organisms within an ecosystem into primary producers and consumers is well known. The primary production generally occurs by means of photosynthesis creating organic matter for the consumers of the ecosystem. However, photosynthesis does not occur above the temperature of 73 °C, which can therefore be defined as the photosynthetic boundary. The primary production above this limit is provided solely by chemolithoautotrophs, which can use inorganic chemicals as energy sources. These organisms often form conspicuous microbial filaments, streamers or mats similar to photosynthetic mats at lower temperatures (Fig. 3.9).

When different thermal habitats are analysed in relation to available energy sources and metabolic types of the organisms present, a characteristic pattern of community structures is found. Clearly the bacterial diversity in geothermal sites is established along environmental gradients of temperature, pH and available electron donors and acceptors with similar communities found in similar conditions in far apart regions (Menzel et al. 2015; Inskeep et al. 2010; Miller et al. 2009).

Terrestrial chemolithoautotrophic species of the *Aquificales* often form conspicuous thick microbial mats at high temperatures. The particular species composition depends on the physicochemical conditions in the hot spring, e.g. temperature, pH and available chemical energy sources (Takacs-Vesbach et al. 2013; Reysenbach et al. 2005; Hjorleifsdottir et al. 2001; Skirnisdottir et al. 2000). The most important electron donors appear to be the geothermal gases H_2S or H_2, and their absolute concentrations and ratios may have marked effects on the microbial diversity. For

Table 3.1 Energy sources and representative organisms in freshwater, alkaline hot springs

Energy source	Representative organisms	T_{max}	T_{opt}	pH_{opt}
Primary producers				
H_2O/light	*Synechococcus lividus*	73	65	8.0
H_2S/light/org	*Chloroflexus aurantiacus*	70	56	8.0
H_2/O_2	*Hydrogenobacter thermophilus*	77	72	6.8
H_2S/O_2	*Thermocrinis ruber*	89	80	7–8.5
H_2/CO_2	*Methanobacterium thermoauotrophicum*	75	65	7.4
H_2/SO_4^{2-}	*Thermodesulfobacterium thermophilum*	85	65	7.5
Consumers				
Org.m/O_2	*Thermus* sp.	79	65–70	7.0
Org.m/NO_3	*Thermus* sp.	79	65–70	7.0
Org.m/O_2	*Geobacillus* sp.	80	50–60	7.0
Org.m/org.m	*Clostridium* sp.	80	60–70	6.0
Org.m/org.m	*Thermoanaerobacter* sp.	78	55–75	6–8.0
Org.m/SO_4^{2-}	*Desulfotomaculum* sp.	70	55–65	7.0

T_{max}, T_{opt} and pH_{opt} refer to maximum temperature for growth, optimum temperature for growth and optimum pH for growth, respectively. Org. m refers to organic material

Table 3.2 Energy sources and representative organisms in acidic solfatara fields

Energy source	Representative organisms	T_{max}	T_{opt}	pH_{opt}
Primary producers (autotrophs)				
S/O_2	*Sulfolobus acidocaldarius*	80	75	2.5
S/O_2	*Acidianus infernus*	96	90	2.0
H_2/S	*Acidianus infernus*	96	90	2.0
Consumers (heterotrophs)				
Org.m/O_2	*Thermoplasma volcanium*	67	60	2.0
Org.m/O_2	*Sulfolobus acidocaldarius*	80	75	2.5

See Table 3.1 for abbreviations

example, microbial mats in alkaline (~pH 8) low-sulfide hot springs at ~80–92 °C are predominately populated by hydrogen-oxidizing *Thermocrinis* species, whereas circumneutral high-sulfide springs around 70 °C are characterized by grey or yellowish microbial mats dominated by *Sulfurihydrogenibium* species (Reysenbach et al. 2005; Hjorleifsdottir et al. 2001; Skirnisdottir et al. 2000). Sulfide as an energy source selects for *Sulfurihydrogenibium* in the latter case, but its toxicity may also select against *Thermocrinis* species (Hjorleifsdottir et al. 2001; Skirnisdottir et al. 2000). Genetically distinct but related *Aquificales* groups appear to occupy similar niches in far apart regions. *Thermocrinis ruber* is the dominant species of the pink filamentous streamers in Yellowstone National Park (Huber et al. 1998), and a related species, *Thermocrinis albus*, dominates the commonly observed grey or bluish threads and streamers in Icelandic hot springs above 80 °C (Skirnisdottir et al. 2000; Eder and Huber 2002).

Examples of organisms that are important representatives of certain metabolic groups in different types of hot springs are listed in Tables 3.1–3.4.

Table 3.3 Energy sources and representative organisms in anaerobic geothermal mud and soil

Energy source	Representative organisms	T_{max}	T_{opt}	pH_{opt}
Primary producers (autotrophs)				
H_2/CO_2	*Methanothermus fervidus*	97	83	6.5
H_2/CO_2	*Methanotorris igneus*	91	88	5.7
H_2/S	*Thermoproteus tenax*	96	90	5.0
CO/S	*Thermoproteus tenax*	96	90	5.0
H_2/S	*Pyrodictium occultum*	110	105	6.5
H_2/SO_4^{2-}	*Archaeoglobus fulgidus*	95	83	7.0
Consumers (heterotrophs)				
Org.m/org.m	*Thermotoga maritima*	90	80	6.5
Org.m/S	*Thermotoga neapolitana*	90	80	7.0
Org.m/S	*Pyrobaculum islandicum*	102	100	6.0
Org.m/SO_4^{2-}	*Archaeoglobus fulgidus*	95	83	7.0
Org.m/S	*Pyrobaculum islandicum*	102	100	6.0
Org.m/org.m[b]	*Pyrococcus furiosus*	104	100	6.0
Org.m/S	*Thermoproteus tenax*	96	90	5.0
Org.m/SO_4^{2-}	*Archaeoglobus profundus*	90	82	6.0

See Table 3.1 for abbreviations

Table 3.4 Energy sources and representative organisms in sulfide-rich hot springs

Energy source	Representative organisms	T_{max}	T_{opt}	pH_{opt}
Primary producers (autotrophs)				
H_2S/O_2	*Sulfurihydrogenibium kristjanssonii*	73	68	6.6
Consumers (heterotrophs)				
Org.m/O_2	*Thermus scotoductus and Thermus oshimai* (*Iceland*), *Thermus aquaticus* (*USA*)	73–79	65–70	6,5

See Table 3.1 for abbreviations

3.4 Biogeography of Thermophiles

Geothermal areas have yielded a large number of highly diverse thermophilic and hyperthermophilic genera of bacteria and archaea. Not surprisingly many of these genera, such as *Thermus*, *Thermoplasma*, *Rhodothermus*, *Bacillus*, *Sulfolobus* and *Hydrogenobacter*, to name just a few, have a worldwide distribution. On the species level, the situation is more complicated, owing to the phenotypic similarity of thermophilic species and also because the genetic structures are in most cases unknown. However, the methods of molecular systematics have started to reveal clear endemic patterns in the distribution of some thermophiles at and below the level of species, and there are cases of a conspicuous absence of a particular species in certain geothermal areas (Hreggvidsson and Kristjansson 2003). It can be expected that culture-independent studies will continue to be of major importance in examining the global distribution of thermophiles.

Geographical isolation might be expected to be a significant factor in causing and enhancing the divergence of microbes. The geographic structure underlying the

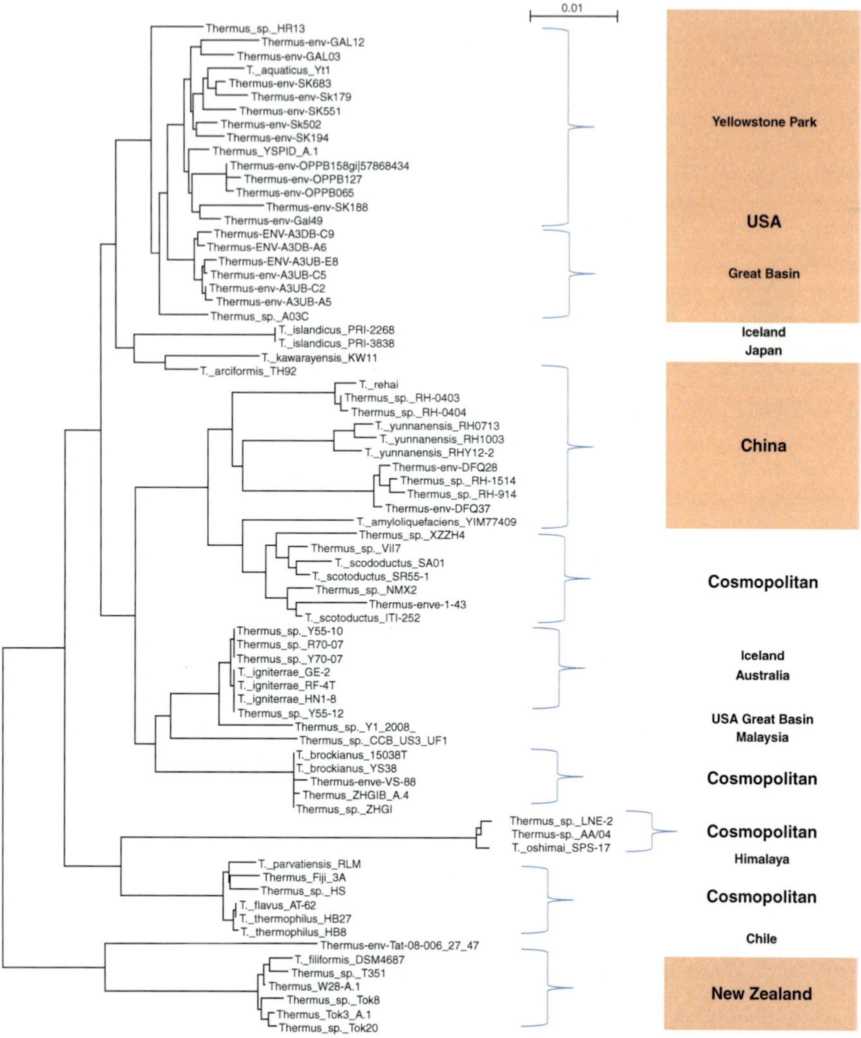

Fig. 3.11 Phylogeography of *Thermus* based on neighbour-joining (NJ) of 16S rRNA gene sequences

thermal biosphere promotes the separation of populations, disrupts the cohesive force and subsequently accelerates genetic divergence by various mechanisms. Founder effects and genetic drift may occur and genetically separate lineages, and different geothermal regions may have different selective pressures leading to local environmental adaptations by periodic selection. Geographical isolation also creates opportunities for isolated evolutionary events, e.g. local lateral gene transfer occurrences opening new niches for separate populations within the species (Hreggvidsson and Kristjansson 2003; Cohan 2001, 2002).

The global geothermal ecosystem might be expected to harbour unique species compositions in far apart habitats. A particularly striking pattern of geographically influenced distribution is found for species of *Thermus* (Fig. 3.11). Distinct differences in species composition seem to exist between widely separated locations, with mixtures of unique endemic lineages and more cosmopolitan species (Hreggvidsson et al. 2006). Another example of a discontinuous global distribution at the species level is found among the thermophilic cyanobacteria. Different clades of *Synechococcus* bacteria are found in far apart geothermal locations around the world (Papke et al. 2003) and some of them appear to represent local environmental adaptations. Thus, high-temperature *Synechococcus* species are abundant and easily visible in North American hot springs, where they form greenish mats together with *Chloroflexus* at temperatures up to 73 °C. These species have not been detected in alkaline and neutral hot springs in Iceland between 65 and 73 °C. This temperature interval is dominated by the anoxygenic photoautotroph *Chloroflexus* that forms instead colourful pink and orange mats (Fig. 3.9). Therefore, there appears to be an unoccupied niche for high-temperature oxygenic photosynthetic bacteria in Iceland. A similar but perhaps less striking distribution pattern is seen for the *Aquificales*. Different *Thermocrinis* species appear to occupy comparable niches of low-sulfide, high-temperature hot springs in Iceland and in Yellowstone National Park. Also, microbial mats in high-sulfide hot springs in Japan, America and Iceland are dominated by distinct *Sulfurihydrogenibium* species (Reysenbach et al. 2005; Hreggvidsson and Kristjansson 2003; Skirnisdottir et al. 2000).

The genetic structure of a species is shaped by geography even on a local scale. Multilocus studies of *Rhodothermus marinus* and *Thermus thermophilus* in coastal hot springs in Iceland clearly revealed that the populations at different sites were evolving independently of each other (Hreggvidsson et al. 2006; Petursdottir et al. 2000). This is despite the fact that both are marine species, and microbial transfer between sites should not be impeded by a completely different physical medium of passage as would be the case with microorganisms in terrestrial hot springs passing through air.

The terrestrial sulfur-oxidizing species *Sulfolobus islandicus* is adapted to and thrives in acidic geothermal springs at around 80 °C. The very low pH (~2) of these sulfuric hot springs is another factor restricting the distribution of the species. A multilocus study by Whitaker and co-workers (2003) clearly showed genetic differences between populations in different regions in the Northern Hemisphere, Kamchatka (Russia), Yellowstone National Park, Alaska and Iceland (separated by 250– ~6000 km) but also between populations within areas separated by as little as 15 km. It could be concluded that *Sulfolobus islandicus* strains clustered by a geographic locale rather than by hot spring character, temperature or pH and that the genetic distances between populations increased proportionally with the geographical distance.

Genome sequencing gives evidence that lateral gene flow across species, phyla and even domain boundaries has occurred to a considerable extent between thermophilic microbial lineages. The concept of a dispensable fraction of a genome in a particular species implies that genes have been lost and gained since the separation from a common ancestor. The genome of the thermophile *Sulfolobus solfataricus* contains an unusually large number of transposable elements and, as a consequence, may have been particularly susceptible to recombination and rearrangements (Redder and Garrett 2006).

Reno et al. (2009) examined the pan-genomic structure of *S. islandicus* isolates from far apart locations in order to test whether barriers to dispersal or ecological selection were primarily responsible for shaping its population structure. Most of the peripheral genes came from viruses and plasmids and about one-third was specific to a geographic location. The viruses and plasmids that had lent their genes to *Sulfolobus* in one site were different from those found in another. Also, much of the variation was found in genes devoted to the microbe's defence system against foreign genetic elements, indicating that *S. islandicus* is evolving largely in response to the assault of local viruses.

3.4.1 Dispersal of Thermophiles

Various factors may cause and maintain a discontinuous distribution pattern of a thermophilic species. Obviously, a large distance between geothermal sites reduces the number of migration events from one area to another. However, the nature of the surroundings may also hinder distribution if it is hostile or harmful to the organisms. For example, marine migration routes may be barred for some terrestrial thermophiles and oxygen may be harmful to obligate thermophilic anaerobes.

Alkaline hot springs in low-temperature regions may be interconnected below the surface by streams, or they may be fed by a single subterranean reservoir. The subsurface temperature of the water is higher than in openings at the surface where the water rapidly cools down. The temperature of the underground water stream may in this case act as a physical dispersal threshold preventing dissemination. Depending on the temperature, distribution of some species, but not all, might be obstructed. In this context it bears mentioning that topological features may play a role as migration follows waterways from watersheds, both above or below the ground. Places at higher altitudes may therefore be different and perhaps less diverse than comparable sites in lowlands where migration events are recurrent.

Geothermal regions may be connected across long distances by hot underwater streams. The presence of apparently purely terrestrial thermophiles in the submarine hydrothermal vent in Eyjafjördur, a fjord on the north coast of Iceland, confirmed the freshwater origin of the water. Surprisingly, no terrestrial *Thermus* strains were isolated from the samples indicating the presence of a high-temperature or hydrological barriers along the way. The water could be traced to high inland mountains located about 100 km south of the cone on the basis of stable oxygen and hydrogen isotopic ratios (Marteinsson et al. 2001).

3.4.1.1 Dispersal Capabilities of Thermophiles

Dispersal mechanisms or capabilities differ between microorganisms. Those that endure desiccation or form metabolically inactive resting bodies resistant to harsh environmental conditions may be scattered by winds all over the world. Thus, spore-forming *Geobacillus* species have a worldwide distribution (Zeigler 2014) in contrast to non-spore-forming species, such as *Thermus*. The cyanobacterial species *Mastigocladus laminosus* lives in alkaline hot spring below 57 °C. It is found in geothermal areas all over the world in contrast to the high-temperature *Synechococcus* species. The former species consists of several genetically different groups. Some of them appear to be endemic such as the population in the Waitangi hot springs in New Zealand, whereas others are widespread. *M. laminosus* is very tolerant to desiccation and freezing, which may facilitate airborne dispersal and explain its phylogeographic structure (Miller et al. 2002). Increased ecological specialization may also be an agent of geographical isolation. High-temperature *Synechococcus* isolates have been reported to be sensitive to freezing and desiccation, factors likely to be important in dispersal (Miller and Castenholz 2000) and possibly a trade-off for the adaptation to higher temperatures. Such sensitivity and a surrounding sea may be sufficient to explain the absence of these strains in Iceland. Similarly, temperature shifts from the high temperature of the habitat to the lower temperature of the surroundings may induce cell cycle arrest and chromosomal DNA degradation as observed in *Sulfolobus* cultures (Hjort and Bernander 1999), thus limiting their dispersal range.

3.4.1.2 Chemical and Biological Barriers

Distinctive geochemical or physical characteristics of a particular region may also exert their influence through biological barriers. A particular hot spring chemistry may be characteristic for a certain geological region affecting the colonization success of migrating species. Local biota or ecotypes of a particular species adapted to the existing physicochemical conditions may outcompete incoming strains and species. For example, high arsenic levels in geothermal springs may influence the species composition and overall diversity both directly and indirectly. Arsenic toxicity is mediated in several ways such as by arsenate inhibiting oxidative phosphorylation and arsenite binding to sulfhydryl groups and inactivating protein function (Oremland and Stolz 2003).

The arsenic-rich Champagne Pool is a large hot spring in the Waiotapu region of the North Island in New Zealand formed by a hydrothermal eruption 900 years ago. The area has the sulfur chemistry features of a high-temperature region, hydrogen sulfide gas discharge, sulfur precipitation and build-up of sulfuric acid that are manifested in solfatara fields, mud pits and acidic hot springs (Giggenbach et al. 1994; Hedenquist 1991). The Champagne Pool has a pH of 5.5 which is relatively constant due to buffering by the high flux of CO_2. Gases other than carbon dioxide (73 %) are nitrogen (16.2 %), methane (6.4 %), hydrogen (2.3 %) and hydrogen sulfide (1.7 %). The pool is fed by chloride water directly from a deep hydrothermal reservoir at 230 ° C, but the temperature in the pool is around 75 °C. It has high concentrations of silica and metalloid ion-sulfide complexes. Precipitates formed within the pool are enriched in arsenic, antimony, thallium and mercury. The pool water contains high concentrations of arsenic ions and compounds which were determined at 2.9–4.2 mg/l, thereof around 2 mg (16 μM) arsenite (Hug et al. 2014).

The Champagne Pool water is apparently a rather toxic cocktail for living organisms. Microbial diversity in the pool has been studied using both culturing and culture-independent methods (Hug et al. 2014; Hetzer et al. 2007). The cell density was low compared to other geothermal springs within New Zealand and the overall diversity likewise. Analyses of environmental DNA indicated the dominance of *Sulfurihydrogenibium*. The phylogenetic relationships indicated primarily hydrogen-oxidizing and sulfur metabolism. Hetzer et al. (2007) proposed that a unique chemical character, the presence of arsenic and other metallic compounds, is the limiting factor for the microbial diversity and biomass and that only metal ion-tolerant or metal ion-resistant microorganisms survive these conditions. A new species, *Venenivibrio stagnispumantis*, was isolated from this hot spring. The species grew in the presence of arsenite, arsenate and antimonite at considerably higher concentrations than found in the Champagne Pool spring water (Hetzer et al. 2008). A recent follow-up study on the Champagne Pool microbiome included phylogenetic analysis of 16S rRNA genes from metagenomes and supported the above findings for the Champagne Pool. The dominance of *Sulfurihydrogenibium* was confirmed, and the importance of sulfur metabolism was revealed by the presence of several phylotypes closely related to known sulfur and sulfide oxidizers, as well as sulfur and sulfate reducers. The metagenome analysis revealed genes encoding for arsenate reductase (Hug et al. 2014). Interestingly, arsenite oxidase genes (*aio*) needed for dissimilatory utilization or detoxification were not found in the metagenomic data set suggesting tolerance is the main selective force acting on the microbiome.

It is interesting to compare the Champagne Pool with arsenic-rich hot springs in different geographic locations and geological settings, the Alvord Desert and Yellowstone National Park geothermal systems in North America. The Alvord Desert is located in the Great Basin where the highest temperatures below ground reach levels which are not much higher than 200 °C. The geothermal water is not heated up by magmatic heat as in volcanic areas, but rather the geothermal activity reflects loci of shear transfer between fault systems in the geological strata (Costa et al. 2009, Faulds et al. 2004). This accounts for relatively low levels of sulfide in the water and the absence of sulfuric acid-buffered springs in the Great Basin (Costa et al. 2009). The microbiology and geochemistry of a typical arsenic-rich hot spring system in the Alvord Desert was studied by Connon and co-workers (2008). The total arsenic concentration in the spring was relatively stable at 60 μM, while the ratio of As(III) to As(V) decreased down the efflux channel. The pH was also relatively narrow, in the range of 6.77–6.81, and increased along the flow. Temperature was close to 80 °C at the source. The thermophilic biota was manifested in chemolithoautotrophically driven microbial communities of a quite different visual character along the temperature gradient. Close to the source at 74–78 °C, where arsenite concentration was highest, the microbial diversity was low and displayed in a thin colourless biofilm on rock pebbles. It was dominated by *Sulfurihydrogenibium* and a particular *Thermus* clade, the *T. aquaticus* clade, and *Thermocrinis* was relatively abundant. A profuse orange microbial mat at a lower sampling temperature was more diverse. Metabolic activities construed from the phylogenetic analysis indicated that hydrogen and the available sulfide were the energy sources for the chemolithoautotrophic *Aquificales* species at the spring source. The relatively high pH, the substantial presence of *Thermocrinis* and *Thermus* and the absence of thermoacidophilic species such as *Sulfolobus* and *Hydrogenobaculum* in the Alvord system further indicated limited relevance of sulfide for this particular ecosystem.

The geology of the Yellowstone National Park differs from the Alvord system. It is a volcanic high-temperature area, and the chemistry is dominated by sulfur, but hot springs rich in arsenic are widespread. A culture-independent study of arsenite-oxidizing hot springs in Yellowstone Park by Hamamura et al. (2009) showed that the distribution of different *Aquificales* was determined by pH, which was consistent with prior studies (Reysenbach et al., 2005; Hjorleifsdottir et al. 2001; Skirnisdottir et al. 2000). The pH in turn relates to sulfide concentration in the source water and the extent of subsequent oxidation and accumulation of sulfuric acid in the spring. Arsenic concentrations were relatively high (15–134 μM) in sampling sites of pH 3.1–8. The analysis showed that *Hydrogenobaculum* was dominant in the acidic hot springs (pH 2.6–3.6), *Sulfurihydrogenibium* in the slightly acidic hot springs of high-sulfide content (pH 6.1) and *Thermocrinis* in the alkaline hot springs of low levels of dissolved sulfide. The *Thermus* strains detected in the hot springs of the higher pH of 6.1 and 7.0 fell all within the *T. aquaticus* clade, which is the same *Thermus* clade as observed in high abundance in the Alvord high arsenic hot springs. Taken together, these and other studies demonstrate very clearly differences in bacterial diversity along geochemical gradients of temperature, pH, of the energy sources sulfide and hydrogen as well as of the toxic compounds arsenite and antimonite.

The presence or absence of compounds such as sulfur and arsenic compounds may be important selective factors acting on the microbiome, as energy sources or as toxic elements in high concentrations delimiting distribution of certain taxa. Arsenite oxidase found in many thermophilic species could have a role in a detoxification process as may be the case for *Thermus* and *Hydrogenobaculum* spp. (Connon et al. 2008; Donahoe-Christiansen et al. 2004; Gihring and Banfield 2001) but may also play a part in a dissimilatory pathway supporting growth as has been reported for *Thermocrinis ruber (*Härtig et al. 2014). It is not clear if *Sulfurihydrogenibium* species can utilize arsenite as an energy source for growth (Hamamura et al. 2009; Aguiar et al. 2004). However, arsenate as a terminal electron acceptor with H_2 and H_2S as electron donors does support chemoautotrophic growth of different species of the genus (Nakagawa et al. 2005). The relatively large abundance of *Thermus* of the *T. aquaticus* clade in arsenic-rich hot springs in Yellowstone Park and the Alvord system is of a particular interest and indicates an important role for this clade in arsenite detoxification. Still, the *T. aquatics* bacteria in the Alvord hot spring system form a phylogenetic cluster, separate from their *T. aquaticus* relatives in the Yellowstone Park (Fig. 3.11). This raises the question if the distance keeps these groups genetically apart or if the populations adapted to their respective environments play an 'active' part in restricting migration between the areas by outcompeting incoming strains.

The presence or absence of arsenite oxidase may relate to the source of the isolate. None were detected in *Sulfurihydrogenibium* strains from Iceland – a low arsenic geological setting – but have been found in *Sulfurihydrogenibium* and apparently acquired in one strain by lateral gene transfer from *Thermus* (Hamamura et al. 2010). Also, in an ongoing pan-genomic study of *Thermus* by the authors, arsenite oxidase was found to be present in all species examined except for *T.*

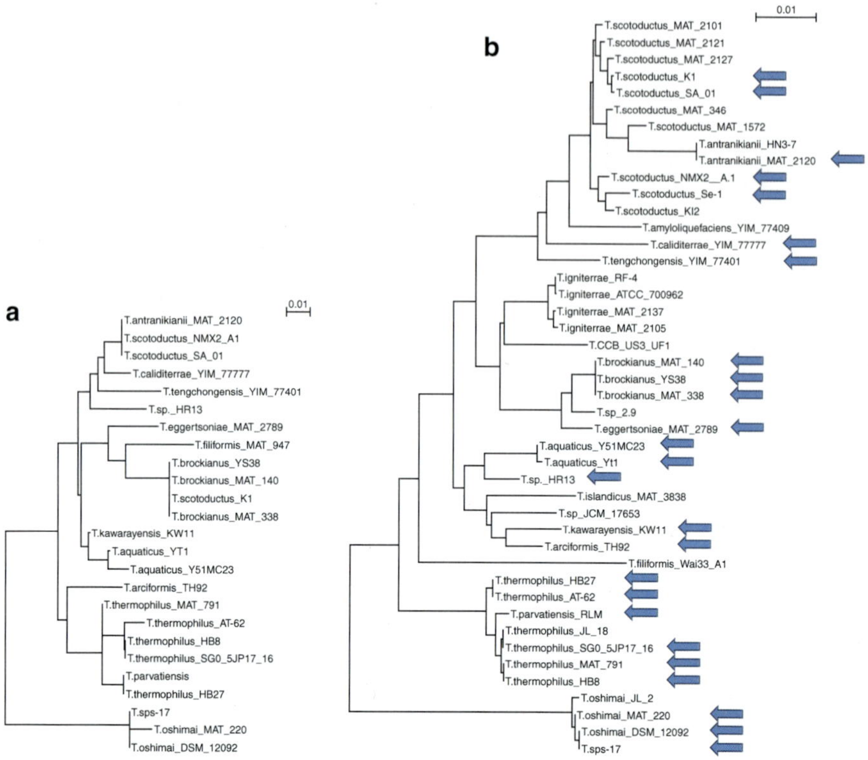

Fig. 3.12 Neighbour-joining (NJ) trees of (**a**) 16S rRNA gene sequences and (**b**) deduced amino acid sequences encoded by *aio*-like genes. Sequences derived from available whole-genome sequences of *Thermus* strains. Arrows point to those found to harbour *aio*-like genes. Strains indicated by MAT numbers were obtained and genome sequenced at Matís

igniterrae and *T. islandicus* from Iceland (Fig. 3.12). The arsenite oxidase sequences formed species-specific phylogenetic groups. However, the gene was not present in all strains of a particular *Thermus* species, indicating gene loss in strains missing the genes. A clear evidence of gene gain could also be found in a strain of *T. scotoductus* from Armenia, the gene apparently originating from *T. brockianus* by lateral gene transfer. Of a particular notice is the absence of the gene in all the Icelandic *T. scotoductus* strains but presence in all other *T. scotoductus* strains. Also, noteworthy is the absence of arsenite oxidase genes in the genomes of *Thermus* species from the Great Boiling Spring, a low arsenic hot spring in the Great Basin. It is missing in both *T. thermophilus* JL-18, and *T. oshimai* JL-2, but present in all other currently sequenced genomes of these two species. Also, the major *Thermus* clades detected in this hot spring belong to the *T. thermophilus* and the *T. scotoductus* clades (Costa et al. 2009), not to the *T. aquaticus* clade that was observed to be dominant in the arsenite-rich hot springs of the Alvord Desert. These results on *Thermus* and the results of Hamamura et al. (2009, 2010) on *Aquificales* strains indicate that the presence of *aio*-like genes can vary between species and even between strains within a species – they may be gained and they may be lost. This probably reflects different environmental factors acting on different subgroups and is realized in ecotypes of different adaptive advantages.

3.4.1.3 Historical Barriers: Time

Those factors where time plays a role in influencing distributions may be termed historical. Successful colonization from an incoming species depends both on the abundance and frequency of migration events, factors that depend on geographical distance and time. Measured on a geological timescale "*everything may be everywhere*" in the sense that a particular species or strain migration may have occurred at some point in time but then not 'intensively enough' to overcome biological or physicochemical barriers. Also, the geological period when a successful migration event takes place may be sufficiently long to establish a distinct genetic difference between the colony and the source populations.

Major geological events such as rare major crustal upheavals or other momentous geological events can also be termed historic on both local and global scales. They may result in the creation of new faults in the geological strata as conduits of heated water and routes of dissemination. Also, new geothermal habitats may be created that need time to develop to their full extent. Conversely, volcanic eruptions may also spread microbes from existing habitats or from the subsurface over long distances in a sufficient magnitude to enable colonization of distant regions. The examples given below highlight in some way the historical dimension of dispersal.

A volcanic eruption and its aftermath may have a significant effect on the dispersal and geographic distribution patterns of hyperthermophiles in the sea by releasing site bound subsurface dwellers. The presence of hyperthermophilic archaea in low-temperature hydrothermal fluids from the Juan de Fuca Ridge that were not detected in the ambient seawater was reported by Holden et al. (1998). This suggested that they had grown below the seafloor at permissive temperatures. Similarly, hyperthermophiles have been enriched from cold plume waters shortly after an eruption (Huber et al. 1990; Delaney et al. 1998). A historic event of this kind may explain the distribution of a particular microbial lineage and theoretically at least the place of origin and approximate time since the 'scattering' may be inferred from gene histories and phylogeographic distributions.

Miller et al. (2002) have speculated on various historical aspects of the distribution of *Mastigocladus laminosus*. On the basis of 16S rRNA and metabolic gene divergence analysis, they attempted to reconstruct the evolutionary history of the species and the timing of diversification events and to identify the site of origin of a relatively recent expansion of a particular subgroup of the species. They discuss how this history may relate to geological events.

Historical reasons have also been suggested to explain the distribution of microbes belonging to the *Aquificales* in the Yellowstone Park (Takacs-Vesbach et al. 2008). Molecular phylogenetic approaches and dispersal-vicariance analyses combined with environmental data were used to examine the distribution of the members of *Sulfurihydrogenibium* in thermal springs in the region. A clear pattern of geographically isolated microbial populations was found. The distribution correlated with the boundary of Yellowstone's calderas (or volcanic craters) and suggested that volcanic eruptions of the past 2 million years explained more of the DNA sequence divergence than contemporary factors, such as habitat preferences or geographical distance.

3.4.2 Biogeography of *Thermus*

Presently, 14 *Thermus* species are validly named. These are *Thermus aquaticus* (Brock and Freeze 1969), *T. thermophilus* (Manaia et al. 1995; Oshima and Imahori 1974), *T. filiformis* (Hudson et al. 1987), *T. scotoductus* (Kristjansson et al. 1994), *T. brockianus* (Williams et al. 1995), *T. oshimai* (Williams et al. 1996), *T. igniterra* and *T. antranikianii* (Chung et al. 2000), *T. islandicus* (Björnsdottir et al. 2009), *T. arciformis* (Zhang et al. 2010), *T. composti* (Vajna et al. 2012), *T. tenchongensis* (Yu et al. 2013), *T. calidíterrae* (Ming et al. 2014) and *T. amyloliquefaciens* (Yu et al. 2015). A number of putative species remain to be described, including *T. eggertssoniae*, which seems to be abundant in Iceland (unpublished).

Geographical isolation may be an important factor in species divergence within the genus *Thermus* (Fig. 3.11). Thus, the *T. aquaticus* lineage has only been found in the North American continent, and the linage shows phylogeographical clustering with genetically different populations in the Great Basin and the Yellowstone Park. The phylogenetic depth is relatively large, indicating that the lineage may have a guild structure consisting of different ecotypes. This could be an example of adaptive radiation of a particular lineage into unoccupied niches in this geothermal area, whereas similar niches in Iceland are populated by different *Thermus* species.

Another example of apparently restricted distribution is the *T. filiformis* clade. It forms a highly distinct lineage in the *Thermus* group and is the only known terrestrial species in New Zealand. This species has not been reported from anywhere else. Now, another species belonging to this lineage but clearly different and apparently endemic to hot springs in Chile has recently been detected by 16S rRNA analysis of the El Tatio hot springs (Fig. 3.11). Other examples of apparently endemic lineages are represented by recently reported species from Tibet and China, *T. tenchongensis* (Yu et al. 2013) and *T. calididiterrae* (Ming et al. 2014) as well as *T. yunnanensis* (Gong et al. 2005) and *T. rehai* (Lin et al. 2002). One species, *T. thermophilus*, seems to be adapted to marine environments and is ubiquitous in submarine and coastal springs worldwide (Kristjansson et al. 1986, 2000; Williams and Sharp 1995). *T. thermophilus* strains have also been reported from nonvolcanic habitats with high content of organics such as compost (Kim et al. 2014; Lyon et al. 2000) and oil fields (Ramos-Padron et al. 2011; Hao et al. 2004), and it is the main *Thermus* species observed at higher temperatures in the Great Boiling hot spring in the nonvolcanic geothermal area of the Great Basin, USA. *T. thermophilus* has not been detected in Icelandic terrestrial geothermal habitats despite intensive sampling efforts, including recent exhaustive massive 16S rRNA amplicon analysis of a wide range of hot springs (unpublished results, Fig. 3.13). Other *Thermus* species such as *T. scotoductus* and *T. brockianus* show a cosmopolitan distribution. Nearly identical

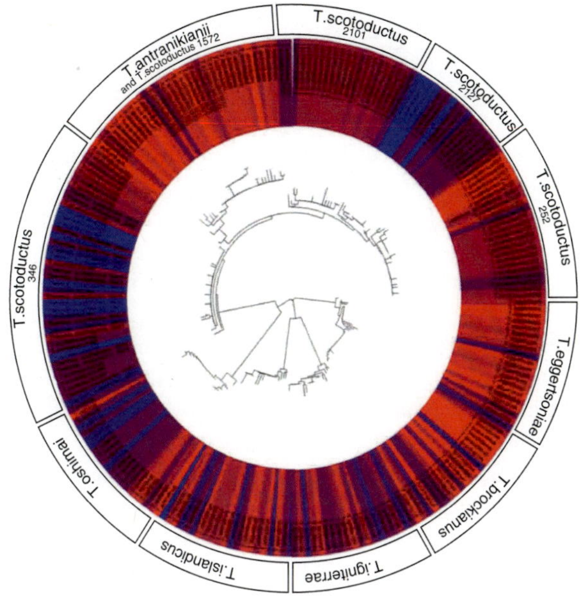

Fig. 3.13 A phylogenetic tree derived from an unpublished study by the authors (2016), where the distribution of *Thermus* in 25 hot springs in Iceland was mapped by massive parallel sequencing. Tens of thousands of partial 16S sequences were obtained and clustered. The clusters were pruned with respect to similarity within hot springs, and each branch-end represents numerous near-identical sequences. Colours indicate temperatures from the coldest of 54° (*blue*) to the warmest of 88 °C (*red*)

16S rRNA sequences of *T. igniterrae* first described from Icelandic geothermal areas have been detected in strains from Australia (Hreggvidsson et al. 2006).

Species boundaries should ideally circumscribe ecologically distinct populations. This is not clear for any of the genotypically different *Thermus* species. The grouping of *Thermus* isolates from distant locations on the basis of phenotypic traits reflects adaptations to conditions at the geographic site rather than phylogenetic relationships (Hudson et al. 1989; Santos et al. 1989). A particular *Thermus* genospecies therefore appears to consist of different ecotypes or local adaptations to conditions of a particular geographical area, which may furthermore be influenced by the presence or absence of other species or strains. This is clearly evident in a recent study in Iceland by the authors where different ecotypes can be seen within the *T. scotoductus* clade in relation to the temperature of the sampling site. Some groups were preferentially found in high-temperature hot springs and others at lower temperatures (unpublished, Fig. 3.13). Also, the geographic distribution of arsenite oxidase genes in genomes of *Thermus* species and strains appears to be discontinuous and reflecting geographic locations and local arsenite environmental concentrations (discussed above).

3.4.3 Ecological Adaptations of *Thermus*

Perhaps *T. thermophilus* comes closest to representing an ecologically distinct population due to its apparent preference of marine geothermal habitats. All other *Thermus* species are found in terrestrial hot springs, and their niches or habitat preferences have not been determined. The presence of different ecotypes within a species also confounds the matter.

Today we have a rather fragmented picture of the ecology of *Thermus* in terrestrial hot springs. Under laboratory conditions the different species are similar in phenotypic properties, such as in pH and temperature optima as well as in carbon utilization patterns. Frequently, we isolated different *Thermus* species from the same hot spring. With reference to the competitive exclusion principle, this raises the question of how phenotypically and physiologically very similar *Thermus* species can coexist in a particular hot spring. However, we cannot really distinguish between representatives of stable populations and transient species, using traditional isolation techniques. The observed species diversity may be partly maintained by migration events from microhabitats within the hot spring, sediments or microbial mats, from different patches of environmental gradients, by periodic disturbances or by environmental fluctuations.

Adaptive traits important for the habitat preferences of many thermophilic species may involve small physiological differences or marginally different responses to physicochemical adverse conditions and may therefore be difficult to evaluate. Also, a particular species could have a slight competitive edge for limiting resources under oligotrophic conditions, or there could be a small but significant difference in growth rates between species at the limits of their temperature or pH growth ranges. Subtle combinations of such traits might determine the niche position of a *Thermus* species and consequently influence the diversity and species abundance in a particular hot spring. It follows that such differences between species would be difficult to detect and evaluate in the laboratory.

Two important niche parameters that may explain growth in a particular habitat are pH and temperature. In studies based on cultivated isolates, we observed that the specific temperatures of the isolation sites did not relate to any of the *Thermus* species lineages examined (Hreggvidsson et al. 2006). However, the pH of the isolation sites indicated a relationship to lineage formation. *T. igniterrae* and *T. brockianus* were generally isolated from hot springs with pH > 8.0, while *T. scotoductus* was more commonly isolated from hot springs of neutral and lower pH. Furthermore, *T. igniterrae* appeared to be very sensitive to ionic strength. No clear phenotype could be attributed to the genetic lineages based on carbon utilization abilities. A number of isolates from each *Thermus* species were also examined, and the species were shown to be genotypically tight, except for *T. scotoductus* that formed several deep subclusters, which perhaps indicated different ecotypes (Hreggvidsson et al. 2006). In other studies we established that at least some of the *Thermus* species may harness energy by oxidizing sulfur compounds (Skirnisdottir et al. 2001; Björnsdottir et al. 2009), and there are reports of *T. scotoductus* strains using nitrate, Fe(III), Mn(IV) or S^O as terminal electron acceptors in the respiratory chain (Balkwill et al. 2004).

Thermocrinis ruber has been reported to use arsenite as a chemolithotrophic energy source for growth (Härtig et al. 2014). The isolate *Thermus* HR13 from arsenite-rich hot spring in California has been reported to oxidize arsenite without energy gain, but it is able to use arsenate as a terminal electron acceptor in place of oxygen (Gihring and Banfield 2001). Different abilities to tolerate, detoxify or exploit these compounds in their electron transport chain may explain differential distribution of some *Thermus* species. Water activity in the habitats may also be important for the distribution, e.g. the requirement, ability or inability to tolerate an ionic strength above a certain level. Differences in the tolerance to salt concentration have been observed between the different genospecies. The highest salt tolerance was found among strains of the *T. thermophilus* lineage, which is explained by its habitat of marine and coastal hot springs. The opposite was found for the *T. igniterrae* lineage that appeared to be very sensitive to ionic strength (Hreggvidsson et al. 2006). *T. igniterrae* shows a discontinuous and unexpected distribution. So far it has only been found in Iceland and north-eastern Australia. In Australia it was detected and isolated as the dominant *Thermus* species at the higher temperatures in a nonvolcanic hot runoff from a borehole in the Great Artesian Basin aquifer in Central Queensland (Spanevello and Patel 2004). The physicochemical characteristic of this habitat of *T. igniterrae* in Australia agrees well with our findings, which indicate that the species prefers high-temperature and alkaline hot springs low in mineral salt content and is sensitive to NaCl, bivalent cations and sulfur compounds. In this case apparently, *the environment selects.*

The conclusion is that the genospecies of *Thermus* consist of a number of ecologically distinct populations distinguished only by few adaptive traits to physicochemical conditions and that their distribution ranges are dictated by both geographical distances and environmental variables. It remains to be seen if adaptive traits observed in one region for many genospecies are universal. The niche space may have been defined for some of them at least in one or few dimensions, e.g. by tolerance and intolerance to salts for *T. thermophilus* and *T. igniterrae*, respectively, by species- and strain-specific tolerance to toxic compounds such as to arsenite or mercury or by adaptation to different temperature zones as evident within *T. scotoductus* or to high alkalinity of *T. brockianus* and *T. igniterrae*. An ecological approach in conjunction with genomic analysis and multilocus population studies will definitely reveal more about the different adaptations of *Thermus*, the evolutionary history of the genus, and possibly enable the positioning and tracing of ecological radiations of different species in time and space.

Acknowledgements We thank Haukur Jóhannesson for providing photographs as well as for reviewing a part of the text.

References

Aguiar P, Beveridge TJ, Reysenbach AL (2004) *Sulfurihydrogenibium azorense,* sp. nov., a thermophilic hydrogen-oxidizing microaerophile from terrestrial hot springs in the Azores. Int J Syst Evol Microbiol 54:33–39

Arnórsson S (2003) Arsenic in surface- and up to 90 °C ground waters in a basalt area, N-Iceland: processes controlling its mobility. Appl Geochem 18:1297–1312

Balkwill DL, Kieft TL, Tsukuda T, Kostandarithes HM, Onstott TC, Macnaughton S, Bownas J, Fredrickson JK (2004) Identification of iron-reducing *Thermus* strains as *Thermus scotoductus*. Extremophiles 8:37–44

Björnsdottir SH, Petursdottir SK, Hreggvidsson GO, Skirnisdottir S, Hjorleifsdottir S, Arnfinnsson J, Kristjansson JK (2009) *Thermus islandicus* sp. nov., a mixotrophic sulfur-oxidizing member of the genus *Thermus*. Int J Syst Evol Microbiol 59:2962–2966

Bödvarsson G (1961) Physical characteristics of natural heat resources in Iceland. Jökull 11:29–38

Brock TD (1986) An overview of the thermophiles. In: Thermophiles: general, molecular and applied microbiology. John Wiley & Sons, New York, pp. 1–16

Brock TD, Freeze H (1969) *Thermus aquaticus* gen. nov. and sp. nov., a non-sporulating extreme thermophile. J Bacteriol 98:289–297

Chung AP, Rainey FA, Valente M, Nobre MF, da Costa MS (2000) *Thermus igniterrae* sp. nov. and *Thermus antranikianii* sp. nov., two new species from Iceland. Int J Syst Evol Microbiol 50:209–217

Cohan FM (2001) Bacterial species and speciation. Syst Biol 50:513–524

Cohan FM (2002) What are bacterial species? Annu Rev Microbiol 56:457–487

Cohan FM (2006) Towards a conceptual and operational union of bacterial systematics, ecology, and evolution. Philos Trans R Soc Lond B Biol Sci 361:1985–1996

Connon SA, Koski AK, Neal AL, Wood SA, Magnuson TS (2008) Ecophysiology and geochemistry of microbial arsenic oxidation within a high arsenic, circumneutral hot spring system of the Alvord Desert. FEMS Microbiol Ecol 64:117–128

Costa KC, Navarro JB, Shock EL, Zhang CL, Soukup D, Hedlund BP (2009) Microbiology and geochemistry of great boiling and mud hot springs in the United States. Extremophiles 13:447–459

Delaney JR, Kelley DS, Lilley MD, Butterfield DA, Baross JA, Wilcock WSD, Embley RW, Summit M (1998) The quantum event of oceanic crustal accretion: impacts of diking at mid-ocean ridges. Science 281:222–230

Donahoe-Christiansen J, D'Imperio S, Jackson CK, Inskeep WP, McDermott TR (2004) Arsenite-oxidizing *Hydrogenobaculum* strain isolated from an acid-sulfate-chloride geothermal spring in Yellowstone National Park. Appl Environ Microbiol 70:1865–1868

Eder W, Huber R (2002) New isolates and physiological properties of the Aquificales and description of *Thermocrinis albus* sp. nov. Extremophiles 6:309–318

Faulds JE, Coolbaugh M, Blewitt G, Henry CD (2004) Why is Nevada in hot water? Structural controls and tectonic model of geothermal systems in the Northwestern Great Basin. Geoth Res T 28:649–654

Giggenbach WF, Sheppard DS, Robinson BW, Stewart MK, Lyon GL (1994) Geochemical structure and position of the Waiotapu geothermal field, New Zealand. Geothermics 23:599–644

Gihring TM, Banfield JF (2001) Arsenite oxidation and arsenate respiration by a new *Thermus* isolate. FEMS Microbiol Lett 204:335–340

Gogarten JP, Doolittle WF, Lawrence JG (2002) Prokaryotic evolution in light of gene transfer. Mol Biol Evol 19:2226–2238

Gong N, Chen C, Xie L, Chen H, Lin X, Xhang R (2005) Characterization of a thermostable alkaline phosphatase from a novel species *Thermus yunnanensis* sp. nov. and investigation of its cobalt activation at high temperature. Biochim Biophys Acta 1750:103–111

Hamamura N, Macur RE, Korf S, Ackerman G, Taylor WP, Kozubal M, Reysenbach AL, Inskeep WP (2009) Linking microbial oxidation of arsenic with detection and phylogenetic analysis of arsenite oxidase genes in diverse geothermal environments. Environ Microbiol 11:421–431

Hamamura N, Macur RE, Liu Y, Inskeep WP, Reysenbach AL (2010) Distribution of aerobic arsenite oxidase genes within the Aquificales. In: Hamamura N, Suzuki S, Mendo S, Barroso CM, Iwata H, Tanabe S (eds) Interdisciplinary studies on environmental chemistry – biological responses to contaminants. TERRAPUB, Tokyo, pp. 47–55

Hao R, Lu A, Wang G (2004) Crude-oil-degrading thermophilic bacterium isolated from an oil field. Can J Microbiol 50:175–182

Hedenquist JW (1991) Boiling and dilution in the shallow portion of the Waiotapu geothermal system, New Zealand. Geochim Cosmochim Acta 55:2753–2765

Hetzer A, Morgan HW, McDonald IR, Daughney CJ (2007) Microbial life in Champagne Pool, a geothermal spring in Waiotapu, New Zealand. Extremophiles 11:605–614

Hetzer A, McDonald IR, Morgan HW (2008) *Venenivibrio stagnispumantis* gen. nov., sp. nov., a thermophilic hydrogen-oxidizing bacterium isolated from Champagne Pool, Waiotapu, New Zealand. Int J Syst Evol Microbiol 58:398–403

Hjorleifsdottir S, Skirnisdottir S, Hreggvidsson GO, Holst O, Kristjansson JK (2001) Species composition of cultivated and non-cultivated bacteria from short filaments in an Icelandic hot spring at 88 °C. Microb Ecol 42:117–125

Hjort K, Bernander R (1999) Changes in cell size and DNA content in *Sulfolobus* cultures during dilution and temperature shift experiments. J Bacteriol 181:5669–5675

Holden JF, Summit M, Baross JA (1998) Thermophilic and hyperthermophilic microorganisms in 3–30 °C hydrothermal fluids following a deep-sea volcanic eruption. FEMS Microbiol Ecol 25:33–41

Hreggvidsson GO, Kristjansson JK (2003) Thermophily. In: Gerday C, Glansdorff N (eds) Extremophiles – Encyclopedia of Life Support Systems (EOLSS) Developed under the Auspices of the UNESCO. Eolss Publishers, Oxford , http://www.eolss.net

Hreggvidsson GO, Skirnisdottir S, Smit B, Hjorleifsdottir S, Marteinsson VT, Petursdottir SK, Kristjansson JK (2006) Polyphasic analysis of *Thermus* isolates from geothermal areas in Iceland. Extremophiles 10:563–575

Huber R, Stoffers P, Cheminee JL, Richnow HH, Stetter KO (1990) Hyperthermophilic archaebacteria within the crater and open-sea plume of erupting Macdonald Seamount. Nature 345: 179–181

Huber R, Eder W, Heldwein S, Wanner G, Huber H, Rachel R, Stetter KO (1998) *Thermocrinis ruber* gen. nov., sp. nov., a pink-filament-forming hyperthermophilic bacterium isolated from Yellowstone National Park. Appl Environ Microbiol 64:3576–3583

Hudson JA, Morgan HW, Daniel RM (1987) *Thermus filiformis* sp. nov., a filamentous caldoactive bacterium. Int J Syst Bacteriol 37:431–436

Hudson JA, Morgan HW, Daniel RM (1989) Numerical classification of *Thermus* isolates from globally distributed hot springs. Syst Appl Microbiol 11:250–256

Hug K, Maher WA, Stott MB, Krikowa F, Foster S, Moreau JW (2014) Microbial contributions to coupled arsenic and sulfur cycling in the acid-sulfide hot spring Champagne Pool, New Zealand. Front Microbiol 5:569

Härtig C, Lohmayer K, Kolb S, Horn MA, Inskeep WP, Planer-Friedrich B (2014) Chemolithotrophic growth of the aerobic hyperthermophilic bacterium *Thermocrinis ruber* OC 14/7/2 on monothioarsenate and arsenite. FEMS Microbiol Ecol 90:747–760

Inskeep WP, Rusch DB, Jay ZJ et al (2010) Metagenomes from high-temperature chemotrophic systems reveal geochemical controls on microbial community structure and function. PLoS One 5:e9773

Kaasalainen H, Stefansson A (2011) Chemical analysis of sulfur species in geothermal waters. Talanta 85:1897–1903

Kashefi K, Lovley D (2003) Extending the upper temperature limit for life. Science 301:934

Keller NS, Stefansson A, Sigfusson B (2014) Arsenic speciation in natural sulfidic geothermal waters. Geochim Cosmochim Acta 142:15–26

Kim YM, Cho HU, Park KY, Cho KH (2014) Identification of the bacterial community of a pilot scale thermophilic aerobic bioreactor treating sewage sludge. Int Biodeter Biodegr 92: 66–70

Kristjansson JK, Hreggvidsson GO (1995) Ecology and habitats of extremophiles. World J Microbiol Biotechnol 11:17–25

Kristjansson JK, Hreggvidsson GO, Alfredsson GA (1986) Isolation of halotolerant *Thermus* spp. from submarine hot springs in Iceland. Appl Environ Microbiol 52:1313–1316

Kristjansson JK, Hjorleifsdottir S, Marteinsson VT, Alfredsson GA (1994) *Thermus scotoductus*, sp. nov., a pigment-producing thermophilic bacterium from hot tap water in Iceland and including *Thermus* sp. X-1. Syst Appl Microbiol 17:44–50

Kristjansson JK, Hreggvidsson GO, Grant WD (2000) Taxonomy of extremophiles. In: Priest FG, Goodfellow M (eds) Applied microbial systematics. Kluwer Academic Publishers, Dordrecht, pp. 231–292

Langner HW, Jackson CR, McDermott TR, Inskeep WP (2001) Rapid oxidation of arsenite in a hot spring ecosystem, Yellowstone National Park. Environ Sci Technol 35:3302–3309

Lawrence JG (2002) Gene transfer in bacteria: speciation without species? Theor Popul Biol 61:449–460

Levin R (1981) Periodic selection, infectious gene exchange and the genetic structure of *E. coli* populations. Genetics 99:1–23

Lin L, Chen C, Peng Q, Ben K, Zhou Z (2002) *Thermus rehai* sp. nov. isolated from Rehai of Tengchong, Yunnan province, China. J Basic Microbiol 42:337–344

Lyon PF, Beffa T, Blanc M, Auling G, Aragno M (2000) Isolation and characterization of highly thermophilic xylanolytic *Thermus thermophilus* strains from hot compost. Can J Microbiol 46:1029–1035

Manaia CM, Hoste B, Guiterrez MC, Gillis M, Ventosa A (1995) Halotolerant *Thermus* strains from marine and terrestrial hot springs belongs to *Thermus thermophilus* (ex Oshima and Imahori, 1974) nom. rev. emend. Syst Appl Microbiol 17:526–532

Marteinsson VT, Kristjánsson JK, Kristmannsdóttir H, Dahlkvist M, Sæmundsson K, Hannington M, Petursdottir SK, Geptner A, Stoffers P (2001) Discovery and description of giant submarine smectite cones on the seafloor in Eyjafjordur, Northern Iceland, and a novel thermal microbial habitat. Appl Environ Microbiol 67:827–833

Menzel P, Gudbergsdottir SR, Rike AG, Lin L, Zhang Q, Contursi P, Moracci M, Kristjansson JK, Bolduc B, Gavrilov S, Ravin N, Mardanov A, Bonch-Osmolovskaya E, Young M, Krogh A, Peng X (2015) Comparative metagenomes of eight geographically remote terrestrial hot springs. Microb Ecol 70:411–424

Miller SR, Castenholz RW (2000) Evolution of thermotolerance in hot spring cyanobacteria of the genus *Synechococcus*. Appl Environ Microbiol 66:4222–4229

Miller SR, Castenholz RW, Pedersen D (2002) Phylogeography of the thermophilic cyanobacterium *Mastigocladus laminosus*. Appl Environ Microbiol 73:4751–4759

Miller S, Strong A, Jones K, Ungerer M (2009) Bar-coded pyrosequencing reveals the shared bacterial community properties along the temperature gradients of the alkaline hot springs in Yellowstone National Park. Appl Environ Microbiol 75:4565–4572

Ming H, Yin YR, Li S, Nie GX, Yu TT, Zhou EM, Liu L, Dong L, Li WJ (2014) *Thermus caliditerrae* sp. nov., a novel thermophilic species isolated from a geothermal area. Int J Syst Evol Microbiol 64:650–656

Nakagawa S, Shtaih Z, Banta A, Beveridge TJ, Sako Y, Reysenbach AL (2005) *Sulfurihydrogenibium yellowstonense* sp. nov., an extremely thermophilic, facultatively heterotrophic, sulfur-oxidizing bacterium from Yellowstone National Park, and emended description of the genus *Sulfurihydrogenibium, Sulfurihydrogenibium subterraneum* and *Sulfurihydrogenibium azorense*. Int J Syst Evol Microbiol 55:2263–2268

Oremland RS, Stolz JF (2003) The ecology of arsenic. Science 300:939–944

Oshima T, Imahori K (1974) Description of *Thermus thermophilus* (Yoshida and Oshima) comb. nov., a nonsporulating thermophilic bacterium from a Japanese thermal spa. Int J Syst Bacteriol 24:102–112

Palmason G (2005) Jardhiti – Edli og nyting audlindar. Hid islenska bokmenntafelag, Reykjavik (Geothermal energy- its nature and applications Reykjavik, Icelandic Literary Society; 298 pages)

Papke RT, Ramsing NB, Bateson MM, Ward DM (2003) Geographical isolation in hot spring cyanobacteria. Environ Microbiol 5:650–659

Petursdottir SK, Hreggvidsson GO, da Costa MS, Kristjansson JK (2000) Genetic diversity analysis of *Rhodothermus* reflects geographical origin of the isolates. Extremophiles 4:267–274

Ramos-Padron E, Bordenave S, Lin S, Bhaskar IM, Dong X, Sensen CW, Fournier J, Voordouw G, Gieg LM (2011) Carbon and sulfur cycling by microbial communities in a gypsum-treated oil sand tailing pond. Environ Sci Technol 45:439–446

Redder P, Garrett RA (2006) Mutations and rearrangements in the genome of *Sulfolobus solfataricus* P2. J Bacteriol 188:4198–4206

Reno ML, Held NL, Fields CJ, Burke PV, Whitaker RJ (2009) Biogeography of the *Sulfolobus islandicus* pan-genome. Proc Natl Acad Sci U S A 106:8605–8610

Reysenbach AL, Banta A, Civello S, Daly J, Mitchel K, Lalonde S, Konhauser KO, Rodman A, Rusterholtz K, Takacs-Vesbach C (2005) The Aquificales of Yellowstone National Park. In: Inskeep WP, McDermott TR (eds) Geothermal biology and geochemistry in Yellowstone National Park. Thermal Biology Institute, Montana State University, Bozeman, pp. 129–142

Romero L, Alonso H, Campano P, Fanfani L, Cidu R, Dadea C, Keegan T, Thornton I, Farago M (2003) Arsenic enrichment in waters and sediments of the Rio Loa (Second Region, Chile). Appl Geochem 18:1399–1416

Santos MA, Williams RAD, da Costa MS (1989) Numerical taxonomic study of *Thermus* isolates from hot springs in Portugal. Syst Appl Microbiol 12:10–15

Skirnisdottir S, Hreggvidsson GO, Hjorleifsdottir S, Marteinsson VT, Petursdottir SK, Holst O, Kristjansson JK (2000) Influence of sulfide and temperature on species composition and community structure of hot spring microbial mats. Appl Environ Microbiol 66:2835–2841

Skirnisdottir S, Hreggvidsson GO, Holst O, Kristjansson JK (2001) Isolation and characterization of a mixotrophic sulfur oxidizing *Thermus scotoductus*. Extremophiles 5:45–51

Spanevello MD, Patel BKC (2004) The phylogenetic diversity of *Thermus* and *Meiothermus* from microbial mats of an Australian subsurface aquifer runoff channel. FEMS Microbiol Ecol 50:63–73

Stetter KO (1996) Hyperthermophilic procaryotes. FEMS Microbiol Rev 18:149–158

Takacs-Vesbach C, Mitchell K, Jackson-Weaver O, Reysenbach AL (2008) Volcanic calderas delineate biogeographic provinces among Yellowstone thermophiles. Environ Microbiol 10:1681–1689

Takacs-Vesbach C, Inskeep WP, Jay ZJ, Herrgard MJ, Rusch DB, Tringe SG, Kozubal MA, Hamamura N, Macur RE, Fouke BW, Reysenbach AL, McDermott TR, Jennings RM, Hengartner NW, Xie G (2013) Metagenome sequence analysis of filamentous microbial communities obtained from geochemically distinct geothermal channels reveals specialization of three *Aquificales* lineages. Front Microbiol 4:84

Tettelin H, Masignanai V, Cieslewicz MJ et al (2005) Genome analysis of multiple pathogenic isolates of *Streptococcus agalactiae* for the microbial "pan-genome". Proc Natl Acad Sci U S A 102:13950–13955

Vajna B, Kanizsai S, Keki Z, Marialigeti K, Schumann P, Toth EM (2012) *Thermus composti* sp. nov., isolated from oyster mushroom compost. Int J Syst Evol Microbiol 62:1468–1490

Valverde A, Tuffin M, Cowan DA (2012) Biogeography of bacterial communities in hot springs: a focus on the actinobacteria. Extremophiles 16:669–679

Wang S, Hou W, Dong H, Jiang H, Huang L, Wu G, Zhang C, Song Z, Zhang Y, Ren H, Zhang J, Zhang L (2013) Control of temperature on microbial community structure in hot springs of the Tibetan Plateau. PLOS one 8:e62901

Ward DM (1998) A natural species concept for prokaryotes. Curr Opin Microbiol 1:271–277

Ward DM, Cohan FM (2005) Microbial diversity in hot spring cyanobacterial mats: pattern and prediction. In: Inskeep WP, McDermott TR (eds) Geothermal biology and geochemistry in Yellowstone National Park. Thermal Biology Institute, Montana State University, Bozeman, pp. 185–201

Ward DM, Bateson MM, Ferris MJ, Nold SC (1998) A natural view of microbial biodiversity within hot spring cyanobacterial mat communities. Microbiol Mol Biol Rev 62:1353–1370

Ward DM, Cohan FM, Bhaya D, Heidelberg JF, Kühl M, Grossman A (2008) Genomics, environmental genomics and the issue of microbial species. Heredity 100:207–219

Whitaker RJ, Grogan DW, Taylor JW (2003) Geographic barriers isolate endemic populations of hyperthermophilic archaea. Science 301:976–978

Wiegel GJ, Ljungdahl LG (1986) The importance of thermophilic bacteria in biotechnology. Crit Rev Biotechnol 3:39–108

Williams R, Sharp R (1995) The taxonomy and identification of *Thermus*. In: Sharp R, Williams R (eds) *Thermus* species. Plenum, New York, pp. 1–42

Williams RAD, Smith KE, Welch SG, Micallef J, Sharp RJ (1995) DNA relatedness of *Thermus* strains, description of *Thermus brockianus* sp. nov., and proposal to reestablish *Thermus thermophilus* (Oshima and Imahori). Int J Syst Bacteriol 45:795–499

Williams RAD, Smith KE, Welch SG, Micallef J (1996) *Thermus oshimai* sp. nov., isolated from hot springs in Portugal, Iceland, and the Azores and comment on the concept of a limited geographical distribution of *Thermus* species. Int J Syst Bacteriol 46:403–408

Yu TT, Yao JC, Ming H, Yin YR, Zhou EM, Liu MJ, Tang SK, Li WJ (2013) *Thermus tengchongensis* sp. nov., isolated from a geothermally heated soil sample in Tengchong, Yunnan, southwest China. Antonie Van Leeuwenhoek 103:513–518

Yu TT, Ming H, Yao JC, Zhou EM, Park DJ, Hozzein WN, Kim CJ, Wadaan MA, Li WJ (2015) *Thermus liquefaciens* sp. nov., isolated from a hot spring sediment sample. Int J Syst Evol Microbiol 65:2491–2495

Zeigler DR (2014) The *Geobacillus* paradox: why is a thermophilic bacterial genus so prevalent on a mesophilic planet? Microbiology 160:1–11

Zhang XQ, Ying Y, Ye Y, Xu XW, Zhu XF, Wu M (2010) Thermus arciformis sp. nov., a thermophilic species from a geothermal area. Int J Syst Evol Microbiol 60:834–839

Bacterial Adaptation to Hot and Dry Deserts

4

Thierry Heulin, Gilles De Luca, Mohamed Barakat, Maxime Gommeaux, Arjan de Groot, Laurence Blanchard, Philippe Ortet, and Wafa Achouak

Contents

The Moula-moula Bird

The Moula-moula bird is a wheatear of the desert. It is the only living soul that might enliven the desolation of volcanic areas with its two, black and white, clean-cut colours. I often used to follow it with my eyes, as it would tinkle its faint trill above the bald heights of the hillocks. It cannot stay still as its feet would get burned by the rock. The sun, which would elsewhere be life, here is devastating furnace.

Only, there is the Moula-moula bird to pour a note of freshness on the inferno. Its faint thrill furtively gives the illusion of greenness; it falls upon the ardent rocks as drops of dew. One only has to close one's eyes and listen to the crystal shrill then one forgets this bare world reduced to its vertebral column and also forgets the surveying mineral world.

Tahar Djaout (1954–1993) *L'invention du desert* (Éditions du Seuil, 1987) Kindly translated from French by Gaëlle Catois

T. Heulin (✉) • G. De Luca • M. Barakat • P. Ortet • W. Achouak
CEA, CNRS, Aix-Marseille Université, UMR 7265 Biosciences and Biotechnology Institute of Aix-Marseille (BIAM), Laboratory of Microbial Ecology of the Rhizosphere and Extreme Environments (LEMIRE), CEA/Cadarache, 13108 St-Paul-lez-Durance, France
e-mail: thierry.heulin@cea.fr

M. Gommeaux
Université de Reims Champagne-Ardenne,
GEGENAA, EA 3795, 2 esplanade Roland Garros, 51100 Reims, France

A. de Groot • L. Blanchard
CEA, CNRS, Aix-Marseille Université, UMR 7265 Biosciences and Biotechnology Institute of Aix-Marseille (BIAM), Laboratory of Cellular Bioenergetics (LBC), CEA/Cadarache, 13108 St-Paul-lez-Durance, France

H. Stan-Lotter, S. Fendrihan (eds.), *Adaption of Microbial Life to Environmental Extremes*, DOI 10.1007/978-3-319-48327-6_4

4.1 Introduction

Prokaryotic microorganisms are known to be highly adaptable to diverse environmental conditions and to thrive in harsh environments. Halophilic microorganisms (Bacteria and Archaea) tolerate and grow in the presence of salt concentrations ten times higher than seawater, whereas acidophiles stand for a pH of 1, and hyperthermophiles face temperatures above 85 °C. Bacteria are able to sense changing environmental parameters, such as temperature, pressure, pH, ionic strength, solute concentrations, and water availability, and to adapt by protecting biological molecules and adjusting biochemical reactions in response to extreme conditions.

These extreme conditions may be transient or permanent and will greatly influence the various adaptation mechanisms. Actually, four strategies might be used to overcome environmental stresses: compensation, conservation, protection, and damage repair.

Facing transient extreme conditions, compensatory responses seek to restore equilibrium and to maintain normal functions, such as, for instance, up-regulation of efflux pumps to extrude toxic metals when the concentration of heavy metals increases. Compensation can also be engaged for long-term adaptation, as exemplified by *Helicobacter pylori*, the extremely acid-resistant and unique microorganism able to thrive within the human stomach by producing copious amounts of urease (Mobley et al. 1995). Another response of bacteria to stress is to assume 'non-growth' states. When spore-forming bacteria face a strong or prolonged stress, they achieve conservation responses by entering a non-dividing state and shifting to a dormancy state under spore form. It is a reversible state of reduced basal metabolic rate in a unit that maintains viability (Barer 2003). Protective responses are used to maintain the physical integrity of living organisms. Stresses as desiccation may cause water loss resulting in cell-volume collapse and in serious damages of cellular macromolecules such as proteins and nucleic acids. To protect themselves against desiccation, many bacterial cells accumulate solutes, including carbohydrates, amino acids, quaternary amines, and tetrahydropyrimidines (Takagi 2008). Lastly, members of *Deinococcaceae* show an exceptional ability to withstand the lethal effects of DNA-damaging agents and to repair efficiently hundreds of DNA double- and single-strand breaks as well as other types of DNA damages following desiccation or gamma radiation.

Bacterial adaptive responses to permanent harsh environments may include survival at the surface of hot arid desert soils that are exposed to heat (up to 58 °C in summer), desiccation, and intense ultraviolet (UV) radiation. To thrive in such extreme conditions, specific and (or) unusual adaptive mechanisms may be involved.

4.2 Characteristics of Hot and Dry Deserts with Emphasis on the Sahara

Hot and dry deserts can be considered as a paradigm of extreme environment for life because in these conditions the main limiting factor is water combined with drastic and highly contrasted temperatures. In extreme desert environments such as the Sahara, the mean annual rainfall is less than 50 mm, and years without any rainfall event are not an exception (e.g., Le Houérou 1997). During the day, at the surface of the desert, the water is almost missing due to high temperature, and, except some rare rainfall events, water can be present only at the end of some nights when the difference of temperatures between soil and air is suitable for dew. Under these conditions, there is a direct link between light intensity and water availability.

The main characteristics of hot and dry deserts are the following (Le Houérou 1986, 1997):

- Scarce and irregular rainfall with an annual average lower than 100 mm
- Dew at the end of some nights/early in the morning
- Great amplitude of temperatures between night and day
- Wind responsible for 'soil' erosion
- Very low diversity of plants and animals
- Very low organic content of 'soil'

Besides the hot and dry deserts, the polar regions are also considered as cold deserts with the same limiting factor (water/ice). Hot and dry deserts are the largest desert regions mainly located at the tropics of Cancer and Capricorn, representing about 50 million km^2 (one-third of continental surfaces). In Africa, the larger ones are the Namib and Kalahari deserts in the South (Capricorn tropic) and the Sahara desert in the North (Cancer tropic). The Sahara is the largest hot and dry desert, and it extends 9 million km^2 from Mauritania to Egypt.

The climate of Sahara has varied greatly during the past several million years, but the onset of persistent and widespread arid conditions is thought to have occurred approximately 2.5 million years ago during the Pliocene, in conjunction with the onset of major glaciation in the northern hemisphere (Swezey 2009). Since then, arid conditions in the Sahara have waxed and waned as the glaciers have advanced and retreated. During the last glacial maximum at approximately 31–23 thousand years ago (ka), for example, arid conditions were particularly intense and widespread, and desert conditions expanded well beyond their modern range (e.g. Sarnthein 1978). A general chronology for the Sahara during the late Quaternary reveals a pattern of arid conditions and aeolian sediment mobilization from approximately 25–11 ka, relatively humid conditions and aeolian sediment stabilization from approximately 11–5 ka (time of the Holocene 'Green Sahara' or 'African Humid Period'), and a return to arid conditions and aeolian sediment mobilization since 5 ka (Swezey 2001; Kröpelin et al. 2008; Bristow and Armitage 2016-in press). These arid-humid climate oscillations in the Sahara are thought to have had major influences on the evolution and adaptation of numerous species (e.g. Gonçalves et al. 2012; Migliore et al. 2013; Brito et al. 2014; Sow et al. 2014; Leite et al. 2015).

4.3 Bacterial Mechanisms for Desiccation Tolerance

The production of bacterial spores (endospores of *Firmicutes*, exospores of *Actinobacteria*) and akinetes (*Cyanobacteria*) is the most extensively documented mechanism explaining bacterial desiccation tolerance. Other examples of bacterial cell differentiation are cysts of *Azotobacter* (*Gammaproteobacteria*) and myxospores of *Myxococcus* (*Deltaproteobacteria*). A completely different mechanism was more recently described: no cell differentiation involved in *Deinococcus* but a highly efficient and rapid ability to repair DNA damage (Mattimore and Battista 1996). This mechanism for tolerance of desiccation, which is correlated to tolerance of gamma radiation, is discussed in Sect. 6.

Using a range of complementary microbiological approaches (culture-dependent and culture-independent methods), an extensive diversity of bacterial species have been documented in nutrient-poor environments such as deserts. These data confirm that *Firmicutes* (*Bacillus*, *Paenibacillus*, etc.) and *Actinobacteria* (*Arthrobacter*, *Geodermatophilus*, etc.) represent the dominant bacterial communities in deserts such as the Sahara (Chanal et al. 2006; Benzerara et al. 2006; Gommeaux et al. 2010) and Namib (Prestel et al. 2008). These studies also have revealed the abundance and a huge diversity of *Proteobacteria* belonging to the four subgroups (*alpha*, *beta*, *gamma*, and *delta*) and the absence of *Myxococcus* and *Azotobacter* isolates (or 16S rDNA sequences).

Several publications report the presence of rhizobia in deserts, especially strains able to nodulate the legume tree *Acacia* (Zerhari et al. 2000; Khbaya et al. 1998). The mechanism mentioned to explain the adaptation of rhizobia to dryness conditions is the synthesis of osmoprotectants (glycine-betaine, sucrose, ectoine). More recently, it was demonstrated that mannosucrose and trehalose are also involved in the desiccation tolerance of *Rhizobium* (Essendoubi et al. 2007). Exploring the diversity of extracellular polymer-producing bacteria in desert soils of southern Algeria, Kaci et al. (2005) showed that *Rhizobium* strains producing heteroglycan constitute the most abundant population able to colonize plant roots. The synthesis of such extracellular polymers, potentially able to limit water loss, is probably an important mechanism explaining the desiccation tolerance of rhizobia, as shown in terrestrial vegetative cells of cyanobacteria (Potts 1994; Billi and Potts 2002).

The cyanobacteria are classical inhabitants of desert surfaces (Garcia-Pichel and Belnap, 1996; Karnieli et al. 1999). For instance, Garcia-Pichel and Pringault (2001) showed that the filamentous cyanobacterium *Oscillatoria* could migrate to the soil surface in response to wetting events or retreating below in response to drying event. This nomadic behaviour seems to be the main explanation of *Oscillatoria* adaptation to desert life. An interesting work on cyanobacterial diversity of crust formation in the desert of the Colorado Plateau (Utah, USA) showed that the complement of culture-dependent and culture-independent techniques revealed a new cluster named '*Xeronema*' grouping cyanobacteria closely related to *Phormidium* (Garcia-Pichel et al. 2001).

4.4 Counting and Describing Bacterial populations in Desert Environments

Two recent works on this topic were carried out in the frame of a French research project (Treasures of Sahara[1]). The first one was dedicated to the mineral and microbial analysis of sand sampled in Merzouga dunes of Morocco (Gommeaux et al. 2010). The second one was performed in a semiarid region of Tataouine in southern Tunisia (Benzerara et al. 2006) with a special focus on gamma radiation-/desiccation-tolerant bacteria (Chanal et al. 2006).

Considering the Merzouga dunes, the mineral analysis reveals mostly quartz grains, pure or plated with iron oxides, some carbonate grains, and other minor silicate grains (Fig. 4.1; see Gommeaux et al. 2010). The only source of carbon in these conditions is the presence of occasional cyanobacterial crusts. From a series of 160 sand grains randomly chosen from the sand treated with Syto9 (a green fluorescent nucleic acid stain), a number of 10.4 ± 1.0 fluorescent spots per grain were obtained (Fig. 4.2; see Gommeaux et al. 2010). Considering the number of 2.1×10^4 grains per gram of sand, a minimum of $2.2 \pm 0.2 \times 10^5$ bacteria per gram were present on the surface of individual sand grains. An original method based on a 'grain-by-grain' cultivation with direct plating of grains on solid $0.1 \times$ tryptic soy agar (TSA) medium revealed about 14 % of the numbers obtained by Syto9 direct counting, constituting a greatly more efficient method to reveal 'culturable' bacteria compared to the classical method based on suspension dilution and plating (1.6 % of the numbers obtained by Syto9 direct counting; see Gommeaux et al. 2010). Both culture-based and molecular-based (cloning sequencing of 16S rDNA) analyses of bacterial diversity reveal that *Firmicutes*, *Actinobacteria*, *Proteobacteria*, and '*Cytophaga-Flexibacter-Bacteroides*' (CFB) are the most frequent groups. Green non-sulphur (GNS) bacteria *Acidobacteria* and *Planctomycetes* were present, but less frequently. Dividing the fluorescence cell count by the number of operational taxonomic units (OTUs) yielded an estimated 1560 ± 140 cells per OTU per gram of sand. This figure is much lower than estimated by Torsvik et al. (2002) for a broad range of natural and anthropic environments, in which the average number of individuals per dominating taxon per gram or millilitre is in the range of 10^4–10^5. Thus, the very low biomass in Merzouga sand is not correlated with a reduction in bacterial diversity but mainly with a very low number of cells per taxon (Gommeaux et al. 2010).

[1] French laboratories from Cadarache, Marseille, Orsay, and Rennes in collaboration with laboratories in Rabat (Morocco) and in El-Fjé-Medenine (Tunisia). This work was supported by the GEOMEX program grant from the Centre National de la Recherche Scientifique (CNRS).

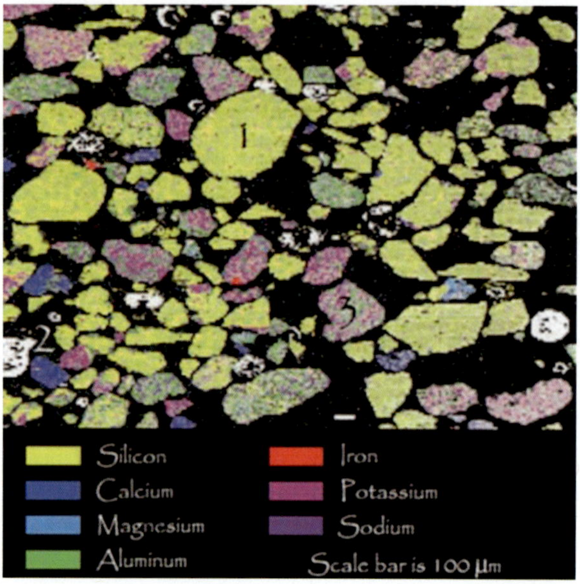

Fig. 4.1 Scanning electron microscope (SEM) observation of a thin section of Merzouga sand. The superimposed colours correspond to energy-dispersive X-ray spectroscopy (EDS) mapping of selected elements. At least three types of grains were distinguished (quartz grains as 1, carbonate as 2, non-quartz silicate as 3), according to the bulk chemistry of the grains (From Gommeaux et al. 2010, with permission)

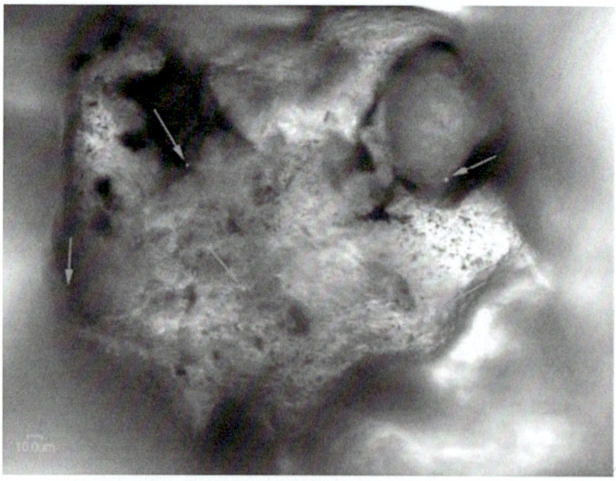

Fig. 4.2 Observation of Syto9-labelled bacteria on the grain surface. Composite image of Syto9 fluorescence spots superimposed on the white-light image using Adobe Photoshop software. The dots are interpreted as individual cells, whereas the spots are interpreted as groups of a few cells or microcolonies (From Gommeaux et al. 2010, with permission)

The genera *Chelatococcus* and *Saccharothrix* are strongly attached to sand grains, considering their exclusive isolation by the 'grain-by-grain' method (Gommeaux et al. 2010). Another important conclusion of this work deals with the correlation between mineral and bacterial diversity. For instance, *Arthrobacter* appears to be well adapted to this harsh environment with a preference for grains other than the dominant mineral quartz.

The analysis of microbial diversity was also performed on Tataouine sand grains (Benzerara et al. 2006; Chanal et al. 2006). The main difference to the Merzouga dunes was the number of cultivated bacteria on 0.1 × TSA medium, which was tenfold higher in the Tataouine sample. This result can be explained by steppic vegetation in Tataouine (semiarid region) and a correlated higher carbon content of the sandy soil. The classical culture-dependent technique reveals that the most frequent isolates belonged to *Firmicutes*, *Actinobacteria*, *Proteobacteria*, and CFB, in agreement with Merzouga data. The culture-independent technique (cloning sequencing of 16S rDNA) revealed clone sequences corresponding to *Proteobacteria*, *Actinobacteria*, and *Acidobacteria*. It was also shown that strains in the Tataouine sample that were tolerant to gamma radiation belong to *Firmicutes* (*Bacillus*), *Deinococcus*, and *Chelatococcus* (*Alphaproteobacteria*) genera. *Chelatococcus* were isolated as desiccation-tolerant strains from the surface soil of a shrub-steppe (Fredrickson et al. 2008) and from the Merzouga sand grains (Gommeaux et al. 2010).

A surprising result was the absence of known thermotolerant bacteria (for example, in both studies no isolate or clone sequence related to *Thermus* was obtained). The evidence of clone sequences of non-thermophile *Crenarchaeota* in the Tataouine sample (Chanal et al. 2006) reinforced the notion that tolerance of desiccation rather than tolerance of temperature is the more important desert-adaptive microbial trait. Water allows bacterial activity and is present only when the temperature of the sand is low (below 20 °C).

Considering the cyanobacterial diversity in the Tataouine samples, sequences of *Oscillatoria*, *Anabaena*, *Nostoc*, and *Symploca* were identified after cloning sequencing of 16S-rDNA (Benzerara et al. 2006). Such large diversity of cyanobacteria in deserts was previously described by Garcia-Pichel et al. (2001).

A more complete list of described bacterial species isolated from desert regions (85 novel species since 2003) is presented in Table 4.1. Among the newly described bacterial genera/species isolated from desert environments since 2011 (47 new species), Klenk and co-workers isolated five new genera (*Blastocatella*, *Desertibacter*, *Oligoflexus*, *Tenggerimyces*, and *Yuhushiella*) and eight new species belonging to the genus *Geodermatophilus* (*G. africanus*, *G. arenarius*, *G. normandii*, *G. sabuli*, *G. saharensis*, *G. siccatus*, *G. telluris*, and *G. tzadiensis*) from the Sahara desert.

4.5 *Ramlibacter* and Its Life Cycle

The *Ramlibacter* story started in 1931 with the fall of a meteorite near Tataouine in southern Tunisia (Lacroix 1931). One important point of this story is the collection the day after the fall of the larger fragments, which were sent to the Muséum National d'Histoire Naturelle (Paris, France) and used as negative control of meteorite alteration studies (Barrat et al. 1998). Several weathered fragments of the meteorite were collected in 1994 in the sandy soil. Scanning electron microscopic observations revealed alteration zones at the surface of the meteorite crystals (pyroxene and chromite) and also secondary calcite crystals resulting from terrestrial weathering (Barrat et al. 1998 and 1999).

A bacterial agent is responsible for the alteration zones and secondary calcite crystals. This bacterial strain (TTB310), which has been isolated and characterized from a meteorite fragment embedded in sandy soil (Gillet et al. 2000), has

Table 4.1 List of new bacterial species isolated from deserts, validly published since 2003

Title	Authors	References
Actinoalloteichus spitiensis sp. nov., a novel actinobacterium isolated from a cold desert of the Indian Himalayas	Singla AK, Mayilraj S, Kudo T, Krishnamurthi S, Prasad GS, Vohra RM	*Int J Syst Evol Microbiol.* 2005 Nov;55(Pt 6):2561–4
Actinomadura namibiensis sp. nov.	Wink J, Kroppenstedt RM, Seibert G, Stackebrandt E	*Int J Syst Evol Microbiol.* 2003 May;53(Pt 3):721–4
Actinophytocola gilvus sp. nov., isolated from desert soil crusts and emended description of the genus *Actinophytocola* Indananda et al. 2010	Sun HM, Zhang T, Yu LY, Lu XX, Mou XZ, Zhang YQ	*Int J Syst Evol Microbiol.* 2014 Sep.;64(Pt 9):3120–5
Actinopolyspora mzabensis sp. nov., a halophilic actinomycete isolated from an Algerian Saharan soil	Meklat A, Bouras N, Zitouni A, Mathieu F, Lebrihi A, Schumann P, Spröer C, Klenk HP, Sabaou N	*Int J Syst Evol Microbiol.* 2013 Oct;63(Pt 10):3787–92
Altererythrobacter xinjiangensis sp. nov., isolated from desert sand and emended description of the genus *Altererythrobacter*	Xue X, Zhang K, Cai F, Dai J, Wang Y, Rahman E, Peng F, Fang C	*Int J Syst Evol Microbiol.* 2012 Jan.;62(Pt 1):28–32
Agrococcus lahaulensis sp. nov., isolated from a cold desert of the Indian Himalayas	Mayilraj S, Suresh K, Schumann P, Kroppenstedt RM, Saini HS	*Int J Syst Evol Microbiol.* 2006 Aug;56(Pt 8):1807–10
Amycolatopsis australiensis sp. nov., an actinomycete isolated from arid soils	Tan GY, Robinson S, Lacey E, Goodfellow M	*Int J Syst Evol Microbiol.* 2006 Oct;56(Pt 10):2297–301
Arthrobacter liuii sp. nov., resuscitated from Xinjiang desert soil	Yu XY, Zhang L, Ren B, Yang N, Liu M, Liu XT, Zhang LX, Ding LX	*Int J Syst Evol Microbiol.* 2015 Mar;65(Pt 3):896–901
Arthrobacter deserti sp. nov., isolated from a desert soil sample	Hu QW, Chu X, Xiao M, Li CT, Yan ZF, Hozzein WN, Kim CJ, Zhi XY, Li WJ	*Int J Syst Evol Microbiol.* 2016 doi: 10.1099/ijsem.0.000986
Bacillus deserti sp. nov., a novel bacterium isolated from the desert of Xinjiang, China	Zhang L, Wu GL, Wang Y, Dai J, Fang CX	*Antonie Van Leeuwenhoek.* 2011 Feb.;99(2):221–9
Blastocatella fastidiosa gen. nov.,sp. nov., isolated from semiarid savanna soil – the first described species of *Acidobacteria* subdivision 4	Foesel BU, Rohde M, Overmann J	*Syst Appl Microbiol.* 2013 Mar;36(2):82–9
Caenispirillum deserti sp. nov., a spheroplast-forming bacterium isolated from a salt desert	Divyasree B, Lakshmi KV, Bharti D, Sasikala Ch, Ramana ChV	*Int J Syst Evol Microbiol.* 2015 Sep.;65(9):3119–24

Table 4.1 (continued)

Title	Authors	References
Caldanaerovirga acetigignens gen. nov., sp. nov., an anaerobic xylanolytic, alkalithermophilic bacterium isolated from Trego Hot Spring, Nevada, USA	Wagner ID, Ahmed S, Zhao W, Zhang CL, Romanek CS, Rohde M, Wiegel J	*Int J Syst Evol Microbiol*. 2009 Nov;59(Pt 11):2685–91
Cesiribacter roseus sp. nov., a pink-pigmented bacterium isolated from desert sand	Liu M, Qi H, Luo X, Dai J, Peng F, Fang C	*Int J Syst Evol Microbiol*. 2012 Jan.;62(Pt 1):96–9
Citricoccus alkalitolerans sp. nov., a novel actinobacterium isolated from a desert soil in Egypt	Li WJ, Chen HH, Zhang YQ, Kim CJ, Park DJ, Lee JC, Xu LH, Jiang CL	*Int J Syst Evol Microbiol*. 2005 Jan.;55(Pt 1):87–90
Corynebacterium deserti sp. nov., isolated from desert sand	Zhou Z, Yuan M, Tang R, Chen M, Lin M, Zhang W	*Int J Syst Evol Microbiol*. 2012 Apr;62(Pt 4):791–4
Delftia deserti sp. nov., isolated from a desert soil sample	Li CT, Yan ZF, Chu X, Hussain F, Xian WD, Yunus Z, Hozzein WN, Abaydulla G, Li WJ	*Antonie Van Leeuwenhoek*. 2015 Jun;107(6):1445–50
Deinococcus deserti sp. nov., a gamma radiation-tolerant bacterium isolated from the Sahara desert	de Groot A, Chapon V, Servant P, Christen R, Saux MF, Sommer S, Heulin T	*Int J Syst Evol Microbiol*. 2005 Nov;55(Pt 6):2441–6.
Deinococcus xinjiangensis sp. nov., isolated from desert soil	Peng F, Zhang L, Luo X, Dai J, An H, Tang Y, Fang C	*Int J Syst Evol Microbiol*. 2009 Apr;59(Pt 4):709–13
Deinococcus peraridilitoris sp. nov., isolated from a coastal desert	Rainey FA, Ferreira M, Nobre MF, Ray K, Bagaley D, Earl AM, Battista JR, Gómez-Silva B, McKay CP, da Costa MS	*Int J Syst Evol Microbiol*. 2007 Jul;57(Pt 7):1408–12
Desertibacter roseus gen. nov., sp. nov., a gamma radiation-resistant bacterium in the family *Rhodospirillaceae*, isolated from desert sand	Liu M, Dai J, Liu Y, Cai F, Wang Y, Rahman E, Fang C	*Int J Syst Evol Microbiol*. 2011 May;61(Pt 5):1109–13
Description of ***Dietzia lutea*** sp. nov., isolated from a desert soil in Egypt	Li J, Chen C, Zhao GZ, Klenk HP, Pukall R, Zhang YQ, Tang SK, Li WJ	*Syst Appl Microbiol*. 2009 Apr;32(2):118–23
Dyadobacter alkalitolerans sp. nov., isolated from desert sand	Tang Y, Dai J, Zhang L, Mo Z, Wang Y, Li Y, Ji S, Fang C, Zheng C	*Int J Syst Evol Microbiol*. 2009 Jan.;59(Pt 1):60–4
Falsirhodobacter deserti sp. nov., isolated from sandy soil	Wang L, Zhou Z, Wu G, Chen M, Lin M, Zhang W, Chen W	*Int J Syst Evol Microbiol*. 2015 Feb.;65(Pt 2):650–5

(continued)

Table 4.1 (continued)

Title	Authors	References
Geodermatophilus africanus sp. nov., a halotolerant actinomycete isolated from Saharan desert sand	Montero-Calasanz Mdel C, Göker M, Pötter G, Rohde M, Spröer C, Schumann P, Gorbushina AA, Klenk HP	*Antonie Van Leeuwenhoek*. 2013 Aug;104(2):207–16
Geodermatophilus arenarius sp. nov., a xerophilic actinomycete isolated from Saharan desert sand in Chad	Montero-Calasanz MC, Göker M, Pötter G, Rohde M, Spröer C, Schumann P, Gorbushina AA, Klenk HP	*Extremophiles*. 2012 Nov;16(6):903–9
Geodermatophilus normandii sp. nov., isolated from Saharan desert sand	Montero-Calasanz Mdel C, Göker M, Pötter G, Rohde M, Spröer C, Schumann P, Gorbushina AA, Klenk HP.	*Int J Syst Evol Microbiol*. 2013 Sep.;63(Pt 9):3437–43.
Geodermatophilus sabuli sp. nov., a gamma radiation-resistant actinobacterium isolated from limestone in the Sahara desert	Hezbri K, Ghodhbane-Gtari F, Montero-Calasanz MD, Sghaier H, Rohde M, Schumann P, Klenk HP, Gtari M	*Int J Syst Evol Microbiol*. 2015 Jul 1. doi: 10.1099/ijsem.0.000422.
Geodermatophilus saharensis sp. nov., isolated from sand of the Saharan desert in Chad	Montero-Calasanz MC, Göker M, Pötter G, Rohde M, Spröer C, Schumann P, Gorbushina AA, Klenk HP	*Arch Microbiol*. 2013 Mar;195(3):153–9
Geodermatophilus siccatus sp. nov., isolated from arid sand of the Saharan desert in Chad	del Carmen Montero-Calasanz M, Göker M, Rohde M, Schumann P, Pötter G, Spröer C, Gorbushina AA, Klenk HP	*Antonie Van Leeuwenhoek*. 2013 Mar;103(3):449–56
Geodermatophilus telluris sp. nov., an actinomycete isolated from Saharan desert sand	Montero-Calasanz Mdel C, Göker M, Pötter G, Rohde M, Spröer C, Schumann P, Klenk HP, Gorbushina AA	*Int J Syst Evol Microbiol*. 2013 Jun;63(Pt 6):2254–9
Geodermatophilus tzadiensis sp. nov., a UV radiation-resistant bacterium isolated from sand of the Saharan desert	Montero-Calasanz Mdel C, Göker M, Broughton WJ, Cattaneo A, Favet J, Pötter G, Rohde M, Spröer C, Schumann P, Klenk HP, Gorbushina AA	*Syst Appl Microbiol*. 2013 May;36(3):177–82

Table 4.1 (continued)

Title	Authors	References
Description of ***Hymenobacter arizonensis*** sp. nov. from the southwestern arid lands of the United States of America	Reddy GS, Garcia-Pichel F	*Antonie Van Leeuwenhoek*. 2013 Feb.;103(2):321–30. Erratum in: *Antonie Van Leeuwenhoek*. 2013 Feb.;103(2):441–2
Hymenobacter deserti sp. nov., isolated from the desert of Xinjiang, China	Zhang L, Dai J, Tang Y, Luo X, Wang Y, An H, Fang C, Zhang C	*Int J Syst Evol Microbiol*. 2009 Jan.;59(Pt 1):77–82
Hymenobacter xinjiangensis sp. nov., a radiation-resistant bacterium isolated from the desert of Xinjiang, China	Zhang Q, Liu C, Tang Y, Zhou G, Shen P, Fang C, Yokota A	*Int J Syst Evol Microbiol*. 2007 Aug;57(Pt 8):1752–6
Jiangella gansuensis gen. Nov., sp. nov., a novel actinomycete from a desert soil in Northwest China.	Song L, Li WJ, Wang QL, Chen GZ, Zhang YS, Xu LH	*Int J Syst Evol Microbiol*. 2005 Mar;55(Pt 2):881–4
Kineococcus xinjiangensis sp. nov., isolated from desert sand	Liu M, Peng F, Wang Y, Zhang K, Chen G, Fang C	*Int J Syst Evol Microbiol*. 2009 May;59(Pt 5):1090–3
Kocuria aegyptia sp. nov., a novel actinobacterium isolated from a saline, alkaline desert soil in Egypt	Li WJ, Zhang YQ, Schumann P, Chen HH, Hozzein WN, Tian XP, Xu LH, Jiang CL	*Int J Syst Evol Microbiol*. 2006 Apr;56(Pt 4):733–7
Lechevalieria atacamensis sp. nov., ***Lechevalieria deserti*** sp. nov., and ***Lechevalieria roselyniae*** sp. nov., isolated from hyperarid soils	Okoro CK, Bull AT, Mutreja A, Rong X, Huang Y, Goodfellow M	*Int J Syst Evol Microbiol* 2010 60 (2): p. 296–300
Lysobacter xinjiangensis sp. nov., a moderately thermotolerant and alkalitolerant bacterium isolated from a gamma-irradiated sand soil sample	Liu M, Liu Y, Wang Y, Luo X, Dai J, Fang C	*Int J Syst Evol Microbiol*. 2011 Feb.;61(Pt 2):433–7
Mesorhizobium gobiense sp. nov. and ***Mesorhizobium tarimense*** sp. nov., isolated from wild legumes growing in desert soils of Xinjiang, China	Han TX, Han LL, Wu LJ, Chen WF, Sui XH, Gu JG, Wang ET, Chen WX	*Int J Syst Evol Microbiol*. 2008 Nov;58(Pt 11):2610–8
Microbacterium radiodurans sp. nov., a UV radiation-resistant bacterium isolated from soil	Zhang W, Zhu HH, Yuan M, Yao Q, Tang R, Lin M, Yang SZ, Li ZK, Chen M	*Int J Syst Evol Microbiol*. 2010 Nov;60(Pt 11):2665–70
Mycetocola manganoxydans sp. nov., an actinobacterium isolated from the Taklamakan desert	Luo X, Wang J, Zeng XC, Wang Y, Zhou L, Nie Y, Dai J, Fang C	*Int J Syst Evol Microbiol*. 2012 Dec;62(Pt 12):2967–70

(continued)

Table 4.1 (continued)

Title	Authors	References
Nesterenkonia rhizosphaerae sp. nov., an alkaliphilic actinobacterium isolated from rhizosphere soil in a saline-alkaline desert	Wang HF, Zhang YG, Chen JY, Hozzein WN, Li L, Wadaan MA, Zhang YM, Li WJ	*Int J Syst Evol Microbiol*. 2014 Dec;64(Pt 12):4021–6
Nocardioides deserti sp. nov., an actinobacterium isolated from desert soil	Tuo L, Dong YP, Habden X, Liu JM, Guo L, Liu XF, Chen L, Jiang ZK, Liu SW, Zhang YB, Zhang YQ, Sun CH	*Int J Syst Evol Microbiol*. 2015 May;65(Pt 5):1604–10
Nocardiopsis alkaliphila sp. nov., a novel alkaliphilic actinomycete isolated from desert soil in Egypt	Hozzein WN, Li WJ, Ali MI, Hammouda O, Mousa AS, Xu LH, Jiang CL	*Int J Syst Evol Microbiol*. 2004 Jan.;54(Pt 1):247–52
Oligoflexus tunisiensis gen. nov., sp. nov., a Gram-negative, aerobic, filamentous bacterium of a novel proteobacterial lineage, and description of ***Oligoflexaceae*** fam. nov., ***Oligoflexales*** ord. nov. and ***Oligoflexia*** classis nov.	Nakai R, Nishijima M, Tazato N, Handa Y, Karray F, Sayadi S, Isoda H, Naganuma T	*Int J Syst Evol Microbiol*. 2014 Oct;64(Pt 10):3353–9.
Paenibacillus harenae sp. nov., isolated from desert sand in China	Jeon CO, Lim JM, Lee SS, Chung BS, Park DJ, Xu LH, Jiang CL, Kim CJ	*Int J Syst Evol Microbiol*. 2009 Jan.;59(Pt 1):13–7
Paenibacillus gansuensis sp. nov., isolated from desert soil of Gansu Province in China	Lim JM, Jeon CO, Lee JC, Xu LH, Jiang CL, Kim CJ	*Int J Syst Evol Microbiol*. 2006 Sep.;56(Pt 9):2131–4
Paenibacillus tarimensis sp. nov., isolated from sand in Xinjiang, China	Wang M, Yang M, Zhou G, Luo X, Zhang L, Tang Y, Fang C	*Int J Syst Evol Microbiol*. 2008 Sep.;58(Pt 9):2081–5
Pedobacter xinjiangensis sp. nov., from desert, Xinjiang	Tang Y, Wang Y, Ji S, Zhang K, Dai J, Zhang L, Peng F, Fang C	*J Microbiol Biotechnol*. 2010 Feb.;20(2):397–402
Planobacterium taklimakanense gen. nov., sp. nov., a member of the family *Flavobacteriaceae* that exhibits swimming motility, isolated from desert soil	Peng F, Liu M, Zhang L, Dai J, Luo X, An H, Fang C	*Int J Syst Evol Microbiol*. 2009 Jul;59(Pt 7):1672–8
Planococcus stackebrandtii sp. nov., isolated from a cold desert of the Himalayas, India	Mayilraj S, Prasad GS, Suresh K, Saini HS, Shivaji S, Chakrabarti T	*Int J Syst Evol Microbiol*. 2005 Jan.;55(Pt 1):91–4

Table 4.1 (continued)

Title	Authors	References
Pontibacter akesuensis sp. nov., isolated from a desert soil in China	Zhou Y, Wang X, Liu H, Zhang KY, Zhang YQ, Lai R, Li WJ	*Int J Syst Evol Microbiol*. 2007 Feb.;57(Pt 2):321–5
Pontibacter diazotrophicus sp. nov., a novel nitrogen-fixing bacterium of the family *Cytophagaceae*	Xu L, Zeng XC, Nie Y, Luo X, Zhou E, Zhou L, Pan Y, Li W	*PLoS One*. 2014 Mar 19;9(3):e92294
Pontibacter korlensis sp. nov., isolated from the desert of Xinjiang, China	Zhang L, Zhang Q, Luo X, Tang Y, Dai J, Li Y, Wang Y, Chen G, Fang C	*Int J Syst Evol Microbiol*. 2008 May;58(Pt 5):1210–4
Pontibacter ruber sp. nov. and ***Pontibacter deserti*** sp. nov., isolated from the desert	Subhash Y, Sasikala Ch, Ramana ChV	*Int J Syst Evol Microbiol*. 2014 Mar;64(Pt 3):1006–11
Pontibacter soli sp. nov., isolated from the soil of a *Populus* rhizosphere in Xinjiang, China	Dai J, Xu M, Peng F, Jiang F, Chen X, Wang Z, Fang C	*Antonie Van Leeuwenhoek*. 2014 Jan.;105(1):65–72.
Pontibacter yuliensis sp. nov., isolated from soil	Cao H, Nie Y, Zeng XC, Xu L, He Z, Luo X, Wu R.	*Int J Syst Evol Microbiol*. 2014 Mar;64(Pt 3):968–72
Prauserella isguenensis sp. nov., a halophilic actinomycete isolated from desert soil	Saker R, Bouras N, Meklat A, Zitouni A, Schumann P, Spröer C, Sabaou N, Klenk HP	*Int J Syst Evol Microbiol*. 2015 May;65(Pt 5):1598–603
Prauserella shujinwangii sp. nov., from a desert environment	Liu M, Zhang L, Ren B, Yang N, Yu X, Wang J, Ding L, Liu X, Liu Z, Goodfellow M, Zhang L	*Int J Syst Evol Microbiol*. 2014 Nov;64(Pt 11):3833–7.
Pseudomonas arsenicoxydans sp. nov., an arsenite-oxidizing strain isolated from the Atacama Desert	Campos VL, Valenzuela C, Yarza P, Kämpfer P, Vidal R, Zaror C, Mondaca MA, Lopez-Lopez A, Rosselló-Móra R	*Syst Appl Microbiol*. 2010 Jun;33(4):193–7
Pseudomonas duriflava sp. nov., isolated from a desert soil	Liu R, Liu H, Feng H, Wang X, Zhang CX, Zhang KY, Lai R	*Int J Syst Evol Microbiol*. 2008 Jun;58(Pt 6):1404–8
Pseudomonas xinjiangensis sp. nov., a moderately thermotolerant bacterium isolated from desert sand	Liu M, Luo X, Zhang L, Dai J, Wang Y, Tang Y, Li J, Sun T, Fang C	*Int J Syst Evol Microbiol*. 2009 Jun;59(Pt 6):1286–9
Ramlibacter tataouinensis gen. nov., sp. nov. and ***Ramlibacter henchirensis*** sp. nov., cyst-producing bacteria isolated from subdesert soil in Tunisia	Heulin T, Barakat M, Christen R, Lesourd M, Sutra L, De Luca G, Achouak W	*Int J Syst Evol Microbiol*. 2003 Mar;53(Pt 2):589–94

(continued)

Table 4.1 (continued)

Title	Authors	References
Rhodococcus kroppenstedtii sp. nov., a novel actinobacterium isolated from a cold desert of the Himalayas, India	Mayilraj S, Krishnamurthi S, Saha P, Saini HS	*Int J Syst Evol Microbiol*. 2006 May;56(Pt 5):979–82
Saccharibacillus deserti sp. nov., isolated from desert soil	Sun JQ, Wang XY, Wang LJ, Xu L, Liu M, Wu XL	*Int J Syst Evol Microbiol*. 2015 Nov 11. doi: 10.1099/ijsem.0.000766
Saccharibacillus kuerlensis sp. nov., isolated from a desert soil	Yang SY, Liu H, Liu R, Zhang KY, Lai R.	*Int J Syst Evol Microbiol*. 2009 May;59(Pt 5):953–7.
Streptomyces atacamensis sp. nov., isolated from an extreme hyperarid soil of the Atacama Desert, Chile	Santhanam R, Okoro CK, Rong X, Huang Y, Bull AT, Weon HY, Andrews BA, Asenjo JA, Goodfellow M	*Int J Syst Evol Microbiol*. 2012 Nov;62(Pt 11):2680–4
Streptomyces deserti sp. nov., isolated from hyperarid Atacama Desert soil	Santhanam R, Okoro CK, Rong X, Huang Y, Bull AT, Andrews BA, Asenjo JA, Weon HY, Goodfellow M	*Antonie Van Leeuwenhoek*. 2012 Mar;101(3):575–81
Saccharopolyspora ghardaiensis sp. nov., an extremely halophilic actinomycete isolated from Algerian Saharan soil	Meklat A, Bouras N, Zitouni A, Sabaou N, Mathieu F, Schumann P, Spröer C, Klenk HP	J Antibiot (Tokyo). 2014 Apr;67(4):299–303
Saccharothrix hoggarensis sp. nov., an actinomycete isolated from Saharan soil	Boubetra D, Zitouni A, Bouras N, Mathieu F, Lebrihi A, Schumann P, Spröer C, Klenk HP, Sabaou N	*Int J Syst Evol Microbiol*. 2013 Feb.;63(Pt 2):549–5.
Salinicoccus luteus sp. nov., isolated from a desert soil	Zhang YQ, Yu LY, Liu HY, Zhang YQ, Xu LH, Li WJ.	*Int J Syst Evol Microbiol*. 2007 Aug;57(Pt 8):1901–5.
Streptosporangium saharense sp. nov., an actinobacterium isolated from Saharan soil	Chaabane Chaouch F, Bouras N, Mokrane S, Zitouni A, Schumann P, Spröer C, Sabaou N, Klenk HP.	*Int J Syst Evol Microbiol*, 08 January, 2016 doi: 10.1099/ijsem.0.000890
Skermanella rubra sp. nov., a bacterium isolated from the desert of Xinjiang, China	Zhang ZY, Gao XH, Zhang YJ, Jia M, Lu XJ, Ma YC, Tian F, Xie Q, Tang SK.	*Antonie Van Leeuwenhoek*. 2015 Sep.;108(3):627–32.
Skermanella xinjiangensis sp. nov., isolated from the desert of Xinjiang, China	An H, Zhang L, Tang Y, Luo X, Sun T, Li Y, Wang Y, Dai J, Fang C.	*Int J Syst Evol Microbiol*. 2009 Jun;59(Pt 6):1531–4.
Sphingobacterium gobiense sp. nov., isolated from soil of the Gobi Desert	Zhao P, Zhou Z, Chen M, Lin W, Zhang W, Wei G	*Int J Syst Evol Microbiol*. 2014 Dec;64(Pt 12):3931–5

Table 4.1 (continued)

Title	Authors	References
Sphingomonas xinjiangensis sp. nov., isolated from desert sand	An H, Xu M, Dai J, Wang Y, Cai F, Qi H, Peng F, Fang C	*Int J Syst Evol Microbiol*. 2011 Aug;61(Pt 8):1865–9
Streptomyces bullii sp. nov., isolated from a hyperarid Atacama Desert soil	Santhanam R, Rong X, Huang Y, Andrews BA, Asenjo JA, Goodfellow M	*Antonie Van Leeuwenhoek*. 2013 Feb.;103(2):367–73
Streptomyces fukangensis sp. nov., a novel alkaliphilic actinomycete isolated from a saline-alkaline soil	Zhang YG, Wang HF, Liu Q, Hozzein WN, Wadaan MA, Cheng J, Chen YJ, Zhang YM, Li WJ	*Antonie Van Leeuwenhoek*. 2013 Dec;104(6):1227–33
Streptosporangium algeriense sp. nov., an actinobacterium isolated from desert soil	Boubetra D, Bouras N, Zitouni A, Schumann P, Spröer C, Sabaou N, Klenk HP	*Int J Syst Evol Microbiol*. 2015 Dec 8. doi: 10.1099/ijsem.0.000829
Tenggerimyces mesophilus gen. nov., sp. nov., a novel member of the family *Nocardioidaceae*	Sun HM, Zhang T, Wei YZ, Liu HY, Yu LY, Zhang YQ	*Int J Syst Evol Microbiol*. 2015 Jul 1. doi: 10.1099/ijsem.0.000421
Yuhushiella deserti gen. nov., sp. nov., a new member of the suborder *Pseudonocardineae*	Mao J, Wang J, Dai HQ, Zhang ZD, Tang QY, Ren B, Yang N, Goodfellow M, Zhang LX, Liu ZH	*Int J Syst Evol Microbiol*. 2011 Mar;61(Pt 3):621–30

been assigned a new genus and a new species, *Ramlibacter tataouinensis*, belonging to the *Betaproteobacteria* (Heulin et al. 2003). In addition, a second species, *Ramlibacter henchirensis*, has been identified with the same morphological characteristics (Heulin et al. 2003). The presence of *Ramlibacter* in the sandy soil of Tataouine was confirmed by Chanal et al. (2006) using a culture-independent method (cloning sequencing of 16S-rDNA). Since the description of this genus in 2003, only one validated new species has been published, *Ramlibacter solisilvae*, isolated from forest soil of Bac Kan Province in Vietnam (Lee et al. 2014).

The presence of *Ramlibacter* genus has been demonstrated in various environments using culture-independent methods (PCR amplification of 16S rDNA followed by cloning or fingerprinting). For instance, *Ramlibacter* was found associated with spores of *Gigaspora* (Long et al. 2008), in a semiarid soil sample (Rutz and Kieft 2004), in soil crusts of the Colorado Plateau (Gundlapally and Garcia-Pichel 2006), in soil from agricultural systems (Jangid et al. 2008), in heavy metal-contaminated acidic waters from zinc mine residues (Almeida et al. 2009), as a potential *m*-xylene degrader (Xie et al. 2010), and during a propane-stimulated bioremediation process in trichloroethene-contaminated groundwater (Connon et al. 2005). Using a classical method, *Ramlibacter* strains were also isolated from soil (Shrestha et al. 2007). More recently, *Ramlibacter* 16S-rDNA gene sequences were

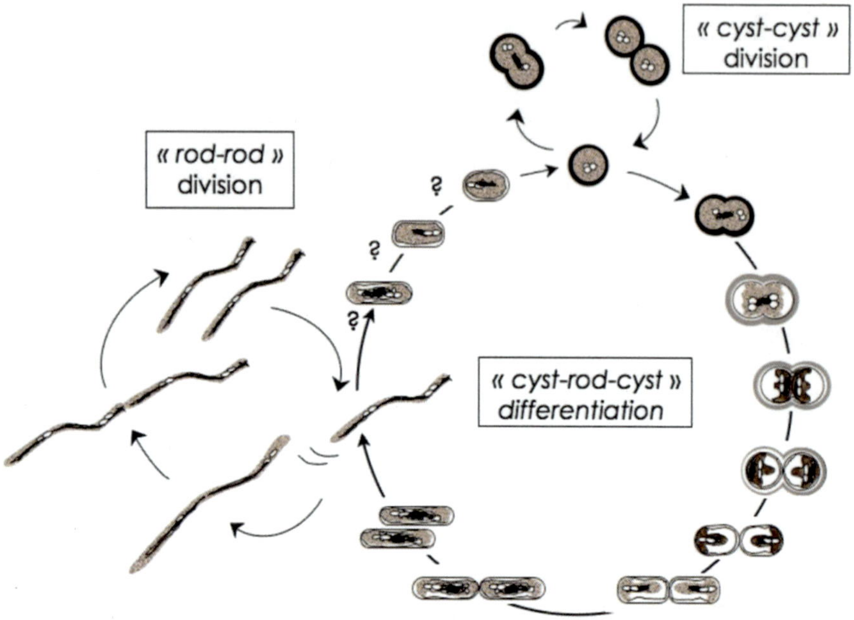

Fig. 4.3 *Ramlibacter tataouinensis* TTB310 cell cycle on nutritive agar. This cell cycle includes the 'cyst-cyst division' step, the 'cyst-rod-cyst differentiation', and the 'rod-rod division' (From De Luca et al. 2011, with permission)

determined by pyrosequencing in murine caecum (Barfod et al. 2013), during micro-aerobic hydrolysis of activated sludges (Zhou et al. 2014), and in the injection water of oil fields in Algeria (Lenchi et al. 2013). All these data (about seventy 16S rDNA sequences and a few isolates) indicate that *Ramlibacter* is adapted to various environments, not only hot and cold deserts.

R. tataouinensis presents an original life cycle characterized by the coexistence of spherical cells (cysts) and rod-shaped cells (Heulin et al. 2003), and three types of cell division occur during the life cycle of this bacterium (Fig. 4.3; see De Luca et al. 2011). The first one consists of a cyst division that generates cysts leading to the formation of bacterial colonies ('*cyst-cyst division*'). The second one, at the border of these colonies, consists of a spherical cell division into rod-shaped cells corresponding to the dissemination form ('*cyst-rod-cyst differentiation*'). The reversion of rod-shaped cells into spherical cells occurs at a distance from original colony (Benzerara et al. 2004a; Gommeaux et al. 2005). The third one consists of a classical rod division ('*rod-rod division*') that probably occurs during the 'social' dissemination of rods. The bacterium *R. tataouinensis* is known to colonize and alter orthopyroxene, which is the major meteorite-forming mineral (Benzerara et al. 2004a). In addition, *R. tataouinensis* can be associated with the biomineralization of calcium phosphate (Benzerara et al. 2004b).

In this way, *R. tataouinensis* is an example of a novel mechanism for desiccation tolerance, illustrated by its ability to divide as a desiccation-tolerant form (cysts). This mechanism is novel when compared to the long-term storage desiccation resistance forms (spores, akinetes, and non-dividing resting cysts) or to the highly efficient DNA repair mechanisms of vegetative cells of *Deinococcus*, resulting in their desiccation tolerance. This desiccation-resistant form resembles mostly the vegetative cells of terrestrial cyanobacteria, which withstand multiple cycles of drying and wetting and/or prolonged desiccation (Potts 1994; Billi and Potts 2002). On the one hand, cysts of *R. tataouinensis* present some traits of *Azotobacter* cysts such as spherical morphotype, absence of motility, cells embedded by a thick capsular material, presence of spherical polyhydroalkanoate lipid granules in the cytoplasm, and long-term resistance to desiccation (Heulin et al. 2003). On the other hand, cysts of *R. tataouinensis* are not resting cells such as the cysts of *Azotobacter* or *Rhodospirillum* (Berleman and Bauer 2004), but are vegetative cells that are able to divide. Cysts have a larger diameter (0.85 μm) and a greater volume (0.34 μm^3) than motile rod-shaped cells (0.24 × 2.9 μm, 0.13 μm^3) with the same cell surface area (2.20 μm^2) (Gommeaux et al. 2005). Therefore, the mechanism of cyst division into motile rod-shaped cells, which takes about 3 h, appears to be very different from the mechanism of cyst germination of *Azotobacter* (Heulin et al. 2003; Gommeaux et al. 2005; Sadoff 1975).

The genome of *R. tataouinensis* TTB310 was completely sequenced by the Genoscope (Genomic Institute, Evry, France) and subsequently annotated by a large consortium of bacteriologists (De Luca et al. 2011). The objectives were to gain insights into possible mechanisms responsible for its unusual lifestyle and its desiccation tolerance. Two-component signalling (TCS) systems are a primary means by which bacteria sense their constantly changing external environment. A thorough analysis of the TCS proteins in the sequenced genome of *R. tataouinensis* TTB310 suggests the following observations: (i) a convergent signalling network due to the higher proportion of histidine kinase sensors (HKs) versus response regulators (RRs); (ii) an intracellular network of signal transduction, since half of the HKs seem to detect intracellular signals; (iii) the involvement of many TCS in post-transcriptional regulation that likely allow a more rapid adaptation compared to transcriptional regulation; and (iv) two chemotaxis systems dedicated to a form of gliding motility. As found in *Caulobacter crescentus* (Curtis and Brun 2010), and suggested in the cyanobacterium *Nostoc punctiforme* ([*Anabaena*] sp. strain PCC 7120) (Kaneko et al. 2001) that both possess a complex program of cell differentiation, a part of these systems could be dedicated to the control of the strain TTB310 cell cycle. The complex network of TCSs found in the genome of *R. tataouinensis* might utilize responses to light (two bacteriophytochromes identified) and a rudimentary circadian hourglass (presence of *kaiC* gene) to anticipate water availability at the dew time, allowing enough time for one division per day. This genome annotation evidenced interesting traits of the strain TTB310 genome that appear to be important for desiccation tolerance, including intermediary metabolism compounds such as trehalose and signal transduction pathways.

4.6 *Deinococcus*, Protein Protection, and DNA Repair

Deinococcus bacteria are famous for their extreme tolerance to gamma and UV radiation, desiccation, and other oxidative stress-generating conditions. This tolerance is likely the result of a combination of different mechanisms, including protection of proteins against oxidation, a compact nucleoid structure, and an efficient stress response allowing repair of massive, stress-induced DNA damage, including hundreds of DNA double-strand breaks (Blasius et al. 2008; Slade and Radman 2011). Natural sources of ionizing radiation on Earth exist at levels much lower than those tolerated by *Deinococcus*, suggesting that radiation tolerance is a consequence of adaptation to natural non-radioactive DNA-damaging conditions such as desiccation (Mattimore and Battista 1996; Tanaka et al. 2004; Cox and Battista, 2005). Indeed, several radiation-tolerant *Deinococcus* strains have been isolated from arid desert soil (de Groot et al. 2005; Rainey et al. 2005; Chanal et al. 2006). Besides factors involved in tolerance to both radiation and desiccation, other components may be more specific for desiccation tolerance such as homologs of desiccation tolerance-associated plant proteins that are found in several *Deinococcus* species, including *Deinococcus deserti* (de Groot et al. 2009). Inactivation of two of these homologs sensitizes *Deinococcus radiodurans* to desiccation, but not to ionizing radiation (Battista et al. 2001).

For *D. radiodurans*, efficient reconstitution of an intact genome is dependent on extended synthesis-dependent strand annealing (ESDSA) and homologous recombination (Zahradka et al. 2006). This DNA repair mechanism requires homologs of proteins that are found in radiation-sensitive bacteria such as RecA and RecFOR, but several *Deinococcus*-specific proteins are probably also involved (Tanaka et al. 2004; Bentchikou et al. 2010). Deinococci possess a highly condensed nucleoid, which may contribute to efficient DNA repair by limiting diffusion of free DNA ends (Levin-Zaidman et al. 2003; Zimmerman and Battista 2005).

Because DNA is damaged by irradiation or desiccation, at least some components of the cell should be protected in order to survive such conditions. Radiation and desiccation tolerance of deinococci and other species has been correlated with protection of proteins from oxidative damage (Daly et al. 2007; Fredrickson et al. 2008). In vitro experiments have shown that protein-free filtrated cell extract of *D. radiodurans* was extremely protective against radiation-induced protein oxidation (Daly et al. 2010). Compared to non-protective cell extracts of radiation-sensitive bacteria (e.g. *E. coli*), cell extracts of *D. radiodurans* are enriched in manganese (Mn^{2+}), phosphate, nucleosides and bases, amino acids, and small peptides. This enrichment is especially true for small peptides, which may result from intracellular protease activities (Daly et al. 2010), and which provide remarkable protein protection (Berlett and Levine 2014).

The first published *Deinococcus* genome sequence was that of *D. radiodurans* (White et al. 1999). Subsequent studies published the genome sequence of *Deinococcus geothermalis* (Makarova et al. 2007) and the genome sequence and proteome analysis of *D. deserti* VCD115 isolated from Sahara surface sand (de Groot et al. 2009). In the Sahara example, *D. deserti* was found to possess several supplementary genes involved in manganese and nutrient import, as well as additional DNA repair genes, including two extra *recA* and three translesion DNA polymerase genes. The three *recA* genes code for two different RecA proteins, both of which contribute to recombinational repair of massive DNA damage (Dulermo et al. 2009). However, only one RecA allows induction of two translesion polymerases that are involved in mutagenic bypass of UV-induced DNA lesions (Dulermo et al. 2009). The supplementary nutrient import and DNA repair genes are probably important for survival and adaptation of *D. deserti* to its hostile environment.

Transcriptomics experiments have shown that expression of many genes, including DNA repair and uncharacterized genes, is induced after exposure of *Deinococcus* to radiation or desiccation (Tanaka et al. 2004; de Groot et al. 2014). The IrrE protein is essential for the induction of many of these genes and also for radiation tolerance (Earl et al. 2002; Makarova et al. 2007; Vujicić-Zagar et al. 2009), showing the importance of an efficient stress response. Recently, it has been demonstrated that IrrE is a metalloprotease that cleaves and inactivates a transcriptional repressor protein called DdrO upon stress exposure, allowing a rapid induction of DNA repair and other genes required for survival (Ludanyi et al. 2014). This novel radiation response mechanism is highly conserved in the *Deinococcaceae*.

Strikingly, the RNA sequencing-based transcriptome analysis of *D. deserti* revealed that an exceptionally high percentage (60 %) of the identified messenger RNAs is leaderless (i.e. lacking a 5′-untranslated region with the ribosome-binding Shine-Dalgarno motif) and proteome data showed that such leaderless mRNAs are efficiently translated in *D. deserti* (de Groot et al. 2014). Moreover, many other identified transcripts are predicted to correspond to leaderless mRNAs encoding small peptides, providing a new explanation for the generation of a cellular pool of small peptides important for protection of proteins against oxidation and thus for radiation/desiccation tolerance and adaptation to harsh environments.

Conclusion

A surprising result from both studies (Merzouga and Tataouine) was the absence of thermotolerant bacteria (e.g. absence of *Thermus*). The evidence of clone sequences of non-thermophile *Crenarchaeota* in the Tataouine sample (Chanal et al. 2006) reinforces the idea that tolerance of desiccation is more important than tolerance of temperature as an adaptive microbial trait for desert conditions. Water allowing bacterial activity is present only when the temperature of the sand is low (below 20 °C).

As demonstrated for other environments, the culture-independent analysis of bacterial diversity in deserts revealed that 70–80 % of 16S-rDNA sequences do not match with described bacterial species, suggesting the presence of a majority of 'new' species and genera to be described. Considering the culture-dependent approach, the grain-by-grain cultivation represents an original and efficient technique allowing the isolation of bacteria strongly attached to a sand grain surface. In the Sahara samples, most of the bacterial isolates or clone sequences belong to phyla and genera for which the mechanisms of adaptation to desiccation are known: *Firmicutes* and *Actinobacteria* (sporulation), and *Deinococcus* and

Rubrobacter (efficient DNA repair). On the other hand, the mechanisms underlying the adaptation to desiccation of most *Proteobacteria* remain to be described. Surprisingly, the very low density of bacteria in desert sand compared to non-arid soils is not due to a much lower diversity but to a very low number of bacteria per taxon. In these hot and dry environments, the number of bacteria is limited by water availability (only few hours per year) and by the very low content in carbon and nutrients (oligotrophic environment). The main conclusion is that numerous bacterial species have been able to adapt in response to extreme conditions prevailing in hot and dry deserts. Many novel mechanisms of desiccation tolerance remain to be unravelled, especially considering proteobacterial species. The ability of these bacteria to tolerate desiccation, UV radiation, heat, oligotrophy, and so many other adverse constraints constitutes a real treasure that may be exploited for biotechnology applications.

Acknowledgments We warmly thank Christopher SWEZEY (US Geological Survey), Frédéric MÉDAIL (Institut Méditerranéen d'Ecologie et de Paléoécologie, Université Paul Cézanne, Aix-Marseille III), and Mylène SALINI for comments and critical reading of the manuscript.

References

Almeida WI, Vieria RP, Machado Cardoso A, Silveira CB, Costa RG, Gonzalez AM, Paranhos R, Medeiros JA, Freitas FA, Albano RM, Martins OB (2009) Archaeal and bacterial communities of heavy metal contaminated acidic waters from zinc mine residues in Sepetiba Bay. Extremophiles 13:263–271

Barer MR (2003) Physiological and molecular aspects of growth, non-growth, culturability and viability in bacteria. In: Coates ARM (ed) Dormancy and low-growth states in microbial disease. Cambridge University Press, Cambridge, pp. 1–35

Barfod KK, Roggenbuck M, Hansen LH, Schjørring S, Larsen ST, Sørensen SH, Krogfelt KA (2013) The murine lung microbiome in relation to the intestinal and vaginal bacterial communities. BMC Microbiol 13:303

Barrat JA, Gillet P, Lécuyer C, Sheppard SM, Lesourd M (1998) Formation of carbonates in the Tatahouine meteorite. Science 280:412–414

Barrat JA, Gillet P, Lesourd M, Blichert-Toft J, Poupeau GR (1999) The Tatahouine diogenite: mineralogical and chemical effects of sixty-three years of terrestrial residence. Meteor Planet Sci 34:91–97

Battista JR, Park MJ, McLemore AE (2001) Inactivation of two homologues of proteins presumed to be involved in the desiccation tolerance of plants sensitizes *Deinococcus radiodurans* R1 to desiccation. Cryobiology 43:133–139

Bentchikou E, Servant P, Coste G, Sommer S (2010) A major role of the RecFOR pathway in DNA double-strand-break repair through ESDSA in *Deinococcus radiodurans*. PLoS Genet 6:e1000774

Benzerara K, Barakat M, Menguy N, Guyot F, De Luca G, Audrain C, Heulin T (2004a) Experimental colonization and alteration of orthopyroxene by the pleomorphic bacteria *Ramlibacter tataouinensis*. Geomicrobiol J 21:341–349

Benzerara K, Menguy N, Guyot F, Skouri F, De Luca G, Barakat M, Heulin T (2004b) Biologically controlled precipitation of calcium phosphate by *Ramlibacter tataouinensis*. Earth Planet Sci Lett 228:439–449

Benzerara K, Chapon V, Moreira D, Lopez-Garcia P, Guyot F, Heulin T (2006) Microbial diversity on the Tatahouine meteorite. Meteor Planet Sci 41:1249–1265

Berleman JE, Bauer CE (2004) Characterization of cyst cell formation in the purple photosynthetic bacterium *Rhodospririllum centenum*. Microbiology(SGM) 150:383–390

Berlett BS, Levine RL (2014) Designing antioxidant peptides. Redox Rep 19:80–86

Billi D, Potts M (2002) Life and death of dried prokaryotes. Res Microbiol 153:7–12

Blasius M, Sommer S, Hübscher U (2008) *Deinococcus radiodurans*: what belongs to the survival kit? Crit Rev Biochem Mol Biol 43:221–238

Bristow CS, Armitage SJ (2016) Dune ages in the sand deserts of the southern Sahara and Sahel. Quat Int 410:46–57

Brito JC, Godinho R, Martínez-Freiría F, Pleguezuelos JM, Rebelo H, Santos X, Vale CG, Velo-Anton G, Boratynski Z, Carvalho SB, Ferreira S, Gonçalves DV, Silva TL, Tarroso P, Campos JC, Leite JV, Nogueira J, Alvares F, Sillero N, Sow AS, Fahd S, Crochet PA, Carranza S (2014) Unravelling biodiversity, evolution and threats to conservation in the Sahara-Sahel. Biol Rev 89:215e231

Chanal A, Chapon V, Benzerara K, Barakat M, Christen R, Achouak W, Barras F, Heulin T (2006) The desert of Tataouine : an extreme environment that hosts a wide diversity of microorganisms and radiotolerant bacteria. Environ Microbiol 8:514–525

Connon SA, Tovanabootr A, Dolan M, Vergin K, Giovannoni SJ, Semprini L (2005) Bacterial community composition determined by culture-independent and –dependent methods during propane-stimulated bioremediation in trichloroethene-contaminated groundwater. Environ Microbiol 7:165–178

Cox MM, Battista JR (2005) *Deinococcus radiodurans*-the consummate survivor. Nat Rev Microbiol 3:882–892

Curtis PD, Brun YV (2010) Getting in the loop: regulation of development in *Caulobacter crescentus*. Microbiol Mol Biol Rev 74:13–41

Daly MJ, Gaidamakova EK, Matrosova VY, Vasilenko A, Zhai M, Leapman RD, Lai B, Ravel B, Li SM, Kemner KM, Fredrickson JK (2007) Protein oxidation implicated as the primary determinant of bacterial radioresistance. PLoS Biol 5:e92

Daly MJ, Gaidamakova EK, Matrosova VY, Kiang JG, Fukumoto R, Lee DY, Wehr NB, Viteri GA, Berlett BS, Levine RL (2010) Small-molecule antioxidant proteome-shields in *Deinococcus radiodurans*. PLoS One 5:e12570

de Groot A, Chapon V, Servant P, Christen R, Fisher-Le Saux M, Sommer S, Heulin T (2005) *Deinococcus deserti* sp. nov., a gamma-radiation tolerant bacterium isolated from the Sahara desert. Int J Syst Evol Microbiol 55:2441–2446

de Groot A, Dulermo R, Ortet P, Blanchard L, Guérin P, Fernandez B, Vacherie B, Dossat C, Jolivet E, Siguier P, Chandler M, Barakat M, Dedieu A, Barbe V, Heulin T, Sommer S, Achouak W, Armengaud J (2009) Alliance of proteomics and genomics to unravel the specificities of Sahara bacterium *Deinococcus deserti*. PLoS Genet 5:e1000434

de Groot A, Roche D, Fernandez B, Ludanyi M, Cruveiller S, Pignol D, Vallenet D, Armengaud J, Blanchard L (2014) RNA sequencing and proteogenomics reveal the importance of leaderless mRNAs in the radiation-tolerant bacterium *Deinococcus deserti*. Genome Biol Evol 6:932–948

De Luca G, Barakat M, Ortet P, Jourlin-Castelli C, Ansaldi M, Py B, Fichant G, Pm C, Voulhoux R, Bastien O, Maréchal E, Henrissat B, Quentin Y, Noirot P, Filloux A, Méjean V, Dubow M, Barras F, Barbe V, Weissenbach J, Mihalcescu I, Verméglio A, Achouak W, Heulin T (2011) The cyst-dividing bacterium *Ramlibacter tataouinensis* genome reveals a well-stocked toolbox for adaptation to a desert environment. PLoS One 6(9):e23784

Dulermo R, Fochesato S, Blanchard L, de Groot A (2009) Mutagenic lesion bypass and two functionally different RecA proteins in *Deinococcus deserti*. Mol Microbiol 74:194–208

Earl AM, Mohundro MM, Mian IS, Battista JR (2002) The IrrE protein of *Deinococcus radiodurans* R1 is a novel regulator of *recA* expression. J Bacteriol 184:6216–6224

Essendoubi M, Brhada F, Eljamali JE, Filali-Maltouf A, Bonnassie S, Georgeault S, Blanco C, Jebbar M (2007) Osmoadaptative responses in the rhizobia nodulating *Acacia* isolated from south-eastern Moroccan Sahara. Environ Microbiol 9:603–611

Fredrickson JK, Li SM, Gaidamakova EK, Matrosova VY, Zhai M, Sulloway HM, Scholten JC, Brown MG, Balkwill DL, Daly MJ (2008) Protein oxidation: key to bacterial desiccation resistance? ISME J 2:393–403

Garcia-Pichel F, Belnap J (1996) Microenvironments and microscale productivity of cyanobacterial desert crusts. J Phycol 32:774–782

Garcia-Pichel F, Pringault O (2001) Cyanobacteria track water in desert soils. Nature 143:380–381

Garcia-Pichel F, López-Cortés A, Nübel U (2001) Phylogenetic and morphological diversity of cyanobacteria in soil desert crusts from the Colorado plateau. Appl Environ Microbiol 67:1902–1910

Gillet P, Barrat JA, Heulin T, Achouak W, Lesourd M, Guyot F, Benzerara K (2000) Bacteria in the Tatahouine meteorite: nanometric-scale life in rocks. Earth Planet Sci Lett 175:161–167

Gommeaux M, Barakat M, Lesourd M, Thiéry J, Heulin T (2005) A morphological transition in the pleiomorphic bacterium *Ramlibacter tataouinensis* TTB310. Res Microbiol 156:1026–1030

Gommeaux M, Barakat M, Montagnac G, Christen R, Guyot F, Heulin T (2010) Mineral and bacterial diversities of desert sand grains from South-East Morocco. Geomicrobiol J 27:76–92

Gonçalves DV, Brito JC, Crochet PA, Geniez P, Padial JM, Harris DJ (2012) Phylogeny of North African *Agama* lizards (Reptilia: Agamidae) and the role of the Sahara desert in vertebrate speciation. Mol Phylogenet Evol 64:582–591

Gundlapally SR, Garcia-Pichel F (2006) The community and phylogenetic diversity of biological soil crusts in the Colorado Plateau studied by molecular fingerprinting and intensive cultivation. Microb Ecol 52:345–357

Heulin T, Barakat M, Christen R, Lesourd M, Sutra L, De Luca G, Achouak W (2003) *Ramlibacter tataouinensis* gen. nov., and *Ramlibacter henchirensis* sp. nov., cyst-producing bacteria isolated from sub-desert soil in Tunisia. Int J Syst Evol Microbiol 53:589–594

Jangid K, Williams MA, Franzluebbers AJ, Sanderlin JS, Reeves JH, Jenkins MB, Endale DM, Coleman DC, Whitman WB (2008) Relative impacts of land-use, management intensity and fertilization upon soil microbial community structure in agricultural systems. Soil Biol Biochem 40:2843–2853

Kaci Y, Heyraud A, Barakat M, Heulin T (2005) Isolation and identification of an EPS-producing *Rhizobium* strain from arid soil (Algeria): characterization of its EPS and the effect of inoculation on wheat rhizosphere soil structure. Res Microbiol 156:522–531

Kaneko T, Nakamura Y, Wolk CP, Kuritz T, Sasamoto S et al (2001) Complete genomic sequence of the filamentous nitrogen-fixing cyanobacterium *Anabaena* sp. strain PCC 7120. DNA Res 8:205–213

Karnieli A, Kidron GJ, Glaesser C, Eyal Ben-Dor E (1999) Spectral characteristics of cyanobacteria soil crust in semiarid environments. Remote Sens Environ 69:67–75

Khbaya B, Neyra M, Normand P, Zerhari K, Filali-Maltouf A (1998) Genetic diversity and phylogeny of rhizobia that nodulate acacia spp. in morocco assessed by analysis of rRNA genes. Appl Environ Microbiol 64:4912–4917

Kröpelin S, Verschuren D, Lézine AM, Eggermont H, Cocquyt C, Francus O, Cazet JP, Fagot M, Ramus B, Russell JM, Conley DJ, Schuster M, von Suchodoletz H, Engstrom DR (2008) Climate-driven ecosystem succession in the Sahara: the past 6000 years. Science 320: 765–768

Lacroix A (1931) Sur la chute récente (27 juin 1931) d'une météorite asidérite dans l'extrême Sud Tunisien. C R Acad Sci Paris 193:305–309

Le Houérou HN (1986) The desert and arid zones of Northern Africa. In: Evenari M, Noy-Meir E, Goodall DW (eds) Hot deserts and arid shrublands. Ecosystems of the world, vol 12B. Elsevier, Amsterdam, pp. 101–147

Le Houérou HN (1997) Climate, flora and fauna changes in the Sahara over the past 500 million years. J Arid Environ 37:619–647

Lee HJ, Lee SH, Lee SS, Lee JS, Kim Y, Kim SC, Jeon CO (2014) *Ramlibacter solisilvae* sp. nov., isolated from forest soil, and emended description of the genus *Ramlibacter*. Int J Syst Evol Microbiol 64(Pt 4):1317–1322

Leite JV, Alvares F, Velo-Anton G, Brito JC, Godhino R (2015) Differentiation of North African foxes and population genetic dynamics in the desert-insights into the evolutionary history of two sister taxa, *Vulpes rueppellii* and *Vulpes vulpes*. Org Divers Evol 15:731–745

Lenchi N, Inceoglu O, Kebbouche-Gana S, Gana ML, Lliros M, Servais P, Tamara Garcia-Armisen T (2013) Diversity of microbial communities in production and injection waters of algerian oilfields revealed by 16S rRNA gene amplicon 454 pyrosequencing. PLoS One 8(6):e66588

Levin-Zaidman S, Englander J, Shimoni E, Sharma AK, Minton KW, Minsky A (2003) Ringlike structure of the *Deinococcus radiodurans* genome: a key to radioresistance? Science 299:254–256

Long LK, Zhu HH, Yao Q, Ai YC (2008) Analysis of bacterial communities associated with spores of *Gigaspora margarita* and *Gigaspora rosea*. Plant and Soil 310:1–9

Ludanyi M, Blanchard L, Dulermo R, Brandelet G, Bellanger L, Pignol D, Lemaire D, de Groot A (2014) Radiation response in *Deinococcus deserti*: IrrE is a metalloprotease that cleaves repressor protein DdrO. Mol Microbiol 94:434–449

Makarova KS, Omelchenko MV, Gaidamakova EK, Matrosova VY, Vasilenko A, Zhai M, Lapidus A, Copeland A, Kim E, Land M, Mavrommatis K, Pitluck S, Richardson PM, Detter C, Brettin T, Saunders E, Lai B, Ravel B, Kemner KM, Wolf YI, Sorokin A, Gerasimova AV, Gelfand MS, Fredrickson JK, Koonin EV, Daly MJ (2007) *Deinococcus geothermalis*: the pool of extreme radiation genes shrinks. PLoS One 2:e995

Mattimore V, Battista JR (1996) Radioresistance of *Deinococcus radiodurans*: functions necessary to survive ionizing radiation are also necessary to survive prolonged desiccation. J Bacteriol 178:633–637

Migliore J, Baumel A, Juin M, Fady B, Roig A, Duong N, Médail F (2013) Surviving in mountain climate refugia: New insights from the genetic diversity and structure of the relict shrub *Myrtus nivellei* (Myrtaceae) in the Sahara desert. PLoS One 8(9):e73795

Mobley HL, Island MD, Hausinger RP (1995) Molecular biology of microbial ureases. Microbiol Rev 59:451–480

Potts M (1994) Desiccation tolerance of prokaryotes. Microbiol Rev 58:755–805

Prestel E, Salamitou S, DuBow MS (2008) An examination of the bacteriophages and bacteria of the Namib desert. J Microbiol 46:364–372

Rainey FA, Rau K, Ferreira M, Gatz BZ, Nobre MF, Bagaley D, Rash BA, Park MJ, Earl AM, Shank NC, Small AM, Henk MC, Battista JR, Kämpfer P, da Costa MS (2005) Extensive diversity of ionizing-radiation-resistant bacteria recovered from Sonoran desert soil and description of nine new species of the genus *Deinococcus* obtained from a single soil sample. Appl Environ Microbiol 71:5225–5235

Rutz BA, Kieft TL (2004) Phylogenetic characterization of dwarf archaea and bacteria from a semiarid soil. Soil Biol Biochem 36:825–833

Sadoff HL (1975) Encystment and germination in *Azotobacter vinelandii*. Bacteriol Rev 39:516–539

Sarnthein M (1978) Sand deserts during glacial maximum and climatic optimum. Nature 272:43–46

Shrestha PM, Noll M, Liesack W (2007) Phylogenetic identity, growth-response time and rRNA operon copy number of soil bacteria indicate different stages of community succession. Environ Microbiol 9:2464–2474

Slade D, Radman M (2011) Oxidative stress resistance in *Deinococcus radiodurans*. Microbiol Mol Biol Rev 75:133–191

Sow AS, Martínez-Freiría F, Dieng H, Fahd S, Brito JC (2014) Biogeographical analysis of the Atlantic Sahara reptiles: Environmental correlates of species distribution and vulnerability to climate change. J Arid Environ 109:65e73

Swezey C (2001) Eolian sediment responses to late Quaternary climate changes; temporal and spatial patterns in the Sahara. Palaeogeogr Palaeoclimatol Palaeoecol 167:119–155

Swezey CS (2009) Cenozoic stratigraphy of the Sahara, Northern Africa. J Afr Earth Sci 53:89–121

Takagi H (2008) Proline as a stress protectant in yeast: physiological functions, metabolic regulations, and biotechnological applications. Appl Microbiol Biotechnol 8:211–223

Tanaka M, Earl AM, Howell HA, Park MJ, Eisen JA, Peterson SN, Battista JR (2004) Analysis of *Deinococcus radiodurans*'s transcriptional response to ionizing radiation and desiccation reveals novel proteins that contribute to radioresistance. Genetics 168:21–33

Torsvik T, Øvreas L, Thingstad TF (2002) Prokaryotic diversity - magnitude, dynamics, and controlling factors. Science 296:1064–1066

Vujicić-Zagar A, Dulermo R, Le Gorrec M, Vannier F, Servant P, Sommer S, de Groot A, Serre L (2009) Crystal structure of the IrrE protein, a central regulator of DNA damage repair in *Deinococcaceae*. J Mol Biol 386:704–716

White O, Eisen JA, Heidelberg JF, Hickey EK, Peterson JD, Dodson RJ, Haft DH, Gwinn ML, Nelson WC, Richardson DL, Moffat KS, Qin H, Jiang L, Pamphile W, Crosby M, Shen M, Vamathevan JJ, Lam P, McDonald L, Utterback T, Zalewski C, Makarova KS, Aravind L, Daly MJ, Minton KW, Fleischmann RD, Ketchum KA, Nelson KE, Salzberg S, Smith HO, Venter JC, Fraser CM (1999) Genome sequence of the radioresistant bacterium *Deinococcus radiodurans* R1. Science 286:1571–1577

Xie S, Sun W, Luo C, Cupples AM (2010) Stable isotope probing identifies novel *m*-xylene degraders in soil microcosm from contaminated and uncontaminated sites. Water Air Soil Pollut 212(1):113–122

Zahradka K, Slade D, Bailone A, Sommer S, Averbeck D, Petranovic M, Lindner AB, Radman M (2006) Reassembly of shattered chromosomes in *Deinococcus radiodurans*. Nature 443:569–573

Zerhari K, Aurag J, Khbaya B, Kharchaf D, Filali-Maltouf A (2000) Phenotypic characteristics of rhizobia isolates nodulating acacia species in the arid and Saharan regions of Morocco. Lett Appl Microbiol 30:351–357

Zhou Z, Qiao W, Xing C, Shen X, Hu D, Wang L (2014) A micro-aerobic hydrolysis process for sludge in situ reduction: Performance and microbial community structure. Bioresour Technol 173:452–456

Zimmerman JM, Battista JR (2005) A ring-like nucleoid is not necessary for radioresistance in the *Deinococcaceae*. BMC Microbiol 5:17

Extremophiles in Antarctica: Life at Low Temperatures

5

David A. Pearce

Contents

D.A. Pearce
Department of Applied Sciences, Faculty of Health and Life Sciences,
Northumbria University, Ellison Building, Newcastle Upon Tyne, NE1 8ST, UK
British Antarctic Survey, Natural Environment Research Council,
High Cross, Madingley Road, Cambridge, CB3 OET, UK
e-mail: david.pearce@northumbria.ac.uk

H. Stan-Lotter, S. Fendrihan (eds.), *Adaption of Microbial Life to Environmental Extremes*, DOI 10.1007/978-3-319-48327-6_5

5.1 Introduction

Extremophiles are microorganisms that survive or thrive in extreme environments. There remains much debate as to what actually constitutes an extreme environment, as evidenced by the coordination action for research activities on life in extreme environments initiative (CAREX http://www.esf.org/hosting-experts/expert-boards-and-committees/space-sciences/activities/carex-project.html) and definitions can vary according to the perspective, i.e. what is extreme to a human isn't necessarily extreme for a microorganism, and also the scale, so, for example, many people view the whole of the Antarctic continent as an extreme environment, when many different types of condition exist. However, extremophiles can thrive in ice, boiling water, acid, the water core of nuclear reactors, salt crystals and toxic waste and in a range of other extreme habitats that were previously thought to be inhospitable for life (Cavicchioli 2002). For the purposes of this discussion, I shall therefore define all microorganisms living in the Antarctic as living in an extreme environment, but will highlight those that are specifically classified as extremophiles. A range of different factors makes the Antarctic environment extreme from the general physical characteristics: extremes of temperature, desiccation and osmotic stress, low-nutrient concentrations, high levels of UVB radiation (under the Antarctic ozone hole) and a highly variable photoperiod (from no light at all to continuous light during a 24 h period).

The continent of Antarctica is large, approximately the size of Europe or the USA, and so there are many additional specific extremes, and these include areas for which specific extremophiles may have adapted – such as regions of volcanic activity, hypersaline lakes, subglacial lakes and even within the ice itself. As a result there are numerous examples of microorganisms that have special adaptations to cope with these often unique combinations of selection pressures, and these lead to a wide range of novel biodiversity, much of which is still yet to be described. Another key feature of the Antarctic ecosystem is the extreme variation within the physical conditions that exist, for example, from freshwater lakes (some of the most oligotrophic environments on Earth) to hypersaline lakes (Laybourn-Parry and Pearce 2016). Microorganisms found under these extreme environmental conditions in the Antarctic are therefore ideal candidates to study the ecophysiological and biochemical adaptations of living at the limits for life, which, surprisingly, are still to be properly defined particularly at low temperatures.

Despite the relative isolation of the Antarctic continent, microorganisms still arrive constantly from around the globe and particularly from the rest of the atmosphere (Pearce et al. 2009, 2016). Whether they remain as permanent members of the Antarctic microbiota or not depends on their rate of arrival, ability to compete with existing populations, potential rate of evolution and rate of removal. Often microorganisms arriving on the Antarctic continent are preadapted to survival in this type of extreme, as they have been pre-filtered through the atmosphere by similar extreme conditions to those found in the Antarctic. Indeed, it has been observed that the harsher the Antarctic environment, the more cosmopolitan the species found (compare studies by Hughes et al. 2004; Pearce et al. 2010). Human impact can also influence this delicate balance. Ah Tow and Cowan (2005) found that the predominant commensal microorganism occurring on human skin, *Staphylococcus epidermidis*, could be detected by PCR in Dry Valley mineral soils collected from heavily impacted areas, but could not be detected in mineral soils collected from low impact and pristine areas from the same region. Cell viability of this non-enteric human commensal appeared to be rapidly lost in Dry Valley mineral soil. However, *S. epidermidis* can persist for long periods in Dry Valley mineral soil as non-viable cells and/or naked DNA.

The Antarctic continent is also one of the most remote places on Earth. It is isolated by distance (the tip of the Antarctic Peninsula is >1000 km from the Southern tip of South America), air movement (the prevailing movement is towards the coast), the Antarctic circumpolar current which moves in an anticlockwise direction around the pole, the polar front where a significant temperature difference prevents free mixing and a relative lack of human activity to introduce new colonists. For this reason, the microorganisms present there are also relatively isolated from the rest of the biosphere. The observed balance between mesophiles growing suboptimally and true psychrophiles may reflect in some way the balance between indigenous microorganisms which have evolved in situ and those which have subsequently colonized the area, however, as yet there is no data to support this theory. There is also the issue of psychrotolerant (organisms which are tolerant of low growth temperatures from −15 °C to +10 °C) vs. psychrophilic (organisms with a requirement for low growth temperatures) isolates. For example, Clocksin et al. (2007) isolated eight strains of chemoorganotrophic bacteria from the water column of Lake Hoare, in the McMurdo Dry Valleys of Antarctica, using cold enrichment temperatures. All isolates grew at 0 °C, and all but one grew at subzero temperatures characteristic of the water column of Lake Hoare. However, the growth temperature optima varied among isolates, with the majority showing optima near 15 °C, indicative of cold-active phenotypes. However, only one isolate was truly psychrophilic, growing optimally around 10 °C and not above 20 °C.

In contrast to the terrestrial realm, the seas around Antarctica are extremely productive and diverse. One of the reasons the seas around Antarctica support so many living things, despite the extreme cold, is the abundance of nutrients in the water. Within the Antarctic environment, it is the marine ecosystem which is the most stable, and this is reflected in the biodiversity of the organisms found there. The Antarctic marine ecosystem bucks the trend of decreasing biodiversity with increasing latitude, so often observed for multicellular organisms. It can also be said to harbour truly psychrophilic microorganisms as the temperature is consistently low. However, even within the marine system, niches exist for extremophiles, such as hydrothermal vent systems, mud volcanoes and cold seeps, and in the relatively unexplored deep-sea environment in general, there is also very high pressure.

Understanding the biodiversity of marine bacterioplankton is crucial for predicting changes in other organisms at the ends of marine food chains. Most bacteria live in microenvironments below the scale of millimetres such as across physical discontinuities of the sea-ice or seawater column or on plant, animal or detritus surfaces. Heterogeneity exists at microscale levels, and changes in microorganisms will produce a cascade of changes through the marine food chain.

Some residual controversy exists over the 'everything is everywhere but the environment selects' debate; however, data are increasingly becoming available, particularly for the cyanobacteria, which suggest that endemic Antarctic forms occur which have not been found elsewhere, and there may indeed be an endemic Antarctic microflora. In contrast to the rather limited diversity of plants and animals to be found in the Antarctic, the microbial diversity of this continent has been shown to be surprisingly diverse (Aguirre de Cárcer et al. 2015; Boetius et al. 2015; Chong et al. 2015; Ghiglione et al. 2012; Terauds et al. 2012; Taton et al. 2006; Tytgat et al. 2014, 2016; Tindall 2004). There have been and are an increasing number of broad-scale microbial biodiversity studies such as the Global Ocean Sampling Project, the Census of Marine Microbes, the Earth Microbiome Project and TARA Oceans which have sought to understand total microbial biodiversity in a range of global environments. However, exploring the biodiversity of Antarctic extreme environments is often targeted, particularly the unusual and less explored more extreme environments.

5.2 Environmental Extremes Associated with the Antarctic

Many of the environmental extremes in Antarctica have the same characteristics and features as extreme environments that are found elsewhere, but which in addition tend to be much colder. Key characteristics are:

Strong winds These dry the Antarctic environments, and they also increase the effects of the cold.

Temperature Antarctica contains the coldest environments on Earth. Environmental temperature has a profound impact on species distributions, abundance and survival, and the importance of temperature in limiting species distributions can be ascribed in part to its powerful effects on subcellular and molecular systems (Hochachka and Somero 2002). Indeed, temperature is a critical environmental factor controlling the evolution and biodiversity of life on earth. The majority of the Earth's biosphere is permanently below 5 °C, dominated by ocean depths, glaciers, alpine and polar regions (Feller and Gerday 2003). Surprisingly, however, although Antarctic temperatures are predominantly cold, in the austral summer, on a sunny day temperatures can reach in excess of 30 °C on dark surfaces, and these alternate with extremes of cold in the winter.

Desiccation Although the Antarctic continent contains a large percentage of the Earths' freshwater, the vast majority of this water is in a solid form and therefore unavailable to organisms. All but around 2 % of the continent is covered with ice, and so the Antarctic is very much a polar desert. However, despite these challenging low temperatures, as long as liquid water is present, life is possible. Even the coldest environments on Earth have enough liquid water to sustain life (Laybourn-Parry 2009).

The physical properties of water They lead to a mechanical impact of external ice (water expands by about 4 % on freezing); this affects biochemical processes, can directly damage cells and membranes and disrupt water relations (controlled by water status and availability).

Limited nutrients Chemical analysis has shown that Antarctic soils are relatively low in nutrient content: 1.95–0.33 % carbon, 0.20–0.03 % nitrogen, 8.00–0.06 % phosphorus and 0.22–0.20 % potassium (Newsham et al. 2010), with an organic matter content of ~ 2.56 % (Lawley et al. 2004).

Patchiness of high nutrients This can lead to localized extremes, for example, exposure to high acidity can occur due to immersion in penguin guano (Fig. 5.1).

Fig. 5.1 Ornithigenic soils on South Thule Island, South Sandwich Islands (Photo by Dr. D. Pearce, British Antarctic Survey)

Significant chemical gradients Such patchiness of nutrients and stressors in general can lead to strong gradients in physicochemical parameters at a wide range of spatial scales – of the order of metres (Chong et al. 2009), kilometres (Chong et al. 2010) or hundreds of kilometres (Yergeau et al. 2008).

High salinity Many of the saline lakes of coastal oases such as the Vestfold Hills are marine derived, formed from pockets of seawater trapped in closed basins when the land rose during isostatic rebound flowing the last glaciation. This suite of lakes ranges from slightly brackish (approx. 4–5 ‰) to hypersaline (240 ‰) (Laybourn-Parry and Pearce 2007, 2016). There is also the potential for alternate immersion in both salt and freshwater due to weather and tides at the coast.

Adverse solar radiation Intense ultraviolet light radiation can occur as a result of the hole in the ozone layer, which is now starting to close. The photoperiod also varies from months of total darkness to total sunlight.

Low biochemical activity Due to low temperatures.

Instability and variability In terms of severity, seasonality and the changes in seasonality with latitude, these physical and chemical conditions can change annually, and these can be exacerbated by climate change (Pearce 2008). In Antarctic freshwater lakes, bacterial community structures have been shown to be influenced by temperature, at temperatures lower than that already experienced as a result of climate change (Pearce 2005). This means that Antarctic environments can be variable or unpredictable, and this in itself imposes a further stress.

Biotic interactions Competition and predation can produce stress. Duarte et al. (2005) showed that bacteria–chlorophyll and bacterial production–primary production relationships in the Southern Ocean differed from the typical relationships applicable to aquatic ecosystems elsewhere. Bacteria responded to phytoplankton blooms, but they responded so weakly that the bacterial production represented a small percentage of primary production (1–10 %). Although other mechanisms might also contribute to the weak bacterial response to phytoplankton blooms, they demonstrated that the reason was likely to be the tight control of bacterial populations by their predators.

Combined stresses Many Antarctic environments such as terrestrial, freshwater, aerial and ice combine more than one of the above stresses. For example, Antarctic soils can have high salinity, low nutrients, low temperature and adverse solar radiation, and these may result in a unique suite of selection pressures.

5.3 Extreme Environments that Also Occur Within the Antarctic

So what type of more extreme environment occurs in the Antarctic? Within the Antarctic continent, there are specific locations in which adaptation to a variety of potential extremes is possible. For this reason, many have been specifically targeted for study in detail.

Aerial The aerial environment is very similar to the Antarctic environment in terms of selection pressures, with temperatures around 0 °C in the stratosphere, where life has been shown to exist (Sattler et al. 2001). It is also low nutrient, desiccating and with high UVB radiation. As a result resistant spore-forming bacteria are often found here (Fig. 5.2) (Bottos et al. 2014).

Volcanic A number of regions of volcanic activity are found both in and around the Antarctic environment; these include Mount Erebus, Deception Island, Montague Island and South Thule (Fig. 5.3).

Hydrothermal The discovery of the Indian Ocean hydrothermal vent communities raised many questions about the biogeography of these ecosystems, for example, how has the distribution of the vent fauna evolved over geological time and how has this been influenced by continental drift and past climate change. One of the obvious routes for dispersal of hydrothermal vent organisms between the Atlantic, Pacific and Indian Ocean is via the Southern Ocean. To address this problem a Natural Environment Research Council (NERC) funded Chemosynthetic Communities of the Southern Ocean (CHESSO) Consortium has been exploring areas of the Southern Ocean where chemical signals in the ocean and the shape of the seabed indicate that hydrothermal vents might be present (Rogers et al. 2012).

Sea ice Antarctic sea ice is highly variable and can triple the size of the Antarctic continent in winter (Fig. 5.4). The sea ice itself is extremely productive with a whole

Fig. 5.2 Aerial transfer of microorganisms into the Antarctic (Photo by Dr. Pearce, British Antarctic Survey)

Fig. 5.3 Deception Island, a volcano to the North of the Antarctic Peninsula (62°58'S 60°39'W) which erupted in 1987 (Photo by Dr. Pearce, British Antarctic Survey)

Fig. 5.4 Sea ice showing coloration due to the presence of associated microorganisms (Photo by Dr. Pearce, British Antarctic Survey)

trophic cascade based around the algae that grow on or under the surface of the ice. Beneath the snow lies a unique habitat for a group of bacteria and microscopic plants and animals that are encased in an ice matrix at low temperatures and light levels, with the only liquid being pockets of concentrated brines. Survival in these conditions requires a complex suite of physiological and metabolic adaptations, but sea-ice organisms thrive in the ice (Thomas and Dieckmann 2002). In the Sea Ice Microbial Communities (SIMCO) study, an extensive array of bacterial isolates were obtained from bottom ice assemblages as well as from algae-free 'clean' ice and under-ice seawater to determine the influence diatom blooms have on the bacterial inhabitants. Previous results had indicated that biodiversity increased three- to sevenfold between ice free of algae and under-ice seawater to ice samples rich in algae (>100 mg chlorophyll m3). From this, 35–95 % of the bacterial biomass from diatom assemblages was estimated to be psychrophilic. By comparison ice lacking algae and under-ice seawater samples are nearly devoid of psychrophiles which were almost always isolates of Psychrobacter glacincola.

Fast ice Psychrobacter salsus sp. nov. and Psychrobacter adeliensis sp. nov. have also been isolated from fast ice from Adelie Land, Antarctica (Shivaji et al. 2004).

Radiation On the snow surface of glaciers and polar caps, psychrophiles are exposed to strong ultraviolet radiation (Fig. 5.5). Antarctic lakes are unproductive, with typical levels of photosynthesis in the region of 0.5–30 μg l-1 day-1. This results from low annual levels of photosynthetically active radiation and ice

Fig. 5.5 Microorganisms growing on the surface of Antarctic snow (Photo by Dr. P. Bucktrout, British Antarctic Survey)

covers that attenuate light to the water column, continuous low temperatures and the lack of any significant input of inorganic nutrients and organic carbon (Laybourn-Parry 2002).

Microbial mats *Planococcus antarcticus* and *Planococcus psychrophilus* sp. nov. were isolated from cyanobacterial mat samples collected from ponds in Antarctica (Reddy et al. 2002) along with psychrophilic *Planococcus maitriensis* sp. nov. and an orange-pigmented bacterium, which was isolated from a cyanobacterial mat sample collected in the vicinity of Schirmacher Oasis, Maitri, the Indian station, in Antarctica (Alam et al. 2003). Van Trappen et al. (2002) isolated 746 heterotrophic bacteria from within microbial mats from ten Antarctic lakes. The frequency of such isolations suggests that this is a habitat conducive to the growth of microorganisms.

Permanently cold marine sediments A 16S ribosomal DNA (rDNA) clone library from permanently cold Antarctic marine sediments was established (Ravenschlag et al. 1999). Screening 353 clones by dot blot hybridization with group-specific oligonucleotide probes suggested a predominance of sequences related to bacteria of the sulfur cycle (43.4 % potential sulfate reducers). Within this fraction, the major cluster (19.0 %) was affiliated with *Desulfotalea* sp. and other closely related psychrophilic sulfate reducers isolated from the same habitat. The cloned sequences showed between 93 and 100 % similarity to these bacteria. Two additional groups were frequently encountered: 13 % of the clones were related to *Desulfuromonas palmitatis* and a second group was affiliated with *Myxobacteria* spp. and *Bdellovibrio* spp. Many clones (18.1 %) belonged to the gamma subclass of the class *Proteobacteria* and were closest to symbiotic or free-living sulfur oxidizers. Rarefaction analysis suggested that the total diversity assessed by 16S rDNA analysis was very high in these permanently cold sediments and was only partially revealed by screening of 353 clones. As more comparative community structures in the marine sediments are published (e.g. Robador et al. 2016) as revealed by pyrosequencing of bacterial 16S rRNA gene amplicons, our understanding of bacterial biodiversity will increase considerably.

Lakes In Antarctic lakes, organisms are confronted by continuous low temperatures as well as a poor light climate and nutrient limitation. Such extreme environments support truncated food webs with no fish, few metazoans and a dominance of microbial plankton (Laybourn-Parry 2002). The key to success lies in entering the short Antarctic summer with actively growing populations. In many cases, the most successful organisms continue to function throughout the year (Laybourn-Parry and Pearce 2007, 2016).

Ornithogenic soils The amount of penguin guano deposited by a breeding population on the shores of Admiralty Bay is about 6.4 tons (dry weight) per day (Rakusa-Suszczewski 1980), causing a significant impact on the local budgets of carbon, nitrogen, phosphorus and other minerals (Tatur 2002); such instances are seen wherever significant penguin colonies exist.

Fig. 5.6 An endolithic cyanobacterial community (Photo by Dr. C. Gilbert, British Antarctic Survey)

Endolithic, sublithic and chasmolithic The endolithic microbial communities that are found in both rocks of the Antarctic dry deserts and the Arctic (Cockell et al. 2002), which comprise lichens, yeasts, cyanobacteria and heterotrophic bacteria, survive low water and nutrient availability (Fig. 5.6). Pointing et al. (2009) demonstrated that life has adapted to form highly specialized communities in distinct lithic niches. Endoliths and chasmoliths in sandstone displayed greatest diversity, whereas soil was relatively depauperate and lacked a significant photoautotrophic component, apart from isolated islands of hypolithic cyanobacterial colonization on quartz rocks in soil contact (Pointing et al. 2009). The findings showed that biodiversity near the cold-arid limit for life is more complex than previously appreciated, but communities lack variability probably due to the high selective pressures of this extreme environment.

Cryoconite holes Cryoconite holes are water-filled cylindrical melt holes on glacial ice surface (Christner et al. 2003; Edwards et al. 2011; Cameron et al. 2012; Telling et al. 2014; Hodson et al. 2015).

Lake sediments A bacterial community has been described from along a historic lake sediment core of Ardley Island, West Antarctica (Li et al. 2006). There are clearly many such examples from around the continent, so the Antarctic harbours a diverse range of highly specialized niches for extremophile growth.

5.4 Extreme Environments Particular to the Antarctic

So what are the extreme environments particular to the Antarctic?

Antarctic subglacial lakes These are one of the few remaining little explored environments on Earth and, from a microbiological perspective, perhaps one of the most interesting. They have the potential to be one of the most extreme environments on Earth and are certainly one of the least accessible, with combined stresses of high pressure, low but probably stable temperature, permanent darkness, predominantly low-nutrient availability (potentially one of the most oligotrophic environments on Earth, particularly in the absence of geothermal activity) and highly variable oxygen concentrations (derived from the ice that provided the original melt water). Indeed, the predominant mode of nutrition is likely to be chemoautotrophic (Siegert et al. 2003). Lake Vostok, the largest subglacial lake found, has shown encouraging signs of containing life. All of the Vostok accreted ice samples between 3541 and 3611 m have been found to contain both prokaryotic and eukaryotic microorganisms (Poglazova et al. 2001). The identification of significant subglacial bacterial activity (Sharp et al. 1999) as well as the work on permafrost communities (Gilichinsky et al. 1995) suggests that life can survive and potentially thrive at low temperatures. Until recently, microbes had only been detected in the two Antarctic subglacial lakes sampled to date, accreted ice from subglacial Lake Vostok in East Antarctica and saturated till from beneath ice streams draining the West Antarctic Ice Sheet (e.g. Karl et al. 1999; Priscu et al. 1999; Christner et al. 2006; Bulat et al. 2009) and saturated till from beneath the Kamb Ice Stream, West Antarctica (Priscu et al. 2005; Lanoil et al. 2009). Data from the accretion ice support the working hypothesis that a sustained microbial ecosystem is present in this subglacial lake environment, despite high pressure, constant cold, low-nutrient input, potentially high oxygen concentrations and an absence of sunlight (Christner et al. 2006). The small numbers of microbes found so far within the accreted ice have DNA profiles similar to those of contemporary surface microbes. In 2013, Pearce et al. reported the first microbiological analysis of a sample taken from a former subglacial lake sediment in Antarctica (Lake Hodgson, on the Antarctic Peninsula). This was one of a number of subglacial lakes just emerging at the margins of the Antarctic ice sheet due to the renewed onset of deglaciation. More recently, Christner et al (2014) published the first geomicrobiological description of water and surficial sediments obtained from direct sampling of a subglacial Antarctic lake. Subglacial Lake Whillans (SLW) which lies beneath approximately 800 m of ice on the lower portion of the Whillans Ice Stream (WIS) in West Antarctica. They found that it was a chemosynthetically driven ecosystem inhabited by a diverse assemblage of bacteria and archaea.

The arid soils of the Antarctic Dry Valleys These constitute some of the oldest, coldest, driest and most oligotrophic soils on Earth. Recent applications of molecular methods have revealed a very wide diversity of microbial taxa, many of which are uncultured and taxonomically unique, and a community that seems to be structured solely by abiotic processes (Cary et al. 2010).

Marine hypersaline deeps Habitats on the seafloor include sediments of varying geology, mineral nodule fields, carbonate mounds, cold seeps, hydrocarbon seeps, saturated brines and hydrothermal vents. Below sea subfloor and deep biosphere, marine sediments merge into the below sea subfloor and deep biosphere.

Ice cave environments Such as those on Mount Erebus.

Permafrost (though this has mainly been studied in the Northern hemisphere) In a study of permafrost, Vishnivetskaya et al. (2007) found that half of the isolates were spore-forming bacteria unable to grow or metabolize at subzero temperatures. Other Gram-positive isolates metabolized, but never exhibited any growth at –10 °C. One Gram-negative isolate metabolized and grew at –10 °C, with a measured doubling time of 39 days. Metabolic studies of several isolates suggested that as temperature decreased below + 4 °C, the partitioning of energy changed with much more energy being used for cell maintenance as the temperature decreases. In addition, cells grown at –10 °C exhibited major morphological changes at the ultrastructural level. Bacteria of the genus *Exiguobacterium* have been repeatedly isolated from ancient permafrost sediments of the Kolyma lowland of Northeast Eurasia. Vishnivetskaya et al. (2007) also reported that the Siberian permafrost isolates *Exiguobacterium sibiricum* 255-15, *E. sibiricum* 7-3, *E. undae* 190-11 and *E. sp.* 5138, as well as *Exiguobacterium antarcticum* DSM 14480, isolated from a microbial mat sample of Lake Fryxell (McMurdo Dry Valleys, Antarctica), were able to grow at temperatures ranging from –6 to 40 °C. In comparison to cells grown at 24 °C, the cold-grown cells of these strains tended to be longer and wider. There are also habitats within the permafrost such as cryopegs. Given their relative isolation and unique conditions, extreme environments in Antarctica offer unique opportunities for biosprospecting (Amos et al. 2015; Dickinson et al. 2016; Purves et al. 2016; Lopatina et al. 2016). Although relatively unexplored, the potential exists for new biosubstances such as novel psychrophilic enzymes. However, much of the work to date has focussed on those microorganisms capable of degrading hydrocarbons associated with low-temperature hydrocarbon spills.

5.5 Key Antarctic Extremophiles

Psychrophiles are organisms capable of growth and reproduction at temperatures from –15 to +10 °C. The ability of psychrophiles to survive and proliferate at low temperatures implies that they have overcome key barriers inherent to permanently cold environments. These challenges include: reduced enzyme activity; decreased membrane fluidity; altered transport of nutrients and waste products; decreased rates of transcription, translation and cell division; protein cold denaturation; inappropriate protein folding; and intracellular ice formation (D'Amico et al. 2006). Cold-adapted organisms have successfully evolved features, genotypic and/or phenotypic, to surmount the negative effects of low temperatures and to enable growth in these extreme environments. Psychrophiles are true extremophiles as they are adapted not only to low temperatures, but frequently also to further environmental constraints. They occur throughout the Antarctic continent, but particularly in the Southern Ocean.

Thermophiles are organisms capable of growth and reproduction at temperatures from between 45 and 80° C. They occur in regions of geothermal activity in Antarctica.

Piezophiles are organisms capable of growth and reproduction at high pressures. Microbial life has been shown to function at gigapascal pressures (Nogi et al. 1998; Sharma et al. 2002), and bacteria recovered from the deep ocean at around 4000 m have been shown to retain both structural integrity and metabolic activity (Pearce 2009). They occur in Antarctica in the deep sea and deep within the ice.

Acidophiles are organisms capable of growth and reproduction under highly acidic conditions where pH is less than or equal to 2.0.

Alkaliphiles are organisms capable of growth and reproduction under alkaline conditions of pH > 9.0; both pH extremes occur in Antarctica, for example, lake ecosystems such as Blood Falls (Mikucki and Priscu 2007).

Halophiles are organisms capable of growth and reproduction at high salinities, for example, Halomonas alkaliantarctica sp. nov., isolated from saline lake Cape Russell in Antarctica, an alkalophilic moderately halophilic, exopolysaccharide-producing bacterium (Poli et al. 2006). Also, euryhaline halophilic bacteria have been described from Suribati Ike, a meromictic lake in East Antarctica (Naganuma et al. 2005). Further, the microbial communities that are found in sea ice are exposed to salt concentrations of several molar in brine veins and are therefore halo-psychrophiles (Staley and Gosink 1999).

5.6 Biodiversity

Microbial communities in the Antarctic are characterized by low numbers of dominant species but have been found to have a high diversity of subdominant species. This observation has implications for the total biodiversity of these environments, which could be much higher than was originally thought. A large percentage of microorganisms detected in Antarctic environments are new to science. Up to 80 % of the dominant members of the microbial community can be detected but have yet to be studied as they have not been cultured. A high proportion of uncultivated microbes are common, generally, but particularly so for Antarctic extremophile environments. A bacterial phylogenetic survey of three environmentally distinct Antarctic Dry Valley soil biotopes showed a high proportion of so-called 'uncultured' phylotypes, with a relatively low diversity of identifiable phylotypes (Smith et al. 2006). One thing is certain, namely, that there is generally a much higher biodiversity than was once thought, and this is coming through in increasing numbers of studies. For example, high 16S rDNA bacterial diversity was described in glacial meltwater lake sediment from Bratina Island, Antarctica (Sjöling and Cowan 2003). High-throughput sequencing studies will inevitably increase estimates of total biodiversity (Pearce et al. 2012).

Particularly abundant groups of microorganisms, identified in microbiological studies in the Antarctic, include the cyanobacteria (photosynthetic bacteria), lichens (symbiotic associations between a cyanobacterium or an algae and a fungus – particularly well adapted for life in the most extreme environments as the symbiosis makes them widely tropically independent of substrate type), psychrophilic bacteria, psychrotolerant bacteria and psychrotrophs. Obligate psychrophilic bacteria are unable to grow above 20 °C; facultative psychrophiles are capable of growth at low temperature, but also grow above 20 °C. Among cold-adapted microorganisms, psychrotrophs tend to dominate in environments that undergo thermal fluctuations. True psychrophiles have a more restricted growth range and are found in permanently cold habitats. However, even in permanently cold environments > 50 % of the bacteria are not psychrophilic. In global terms, psychrotrophs are much more widely distributed than psychrophiles. From permanently cold habitats in the polar regions, the psychrophilic alga *Raphidonema nivale* and the fungus *Sclerotinia borealis* have been isolated. Among terrestrial psychrophiles are members of the genera *Pseudomonas*, *Cytophaga* and *Flavobacterium* (each commonly encountered in the freshwater lakes of Signy Island). However, the potential exists for all forms of microbial life to exist in Antarctic ecosystems. Microorganism groups are likely to include:

Viruses It is now well established that viruses are ubiquitous in aquatic ecosystems worldwide (Wilson et al. 2000; Pearce and Wilson 2003), and bacteriophage are common in polar inland waters (Säwström et al. 2007; Aguirre de Cárcer et al. 2015).

Fungi Basidiomycetous yeasts have been isolated from glacial and subglacial waters of northwest Patagonia (Brizzio et al. 2007) and *Penicillium* in Arctic subglacial ice (Sonjak et al. 2006). Recently, Ji et al. (2016) identified a biodiversity hotspot using multiplex 454 pyrosequencing targeting the bacterial 16S rRNA and the fungal ITS genes; and the bacterial and fungal abundances were estimated using quantitative PCR.

Archaea Archaea have been isolated from the basal ice of the Greenland Ice Core (GISP2 core).

Eukaryotes Willerslev et al. (1999), using PCR amplification of fragments of the 18S rRNA gene, identified a diversity of fungi, plants, algae and protists extracted from 2,000- to 4,000-year-old ice core samples from North Greenland. Lawley et al (2004) described the application of molecular biological techniques to estimate eukaryotic diversity (primarily fungi, algae and protists) in Antarctic soils across a latitudinal and environmental gradient between approximately 60 and 87°S. They found very limited overlap between the eukaryotic biota of the different study sites, combined with their generally low relatedness to existing sequence databases, this indicates a high level of Antarctic site isolation and possibly endemicity, a pattern not consistent with similar studies on other continents.

Cyanobacteria These are photosynthetic bacteria which fix nitrogen and carbon in Antarctic soils. Colouring of snow can be also be caused by a variety of photosynthetic eukaryotes such as *Chlamydomonas nivalis*, *Chloromonas* (*Scotiella*),

Ankistrodesmus, *Raphidonema*, *Mycanthococcus* and certain dinoflagellates (Prescott 1978) and represents a well-known illustration of cold adaptation. *Heteromita* is a very common soil microflagellate that had a worldwide distribution with an optimum temperature for growth around 23 °C. Under Antarctic conditions it demonstrates adaptations which permit survival of freeze-thaw cycles (by rapid and temperature-sensitive encystment and excystment) and by optimal utilization of resources during short periods, which allow this temperate species to grow actively at mean temperatures close to zero (Hughes and Smith 1989). Elsewhere, the ciliate *Holosticha* sp. was reported to be unable to divide above –2 °C (Lee and Fenchel 1972).

The common view that 'everything is everywhere', first formulated at the beginning of the last century by the biologist Beijerinck (Brock 1961) and reintroduced by Fenchel and Finlay in a series of publications (Finlay et al. 1999; Finlay and Clarke 1999; Finlay 2002; Fenchel and Finlay 2004) focused on the marine microbial community, support for the ubiquitous distribution of bacterioplankton came from group- or species-specific analyses.

Because of their small size, great abundance and easy dispersal, it is often assumed that marine planktonic microorganisms have a ubiquitous distribution that prevents any structured assembly into local marine bacterioplankton communities (Pommier et al. 2007). They examined coastal communities at nine locations distributed worldwide through the use of comprehensive clone libraries of 16S ribosomal RNA genes and showed that there were marked differences in the composition and richness of OTUs between locations. Remarkably, the global marine bacterioplankton community showed a high degree of endemism and conversely included few cosmopolitan OTUs. However, a number of new Antarctic species are published on an annual basis (Bowman and McCuaig 2003; Busse et al. 2003; Reddy et al. 2003; Sheridan et al. 2003; Spring et al. 2003; Van Trappen et al. 2003, 2004a–c, 2005; Donachie et al. 2004; Hirsch et al. 2004a,b; Jung et al. 2004; Montes et al. 2004; Pocock et al. 2004; Yi et al. 2005a,b; Smith et al. 2006; Yu et al 2008; Labrenz et al. 2009) along with an increasing number of studies from new extremophilic Antarctic environments.

Biodiversity studies are also of increasing relevance due to the effects of climate change (Jamieson et al. 2012, 2013; Chong et al. 2015). As could be shown, significant changes in microbial community structure as a result of environmental change can be detected at levels below that already experienced as a result of climate change (Pearce 2005). The climate around the Antarctic Peninsula is warming at a rate 3x faster than elsewhere (Turner et al. 2002). These communities might therefore act as indicators of potential effects of climate change.

5.7 Methodology

Extremophiles in Antarctica have been investigated using standard techniques, which can be classified into a number of generic categories:

Culture and physiology The growth and activity of microorganisms.

Microscopy Visualization using both fluorescent and electron microscopy (sometimes used in combination with specific gene probes).

Biochemistry (and biogeochemical cycling) many gene probes are available to assay samples for the presence of biogeochemical activities.

Molecular biology Genomic DNA (using gene probes coupled with FISH – fluorescence in situ hybridization), high-throughput sequencing technology and the construction and analysis of metagenomic libraries are used to screen for novel physiologies.

Physical For example, infrared Raman (used to detect biomolecules).

Recently, epifluorescence microscopy and flow cytometry were used to show that an ice sample contained over 6×10^7 cells ml^{-1} from a Greenland ice core that had remained at –9 °C for over 100,000 years. Anaerobic enrichment cultures inoculated with melted ice were also grown and maintained at –2 °C. Genomic DNA extracted from these enrichments was used for the PCR amplification of 16S rRNA genes with bacterial and archaeal primers and also used in the preparation of clone libraries, and this illustrates a standard polyphasic approach to the study of such environments. In addition, contemporary biological studies are being fuelled by the increasing availability of genome sequences and associated functional studies of extremophiles (e.g. Riedel et al. 2012; Kube et al. 2013; Che et al. 2013; Dsouza et al. 2014, 2015). This is leading to the identification of new biomarkers, an accurate assessment of cellular evolution, insight into the ability of microorganisms to survive in meteorites and during periods of global extinction, and knowledge of how to process and examine environmental samples to detect viable life forms (Cavicchioli 2002). More of these aspects is illustrated in Sect. 5.8.1.–5.8.3. on *Pseudoalteromonas haloplanktis* and methanogenic Archaea.

A critical assessment of detection strategies is important in any community analysis, and the description of Antarctic extremophiles can be hampered by low cell numbers, the methodological difficulty involved in culturing and the detection limits of the assays used. Advances in molecular technology have vastly improved life detection limits, such that microscopy and PCR are now capable of detecting individual cells per ml or the DNA itself at 0.1–0.2 ml^{-1}. Adopting a culture-based approach from Antarctic ice cores, 0, 2 and 10 CFU ml^{-1} have been isolated from Dyer Plateau, Siple Station and Taylor Dome, respectively (Christner et al. 2000), and 1–16 CFU ml^{-1} from a Dronning Maud Land ice core (Pearce et al., unpublished data). Radiolabelled substrates can yield uptake rates at the level of several hundred cells (Karl et al. 1999). However, not one approach is likely to provide a complete unbiased picture of the microorganisms residing in a sample or their relative numbers, and the design of specific, clean sampling strategies is extremely important. Stingl et al. (2008) used a dilution-to-extinction culturing technique for psychrotolerant planktonic bacteria from permanently ice-covered lakes in the McMurdo Dry Valleys, Antarctica. PCR theoretically can detect a single cell; realistically PCR detection limit is 2–8 cells ml^{-1} (Bulat et al. 2002). A minimum DNA concentration for PCR is about 17 ng/2.5 ng; 0.1–0.2 ng l^{-1} (i.e. 0.8–1.6 ng DNA) in the starting sample, for whole genome amplification, it can be up to 1–10 ng input DNA (genomiPhi from GE Healthcare produces 4–7 µg in 1.5h); using a *Flavobacterium* estimate of 8.4 fg DNA per cell (bacteria 6–25 fg DNA per cell), one would need 10^{5-6} cells for genomiPhi to work; if 10^3 cells ml^{-1} are present in the sample, then 1 l would be necessary.

5.8 Adaptations

A number of adaptations have been described for coping with the extreme cold, for a range of different types of organisms, and these include anatomical, behavioural, physiological and genetic, which can apply from bacteria to humans and from algae to grasses. The main benefit from adapting to life in the cold is a lack of competition. Most cellular adaptations to low temperatures and the underlying molecular mechanisms are not fully understood and are still being investigated. In multicellular organisms strategies include: modification of metabolism, elimination of nucleators and the accumulation of cryoprotectants to allow supercooling (freezing resistance).

Specific adaptations include:

The heat-shock response The response to heat stress in six yeast species isolated from Antarctica was examined (Deegenaars and Watson 1998). The yeasts were classified into two groups: one psychrophilic, with a maximum growth temperature of 20 °C, and the other psychrotrophic, capable of growth at temperatures above 20 °C. Elsewhere, molecular cloning and expression analysis of a cytosolic Hsp70 gene have been conducted in the Antarctic ice algae *Chlamydomonas* sp. ICE-L (Liu et al. 2010).

Cold-active secondary metabolites Production of a cold-active killer toxin by *Mrakia frigida* 2E00797 isolated from sea sediment in Antarctica was reported (Hua et al. 2010).

Cold-active enzymes Numerous studies have shown that psychrophilic enzymes are generally characterized by high turnover rates and catalytic efficiencies at low temperatures and reduced stability at moderate and high temperatures. Reduction in the activation energy is also achieved, resulting from an increased structural flexibility of either selected residues located at the active site or of the overall protein structure. The reduced heat stability has been correlated with subtle changes of their sequences compared with mesophilic enzymes, such as decreased levels of Pro and Arg residues, increased numbers or clustering of Gly residues, weakening of intramolecular interactions, increased solvent interactions and decreased number of charged residue interactions and of disulfide bonds (Mavromatis et al. 2003). A cold-active DnaK of an Antarctic psychrotroph *Shewanella* sp. Ac10 has been isolated, supporting the growth of dnaK-null mutant of *Escherichia coli* at cold tem-

Fig. 5.7 Dark pigments produced by Antarctic lichens for protection from UV radiation. (Photo by Dr. C. Gilbert, British Antarctic Survey)

peratures (Yoshimune et al. 2005). Siddiqui and Cavicchioli (2006) reported a systematic comparative analysis of 21 psychrophilic enzymes belonging to different structural families from prokaryotic and eukaryotic organisms. The sequences of these enzymes were multiply aligned to 427 homologous proteins from mesophiles and thermophiles. The net flux of amino acid exchanges from meso/thermophilic to psychrophilic enzymes was measured. Other specific studies of psychrophilic enzymes are also reported in the literature, for example, Feller et al. (1997) and de Pascale et al. (2008). To compensate for reduced metabolic activity at low temperatures, it has been suggested that psychrophiles also synthesize elevated levels of enzymes (Herbert 1989). The high production of a key enzyme to counterbalance its poor catalytic efficiency at low temperature could constitute a novel type of adaptive mechanism to cold environments, as, for example, RUBISCO (Devos et al. 1998). However, much of the adaptation is achieved through structural adaptations at the active site.

Pigment production Pigment production (Fig. 5.7) is a common adaptation to high UVB radiation in Antarctic microorganisms (Dieser et al. 2010).

Freeze-Thaw tolerance As mentioned previously, bacteria of the genus *Exiguobacterium* have been repeatedly isolated from ancient permafrost sediments of the Kolyma lowland of Northeast Eurasia, as well as *Exiguobacterium antarcticum* DSM 14480, isolated from a microbial mat sample of Lake Fryxell (McMurdo Dry Valleys, Antarctica). All are able to grow at temperatures ranging from −6 to 40 °C. In comparison to cells grown at 24 °C, the cold-grown cells of these strains tended to be longer and wider. Bacteria grown in broth at 4 °C showed markedly greater survival following freeze-thawing treatments (20 repeated cycles) than bacteria grown in broth at 24 °C. Surprisingly, significant protection to repeated freeze-thawing was also observed when bacteria were grown on agar at either 4 or 24 °C (Vishnivetskaya et al. 2007).

Photosynthesis The stress of low temperatures on life is exacerbated in organisms that rely on photoautrophic production of organic carbon and energy sources. Phototrophic organisms must coordinate temperature-independent reactions of light absorption and photochemistry with temperature-dependent processes of electron transport and utilization of energy sources through growth and metabolism. Photoautotrophs rely on low-temperature acclimative and adaptive strategies that have been described for other low-temperature-adapted heterotrophic organisms, such as cold-active proteins and maintenance of membrane fluidity. Phototrophic organisms must coordinate temperature-independent reactions of light absorption and photochemistry with temperature-dependent processes of electron transport and utilization of energy sources through growth and metabolism. Psychrophilic (organisms tolerant of growth temperatures of −15 to +10 °C) vs. psychrophilic (optimal growth and reproduction at −15 to +10 °C) photoautotrophs rely on low-temperature acclimative and adaptive strategies that have been described for other low-temperature-adapted heterotrophic organisms, such as cold-active proteins and maintenance of membrane fluidity (Morgan-Kiss et al. 2006).

Symbiosis Lichens have been found only 400 km from the South Pole. They recover very slowly from freezing after the winter with photosynthesis and respiration levels not reaching high levels until late spring. Lichens are able to function with less light and water than other plants and have a high concentration of pigments and acids. Last year a New Zealand research team measured a lichen photosynthezising at close to –20 C, the lowest ever recorded (http://www.anta.canterbury.ac.nz/resources/adapt).

Regulation of membrane fluidity Cold adaptation in *M. burtonii* was shown to involve growth-temperature-regulated membrane-lipid unsaturation. The studies on *M. burtonii* and *H. lacusprofundi* indicate that a general feature of cold adaptation in psychrophilic Archaea might be to increase the abundance of unsaturated lipids at low temperature to ensure that membrane fluidity and thereby membrane function is maintained (see Sect. *5.8.2.*).

Cold-shock response The cold-shock response is a specific pattern of gene expression in response to abrupt changes to lower temperatures. One effect of reducing temperature is to block initiation of protein synthesis. Cold-shock proteins can stabilize mRNA and reinitiate protein production. Others are also linked to maintaining the fluidity of the membrane such as inducible desaturases.

Cold-shock proteins and cold acclimation proteins (CAPs) The structure and function of cold-shock proteins in Archaea have been investigated (Giaquinto et al. 2007).

Cold-inducible proteins Proteomic studies of an Antarctic cold-adapted bacterium, *Shewanella livingstonensis* Ac10, have been conducted for global identification of cold-inducible proteins (Kawamoto et al. 2007). Temperature downshift induces antioxidant response in fungi isolated from Antarctica (Gocheva et al. 2009).

Antifreeze proteins These seem to have less of a role in bacteria. Sea-ice bacteria have been shown to be active at temperatures down to −20 °C and motile down to −10 °C. It is thought that cryoprotectants such as antifreeze may be produced to inhibit ice crystal formation. The antifreeze proteins produced by bacteria are possibly part of the large pool of extracellular polymeric substances (EPSs) located on the cell surface.

Ice-active substances (IASs) Produced by sea-ice diatoms; they may be glycoproteins that bind preferentially to ice crystals causing pitting.

Membrane fluidity Changes in membrane composition can lead to increased flexibility at low temperatures, for example, increase in fatty acid unsaturation, decrease in fatty acid chain length, increase in methyl branching of fatty acids or increase in the ratio of anteiso branching relative to iso branching. Overall, changes in temperature tend to alter the balance between protein flexibility and stability; higher temperatures can make a protein overly flexible, reducing substrate affinity by disrupting the active site and ultimately leading to denaturation. Colder temperatures can make an enzyme overly stable, reducing catalytic rates below the range needed to maintain metabolic homeostasis in the cell (Fields 2001). Temperature can influence the response of microorganisms either directly, by its effects on growth rate, enzyme activity, cell composition and nutritional requirements, or indirectly by its effects on the solubility of solute molecules, ion transport and diffusion, osmotic effects on membranes, surface tension and density (Herbert 1986).

We have learned a great deal about the mechanisms of adaptation to cold environments through the studies of the genome sequences and proteomics of three specific organisms and their comparisons. The most significant finding is that both microbes have flexible proteins, which allow their cells to survive cold temperatures and carry out basic cell functions under extreme conditions. These proteins are more rigid and stable in bacteria that live at higher temperatures. In *M. frigidum*, researchers also identified cold-shock proteins that are not found in heat-loving Archaea. Cold-shock proteins are known to help other organisms, such as Bacteria, adapt to cold environments.

5.8.1 *Pseudoalteromonas haloplanktis*

Pseudoalteromonas haloplanktis is a versatile marine bacterium that grows in the Antarctic and was isolated near the French Dumont d'Urville research station on the Antarctic continent. A collaboration between Genoscope in Evry, the University of Hong Kong and researchers from the Universities of Liège, Naples and Stockholm analysed the genome of this interesting bacterium and was able to reconstruct the adaptations and metabolic mechanisms that allow it to flourish in such an extreme environment. They found out how the bacterium was able to adapt to the presence of atmospheric oxygen, which is very soluble in cold water and then becomes a highly toxic reactive element. They discovered how it was able to keep its membrane fluid and prevent it from freezing at low temperatures, through modification of its lipid composition. Specifically, they found concerted changes in the amino acid composition of its proteins. Organisms living at moderate temperatures relatively infrequently use a delicate amino acid, asparagine, in spite of its advantageous properties, because it deteriorates chemically over time and is thus one of the most significant factors in ageing. The *P. haloplanktis* bacterium, protected from ageing by the cold, was found to use asparagine to a larger degree in its proteins. They also found kinetic and structural optimization of catalysis at low temperatures in a psychrophilic cellulase; no structural alteration related to cold activity could be found in the catalytic cleft, whereas several structural factors in the overall structure can explain the weak thermal stability, suggesting that the loss of stability provides the required active-site mobility at low temperatures (Médigue et al. 2005).

5.8.2 *Methanococcoides burtonii*

Methanococcoides burtonii, a cold-adapted archaeon, has been investigated using proteomics (Campanaro et al. 2010). Cellular processes that are important for cold adaptation were derived from studies of cells grown at 4 °C. The genome sequence has also been studied (Allen et al. 2009). Key aspects of cold adaptation were found to be related to transcription, protein folding and metabolism, including specific roles for RNA polymerase subunit E, a response regulator and peptidyl prolyl *cis/trans* isomerase. A heat-shock protein (DnaK) was found to be expressed during growth at optimal temperatures, indicating that growth at such temperatures was stressful for this cold-adapted organism. Thermal regulation in *M. burtonii* was found to be achieved through complex gene expression events involving gene clusters and operons, through to protein modifications.

5.8.3 *Methanogenium frigidum*

Methanogenium frigidum sp. nov. was isolated from the perennially cold, anoxic hypolimnion of Ace Lake in the Vestfold Hills of Antarctica. This was the first report of a psychrophilic methanogen growing by CO_2 reduction (Franzmann et al. 1997). The cells are psychrophilic, exhibiting most rapid growth at 15 °C and no growth at temperatures above 18–20 °C. Comparative genomics revealed trends in amino acid and tRNA composition and structural features of proteins. Proteins from this cold-adapted ARCHAEON were characterized by a higher content of noncharged polar amino acids, particularly Gln and Thr, and a lower content of hydrophobic amino acids, particularly Leu.

A cold-shock domain (CSD) protein (CspA homolog) was also identified in *M. frigidum*, two hypothetical proteins with CSD folds in *M. burtonii* and a unique winged helix DNA-binding domain protein in *M. burtonii*. This suggests that these types of nucleic acid-binding proteins have a critical role in cold-adapted Archaea. Structural analysis of tRNA sequences from the Archaea indicated that GC content is the major factor influencing tRNA stability in hyperthermophiles, but not in the psychrophiles, mesophiles or moderate thermophiles. Below an optimal growth temperature of 60 °C, the GC content in tRNA was largely unchanged, indicating that any requirement for flexibility of tRNA in psychrophiles is mediated by other means. This is the first time that comparisons have been performed with genome data from Archaea spanning the growth temperature extremes from psychrophiles to hyperthermophiles (Saunders et al. 2003).

5.9 Discussion and Future Perspectives

We can learn a lot from the study of microorganisms that have adapted to life in the cold. Many of the stress response mechanisms are universal. There appears to be no one single response to cold, but a combination of responses at different levels, and these differ by organism. Extremophiles in Antarctica include representatives of all three domains (Bacteria, Archaea and Eukarya); however, the majority are microorganisms, and a proportion of these are Archaea. Recently, however, new groups of abundant, uncultivated Archaea have been found to be widespread in more pedestrian biotopes, including marine plankton, terrestrial soils, lakes and marine and freshwater sediments, and in association with metazoa. Hence, more research effort is needed to characterizing the physiology, biochemistry and genetics of these cosmopolitan but poorly understood Archaea. Knowledge of extremophile habitats is expanding the number and types of extraterrestrial locations that may be targeted for exploration (Cavicchioli 2002), and this is certainly true for Antarctica. Extremophiles do not seem to possess any outrageous measures of adaptation to the cold in Antarctica – it appears to be a combination of more stable proteins, modified enzyme activity, more or less fluid membranes, etc. So is the Antarctic really extreme? Probably not for the organisms that have adapted to thrive there.

As the latest techniques are applied to the study of Antarctic extremophiles, for example, the study of cold adaptation in the marine bacterium, *Sphingopyxis alaskensis*, using quantitative proteomics (Ting et al. 2010), there is an increasing realization of the potential for new and radically innovative biotechnology, including a good potential source of novel biochemicals. Despite the remarkable opportunities that these uncommon organisms present for biotechnological applications, only few instances can be reported for actual exploitation. This lack of progress from the research findings at a laboratory-scale to the actual development of pilot- and large-scale production is correlated with the difficulties encountered in the cultivation of extremophiles. Schiraldi and De Rosa (2002) report achievements in the production of biomass and related enzymes and biomolecules from extremophile sources, especially focusing on the application of novel fermentation strategies.

Extremophiles are a potential source for novel enzymes; a particular interest is for low-temperature enzyme activity, cold-shock induction and ice-active substances (Ferrer et al. 2007). Novel physiological adaptations could also suggest evolutionary separation – biofilm formation and synergy may be two physiological strategies for nutrient acquisition in these systems. The low concentration of nutrient has led to nitrogen fixation levels of 1 g m^{-2} y^{-1} in cyanobacterial mats, so nitrogen availability is a key nutritional factor controlling microbial production in Antarctic freshwater habitats (Olson et al. 1998). Low temperatures might also induce the viable but non-culturable (VBNC) state in Antarctic lake microorganisms, and the VBNC state of some bacteria, collected from Antarctic lakes, has been reported (Chattopadhyay 2000).

We may also learn a great deal from analogy to higher organisms. However, care is required in developing generalizations because one element can serve in many different ways. For example, cryoprotectants can have multiple functions: they appear to favour supercooling or disable nucleators, resist or protect against desiccation, protect frozen membranes or other cell constituents directly, modify the freezing process, mitigate cellular damage from ice crystallization or recrystallization while tissues are frozen or while they are thawing, and repair damage after thawing. In the same way, developmental delays can serve multiple purposes: they may conserve energy, protect against adversity, synchronize the feeding stage with food resources, optimize the timing of reproduction, synchronize individuals with one another, prevent a risky generation late in the year, or assist in further life-cycle programming by allowing the environment to be monitored for a longer period (Danks 2002).

The fact that microflora have been found in relative abundance and diversity in subglacial Lake Whillans and in the basal strata of an Antarctic ice core above the Vostok lake is extremely encouraging in the searching for relict forms of life on the Earth and also as a model for solving a number of problems of exobiology, for instance, for development of methods to penetrate into the under-ice sea at Europa–Jupiter's satellite (Abyzov et al. 2001). Mars has extensive regions rich in clay minerals. Even at low near-surface temperatures, a thin layer of unfrozen water almost certainly coats mineral grains (Möhlmann 2003), accounting for the survival of water detected within a metre of the Mars surface despite the high vapour pressure of solid water (Price 2006). Antarctic subglacial lakes offer an excellent opportunity to develop technologies for exploring these new worlds, because of their hydraulic isolation and proximity to Antarctic infrastructure and because their location can provide an analogue to a Martian polar cap (Price et al. 2002). Such potential habitats include Mars and icy satellites in outer space. Titan, Europa and Callisto have all shown evidence of previously unknown bodies of water that might be home to unique life forms. Cold environments are common throughout the galaxy (Reid et al. 2006).

Antarctic extremophiles could therefore provide a rich source of unexpected and unique adaptive mechanisms, particularly the double or triple extremophiles. They provide molecular insights into psychrophilic lifestyles and an understanding of specific adaptive mechanisms. The genomic era has pushed back the frontiers of discovery of Antarctic extremophiles over the last decades, and the study of extremophiles in the field using techniques of molecular biology will undoubtedly generating a great deal of excitement over the coming years.

Acknowledgments This contribution was funded by the Terrestrial and Coastal Work Package of the Ecosystems Programme, Polar Science for Planet Earth, British Antarctic Survey (BAS) and Natural Environment Research Council.

References

Abyzov SS, Mitskevich IN, Poglazova MN, Barkov IN, Lipenkov VY, Bobin NE, Koudryashov BB, Pashkevich VM, Ivanov MV (2001) Microflora in the basal strata at Antarctic ice core above the Vostok lake. Adv Space Res 28:701–706

Aguirre de Cárcer D, López-Bueno A, Pearce DA, Alcamí A (2015) Biodiversity and distribution of polar freshwater DNA viruses. Sci Adv 1(5):e1400127. doi:10.1126/sciadv.1400127 eCollection 2015

Ah Tow L, Cowan DA (2005) Dissemination and survival of non-indigenous bacterial genomes in pristine Antarctic environments. Extremophiles 9:385–389

Alam SI, Singh L, Dube S, Reddy GS, Shivaji S (2003) Psychrophilic *Planococcus maitriensis* sp. nov. from Antarctica. Syst Appl Microbiol 26:505–510

Allen MA, Lauro FM, Williams TJ, Burg D, Siddiqui KS, De Francisci D, Chong KWY, Pilak O, Chew HH, De Maere MZ, Ting L, Katrib M, Ng C, Sowers KR, Galperin MY, Anderson IJ, Ivanova N, Dalin E, Martinez M, Lapidus A, Hauser L, Land M, Thomas T, Cavicchioli R (2009) The genome sequence of the psychrophilic archaeon, *Methanococcoides burtonii*: the role of genome evolution in cold-adaptation. ISME J 3:1012–1035

Amos GC, Borsetto C, Laskaris P, Krsek M, Berry AE, Newsham KK, Calvo-Bado L, Pearce DA, Vallin C, Wellington EM (2015) Designing and implementing an assay for the detection of rare and divergent NRPS and PKS clones in European, Antarctic and Cuban soils. PLoS One 10:e0138327. doi:10.1371/journal.pone.0138327 eCollection 2015

Boetius A, Anesio AM, Deming JW, Mikucki JA, Rapp JZ (2015) Microbial ecology of the cryosphere: sea ice and glacial habitats. Nat Rev Microbiol 13:677–690. doi:10.1038/nrmicro3522

Bottos EM, Woo AC, Zawar-Reza P, Pointing SB, Cary SC (2014) Airborne bacterial populations above desert soils of the McMurdo Dry Valleys, Antarctica. Microb Ecol 67:120–128

Bowman JP, McCuaig RD (2003) Diversity and biogeography of prokaryotes dwelling in Antarctic continental shelf sediment. Appl Environ Microbiol 69:2463–2484

Brizzio S, Turchetti B, de García V, Libkind D, Buzzini P, van Broock M (2007) Extracellular enzymatic activities of basidiomycetous yeasts isolated from glacial and subglacial waters of northwest Patagonia (Argentina). Can J Microbiol 53:519–525

Brock TD (1961) Milestones in microbiology. Prentice-Hall, Englewood Cliffs

Bulat SA, Alekhina IA, Krylenkov VA, Lukin VV (2002) Molecular biological studies of microbiota in subglacial lake Vostok (the Antarctic) (in Russian). Adv Curr Biol 122:211–221

Bulat SA, Alekhina IA, Lipenkov VY, VV L, Marie D, Petit JR (2009) Cellular concentrations of microorganisms in glacial and lake ice of the Vostok ice core, East Antarctica. Microbiology 78:808–810

Busse H-J, Denner EBM, Buczolits S, Salkinoja-Salonen M, Bennasar A, Kämpfer P (2003) *Sphingomonas aurantiaca* sp. nov., *Sphingomonas aerolata* sp. nov. and *Sphingomonas faeni* sp. nov., air- and dustborne and Antarctic, orange-pigmented, psychrotolerant bacteria, and emended description of the genus *Sphingomonas*. Int J Syst Evol Microbiol 53:1253–1260

Cameron KA, Hodson AJ, Osborn AM (2012) Structure and diversity of bacterial, eukaryotic and archaeal communities in glacial cryoconite holes from the Arctic and the Antarctic. FEMS Microbiol Ecol 82:254–267. doi:10.1111/j.1574-6941.2011.01277.x

Campanaro S, Williams TJ, De Francisci D, Treu L, Lauro FM, Cavicchioli R (2010) Temperature-dependent global gene expression in the Antarctic archaeon. *Methanococcoides burtonii*. Environ Microbiol (online Nov 8: doi:10.1111/j.1462-2920.2010.02367.x)

Cary SC, McDonald IR, Barrett JE, Cowan DA (2010) On the rocks: the microbiology of Antarctic Dry Valley soils. Nat Rev Microbiol 8:129–138

Cavicchioli R (2002) Extremophiles and the search for extraterrestrial life. Astrobiology 2:281–292

Chattopadhyay MK (2000) Cold adaptation of Antarctic microorganisms—possible involvement of viable but non culturable state. Polar Biol 23:223–224

Che S, Song L, Song W, Yang M, Liu G, Lin X (2013) Complete genome sequence of Antarctic bacterium *Psychrobacter* sp. strain G.Genome Announc 1(5).pii: e00725-e00713. doi: 10.1128/genomeA.00725-13

Chong CW, Dunn MJ, Convey P, Annie Tan GY, Wong RCS, Tan IKP (2009) Environmental influences on bacterial diversity of soils on Signy Island, maritime Antarctic. Polar Biol 32: 1571–1582

Chong CW, Pearce DA, Convey P, Tan GYA, Wong RCS, Tan IKP (2010) High levels of spatial heterogeneity in the biodiversity of soil prokaryotes on Signy Island, Antarctica. Soil Biol Biogeochem 42:601–610

Chong CW, Pearce DA, Convey P (2015) Emerging spatial patterns in Antarctic prokaryotes. Front Microbiol 6:1058. doi:10.3389/fmicb.2015.01058 eCollection 2015

Christner BC, Mosley-Thompson E, Thompson LG, Zagorodnov V, Sandman K, Reeve JN (2000) Recovery and identification of viable bacteria immured in glacial ice. Icarus 144:479–485

Christner BC, Kvitko BH, Reeve JN (2003) Molecular identification of Bacteria and Eukarya inhabiting an Antarctic cryoconite hole. Extremophiles 7:177–183

Christner BC, Royston-Bishop G, Foreman CM, Arnold BR, Tranter M, Welch KA, Lyons WB, Tsapin AI, Studinger M, Priscu JC (2006) Limnological conditions in subglacial Lake Vostok, Antarctica. Limnol Oceanogr 51:2485–2501

Christner BC, Priscu JC, Achberger AM, Barbante C, Carter SP, Christianson K, Michaud AB, Mikucki JA, Mitchell AC, Skidmore ML, Vick-Majors TJ, WISSARDScience Team (2014) A microbial ecosystem beneath the West Antarctic ice sheet. Nature 512:310–313. doi:10.1038/nature13667

Clocksin KM, Jung DO, Madigan MT (2007) Cold-active chemoorganotrophic bacteria from permanently ice-covered Lake Hoare, McMurdo Dry Valleys, Antarctica. Appl Environ Microbiol 73:3077–3083

Cockell CS, McKay CP, Omelon C (2002) Polar endoliths – an anticorrelation of climatic extremes and microbial biodiversity. Int J Astrobiol 1:305–310

D'Amico S, Collins T, Marx JC, Feller G, Gerday C (2006) Psychrophilic microorganisms: challenges for life. EMBO Rep 7:385–389

Danks HV (2002) Modification of adverse conditions by insects. Oikos 99:10–24

Deegenaars ML, Watson K (1998) Heat shock response in psychrophilic and psychrotrophic yeast from Antarctica. Extremophiles 2:41–49

de Pascale D, Cusano AM, Autore F, Parrilli E, di Prisco G, Marino G, Tutino ML (2008) The cold-active Lip1 lipase from the Antarctic bacterium *Pseudoalteromonas haloplanktis* TAC125 is a member of a new bacterial lipolytic enzyme family. Extremophiles 12:311–323

Devos N, Ingouff M, Loppes R, Matagne RF (1998) RUBISCO adaptation to low temperatures: a comparative study in psychrophilic and mesophilic unicellular algae. J Phycol 34:655–660

Dickinson I, Goodall-Copestake W, Thorne MAS, Schlitt T, Ávila-Jiménez ML, Pearce DA (2016) Extremophiles in an Antarctic marine ecosystem. Microorganisms 4(1):8. doi:10.3390/microorganisms4010008

Dieser M, Greenwood M, Foreman CM (2010) Carotenoid pigmentation in Antarctic heterotrophic bacteria as a strategy to withstand environmental stresses. AAAR 42:396–405

Donachie SP, Bowman JP, Alam M (2004) *Psychroflexus tropicus* sp. nov., an obligately halophilic Cytophaga–Flavobacterium–Bacteroides group bacterium from an Hawaiian hypersaline lake. Int J Syst Evol Microbiol 54:935–940

Dsouza M, Taylor MW, Turner SJ, Aislabie J (2014) Genome-based comparative analyses of Antarctic and temperate species of *Paenibacillus*. PLoS One 9:e108009. doi:10.1371/journal.pone.0108009

Dsouza M, Taylor MW, Turner SJ, Aislabie J (2015) Genomic and phenotypic insights into the ecology of *Arthrobacter* from Antarctic soils. BMC Genomics 16:36. doi:10.1186/s12864-015-1220-2

Duarte CM, Agustí S, Vaqué D, Agawin NSR, Felipe J, Casamayor EO, Gasol JM (2005) Experimental test of bacteria–phytoplankton coupling in the Southern Ocean. Limnol Oceanogr 50:1844–1854

Edwards A, Anesio AM, Rassner SM, Sattler B, Hubbard B, Perkins WT, Young M, Griffith GW (2011) Possible interactions between bacterial diversity, microbial activity and supraglacial hydrology of cryoconite holes in Svalbard. ISME J 5:150–160. doi:10.1038/ismej.2010.100

Feller G, Zekhnini Z, Lamotte-Brasseur J, Gerday C (1997) Enzymes from cold-adapted microorganisms. The class C β-lactamase from the Antarctic psychrophile *Psychrobacter immobilis* A5. Eur J Biochem 244:186–191

Feller G, Gerday C (2003) Psychrophilic enzymes: hot topics in cold adaptation. Nat Rev Microbiol 1:200–208

Fenchel T, Finlay BJ (2004) The ubiquity of small species: patterns of local and global diversity. Bioscience 54:777–784

Ferrer M, Golyshina O, Beloqui A, Golyshin PN (2007) Mining enzymes from extreme environments. Curr Opin Microbiol 10:207–214

Fields PA (2001) Protein function at thermal extremes: balancing stability and flexibility. Comp Biochem Physiol A 129:417–431

Finlay BJ, Clarke KJ (1999) Ubiquitous dispersal of microbial species. Nature 400:828

Finlay BJ (2002) Global dispersal of free-living microbial eukaryote species. Science 296:1061–1063

Finlay BJ, Esteban GF, Olmo JL, Tyler PA (1999) Global distribution of free-living microbial species. Ecography 22:138–144

Franzmann PD, Liu Y, Balkwill DL, Aldrich HC, De Macario EC, Boone DR (1997) *Methanogenium frigidum* sp. nov., a psychrophilic, H_2-using methanogen from Ace Lake, Antarctica. Int J Syst Bacteriol 47:1068–1072

Ghiglione JF, Galand PE, Pommier T, Pedrós-Alió C, Maas EW, Bakker K, Bertilson S, Kirchmanj DL, Lovejoy C, Yager PL, Murray AE (2012) Pole-to-pole biogeography of surface and deep marine bacterial communities. Proc Natl Acad Sci U S A 109:17633–17638. doi:10.1073/pnas.1208160109

Giaquinto L, Curmi PMG, Siddiqui KS, Poljak A, DeLong E, DasSarma S, Cavicchioli R (2007) The structure and function of cold shock proteins in archaea. J Bacteriol 189:5738–5748

Gilichinsky D, Wagener S, Vishnevetskaya T (1995) Permafrost microbiology. Permafrost Periglacial Processes 6:281–291

Gocheva YG, Tosi S, Krumova ET, Slokoska LS, Miteva JG, Vassilev SV, Angelova MB (2009) Temperature downshift induces antioxidant response in fungi isolated from Antarctica. Extremophiles 13:273–281

Herbert RA (1986) The ecology and physiology of psychrophilic microorganisms. In: Herbert RA, Codd GA (eds) Microbes in extreme environments. The Society for General Microbiology. Academic Press, London, pp 1–24

Herbert RA (1989) Microbial growth at low temperature. In: Gould GW (ed) Mechanisms of action of food preservation procedures. Elsevier Applied Science, London, pp. 71–96

Hirsch P, Gallikowski CA, Siebert J, Peissl K, Kroppenstedt R, Schumann P, Stackebrandt E, Anderson R (2004a) *Deinococcus frigens* sp. nov., *Deinococcus saxicola* sp. nov., and *Deinococcus marmoris* sp. nov., low temperature and drought-tolerating, UV resistant bacteria from continental Antarctica. Syst Appl Microbiol 27:636–645

Hirsch P, Mevs U, Kroppenstedt RM, Schumann P, Stackebrandt E (2004b) Cryptoendolithic actinomycetes from Antarctic sandstone rock samples: *Micromonospora endolithica* sp. nov. and two isolates related to *Micromonospora coerulea* Jensen 1932. Syst Appl Microbiol 27:166–174

Hochachka PW, Somero GN (2002) Biochemical adaptation: mechanism and process in physiological evolution. Oxford University Press, New York

Hodson A, Brock B, Pearce DA, Laybourn-Parry J, Tranter M (2015) Cryospheric ecosystems: a synthesis of snowpack and glacial research. Environ Res Lett 10:110201. doi:10.1088/1748-9326/10/11/110201

Hua MX, Chi Z, Liu GL, Buzdar MA, Chi ZM (2010) Production of a novel and cold-active killer toxin by *Mrakia frigida* 2E00797 isolated from sea sediment in Antarctica. Extremophiles 14:515–521

Hughes KA, McCartney HA, Lachlan-Cope TA, Pearce DA (2004) A preliminary study of airborne biodiversity over peninsular Antarctica. Cell Mol Biol 50:537–542

Hughes J, Smith HG (1989) Temperature relations of *Heteromita globosa* Stein in Signy Island fellfields. In: Heywood RB (ed) University Research in Antarctica. Proceedings of British Antarctic Survey Antarctic Special Topic Award Scheme Symposium, 9-10 November 1988. British Antarctic Survey, Natural Environment Research Council, Cambridge, pp. 117–122

Jamieson RE, Rogers AD, Billett D, Smale DA, Pearce DA (2012) Patterns of bacterioplankton biodiversity in the surface waters of the Scotia Arc, Southern Ocean. FEMS Microbiol Ecol 80:452–468. doi:10.1111/j.1574-6941.2012.01313.x

Jamieson RE, Heywood JL, Rogers AD, Billett DSM, Pearce DA (2013) Bacterial biodiversity in deep-sea sediments from two regions of contrasting surface water productivity near the Crozet Islands, Southern Ocean. Deep Sea Research part I Oceanographic Research papers 75:67–77. doi: 10.1016/j.dsr.2012.12.012

Ji M, van Dorst J, Bissett A, Brown MV, Palmer AS, Snape I, Siciliano SD, Ferrari BC (2016) Microbial diversity at Mitchell Peninsula, Eastern Antarctica: a potential biodiversity "hotspot". Polar Biol 39:237–249. doi:10.1007/s00300-015-1776-y

Jung DO, Achenbach LA, Karr EA, Takaichi S, Madigan MT (2004) A gas vesiculate planktonic strain of the purple non-sulfur bacterium *Rhodoferax antarcticus* isolated from Lake Fryxell, Dry Valleys, Antarctica. Arch Microbiol 182:236–243

Karl DM, Bird DF, Björkman K, Houlihan T, Shackelford R, Tupas L (1999) Microorganisms in the accreted ice of Lake Vostok, Antarctica. Science 286:2144–2147

Kawamoto J, Kurihara T, Kitagawa M, Kato I, Esaki N (2007) Proteomic studies of an Antarctic cold-adapted bacterium, *Shewanella livingstonensis* Ac10, for global identification of cold-inducible proteins. Extremophiles 11:819–826

Kube M, Chernikova TN, Al-Ramahi Y, Beloqui A, Lopez-Cortez N, Guazzaroni ME, Heipieper HJ, Klages S, Kotsyurbenko OR, Langer I, Nechitaylo TY, Lünsdorf H, Fernández M, Juárez S, Ciordia S, Singer A, Kagan O, Egorova O, Petit PA, Stogios P, Kim Y, Tchigvintsev A, Flick R, Denaro R, Genovese M, Albar JP, Reva ON, Martínez-Gomariz M, Tran H, Ferrer M, Savchenko A, Yakunin AF, Yakimov MM, Golyshina OV, Reinhardt R, Golyshin PN (2013) Genome sequence and functional genomic analysis of the oil-degrading bacterium *Oleispiraantarctica*. Nat Commun 4:2156. doi:10.1038/ncomms3156

Labrenz M, Lawson PA, Tindall BJ, Hirsch P (2009) *Roseibaca ekhonensis* gen. nov., sp. nov., an alkalitolerant and aerobic bacteriochlorophyll a-producing alphaproteobacterium from hypersaline Ekho Lake. Int J Syst Evol Microbiol 59:1935–1940

Lanoil B, Skidmore M, Priscu JC, Han S, Foo W, Vogel SW, Tulaczyk S, Engelhardt H (2009) Bacteria beneath the West Antarctic Ice Sheet. Environ Microbiol 11:609–615

Lawley B, Ripley S, Bridge P, Convey P (2004) Molecular analysis of geographic patterns of eukaryotic diversity in Antarctic soils. Appl Environ Microbiol 79:5963–5972

Laybourn-Parry J (2002) Survival mechanisms in Antarctic lakes. Philos Trans R Soc Lond B Biol Sci 357(1423):863–869

Laybourn-Parry J (2009) No place too cold. Science 324:1521–1522

Laybourn-Parry J, Pearce DA (2007) The biodiversity and ecology of Antarctic lakes – models for evolution. Philos Trans R Soc Lond B Biol Sci 362(1488):2273–2289

Laybourn-Parry J, Pearce DA (2016) Heterotrophic bacteria in Antarctic lacustrine and glacial environments. Polar Biol. doi:10.1007/s00300-016-2011-1

Lee CC, Fenchel T (1972) Studies on ciliates associated with sea ice from Antarctica. II. Temperature responses and tolerances in ciliates from Antarctic, temperate and tropical habitats. Arch Protistenk 114:237–244

Li S, Xiao X, Yin X, Wang F (2006) Bacterial community along a historic lake sediment core of Ardley Island, west Antarctica. Extremophiles 10:461–467

Liu S, Zhang P, Cong B, Liu C, Lin X, Shen J, Huang X (2010) Molecular cloning and expression analysis of a cytosolic Hsp70 gene from Antarctic ice algae *Chlamydomonas* sp. I CE-L. Extremophiles 14:329–337

Lopatina A, Medvedeva S, Shmakov S, Logacheva MD, Krylenkov V, Severinov K (2016) Metagenomic analysis of bacterial communities of Antarctic surface snow. Front Microbiol 7:398. doi:10.3389/fmicb.2016.00398 eCollection 2016

Mavromatis K, Feller G, Kokkinidis M, Bouriotis V (2003) Cold adaptation of a psychrophilic chitinase: a mutagenesis study. Protein Eng 16:497–503

Médigue C, Krin E, Pascal G, Barbe V, Bernsel A, Bertin PN, Cheung F, Cruveiller S, D'Amico S, Duilio A, Fang G, Feller G, Ho C, Mangenot S, Marino G, Nilsson J, Parrilli E, Rocha EP, Rouy Z, Sekowska A, Tutino ML, Vallenet D, von Heijne G, Danchin A (2005) Coping with cold: The genome of the versatile marine Antarctica bacterium *Pseudoalteromonas haloplanktis* TAC125. Genome Res 15:1325–1335

Mikucki JA, Priscu JC (2007) Bacterial diversity associated with Blood Falls, a subglacial outflow from the Taylor Glacier, Antarctica. Appl Environ Microbiol 73:4029–4039

Montes MJ, Mercade E, Bozal N, Guinea J (2004) *Paenibacillus antarcticus* sp. nov., a novel psychrotolerant organism from the Antarctic environment. Int J Syst Evol Microbiol 54:1521–1526

Möhlmann DTF (2003) Unfrozen subsurface water on Mars: presence and implications. Int J Astrobiol 2:213–216

Morgan-Kiss RM, Priscu JC, Pocock T, Gudynaite-Savitch L, Huner NP (2006) Adaptation and acclimation of photosynthetic microorganisms to permanently cold environments. Microbiol Mol Biol Rev 70:222–252

Naganuma T, Hua PN, Okamoto T, Ban S, Imura S, Kanda H (2005) Depth distribution of euryhaline halophilic bacteria in Suribati Ike, a meromictic lake in East Antarctica. Polar Biol 28:964–970

Newsham KK, Pearce DA, Bridge PD (2010) Minimal influence of water and nutrient content on the bacterial community composition of a maritime Antarctic soil. Microbiol Res 165:523–530

Nogi Y, Kato C, Horikoshi K (1998) Taxonomic studies of deep-sea barophilic *Shewanella* strains and description of *Shewanella violacea* sp. nov. Arch Microbiol 170:331–338I

Olson JB, Steppe TF, Litaker RW, Paerl HW (1998) N_2 fixing microbial consortia associated with the ice cover of Lake Bonney, Antarctica. Microb Ecol 36:231–238

Pearce DA, Wilson WH (2003) Viruses in Antarctic ecosystems. Antarct Sci 15:319–331

Pearce DA (2005) The structure and stability of the bacterioplankton community in Antarctic freshwater lakes, subject to extremely rapid environmental change. FEMS Microbiol Ecol 53:61–72

Pearce DA (2008) Climate change and the microbiology of the Antarctic Peninsula region. Sci Prog 91:203–217

Pearce DA (2009) Antarctic subglacial lakes – a new frontier in microbial ecology. ISME J 3:877–880

Pearce DA, Bridge PD, Hughes K, Sattler B, Psenner R, Russell NJ (2009) Microorganisms in the atmosphere over Antarctica. FEMS Microbiol Ecol 69:143–157

Pearce DA, Hughes KA, ss, Lachlan-Cope TA, Jones AE (2010) Biodiversity of air-borne microorganisms at Halley station, Antarctica. Extremophiles 14:145–159

Pearce DA, Newsham KK, Thorne MA, Calvo-Bado L, Krsek M, Laskaris P, Hodson A, Wellington EM (2012) Metagenomic analysis of a southern maritime antarctic soil. Front Microbiol 3:403. doi:10.3389/fmicb.2012.00403 eCollection 2012

Pearce DA, Hodgson DA, Thorne MAS, Burns G, Cockell CS (2013) Preliminary analysis of life within a former subglacial lake sediment in Antarctica. Diversity 5:680–702. doi:10.3390/d5030680

Pearce DA, Alekhina IA, Terauds A, Wilmotte A, Quesada A, Edwards A, Dommergue A, Sattler B, Adams BJ, Magalhães C, Chu WL, Lau MC, Cary C, Smith DJ, Wall DH, Eguren G, Matcher G, Bradley JA, de Vera JP, Elster J, Hughes KA, Cuthbertson L, Benning LG, Gunde-

Cimerman N, Convey P, Hong SG, Pointing SB, Pellizari VH, Vincent WF (2016) Aerobiology over Antarctica - a new initiative for atmospheric ecology. Front Microbiol 7:16. doi:10.3389/fmicb.2016.00016

Pocock T, Lachance MA, Proschold T, Priscu JC, Kim SS, Huner NPA (2004) Identification of a psychrophilic green alga from Lake Bonney Antarctica: *Chlamydomonas raudensis* Ettl. (UWO 241) *Chlorophyceae*. J Phycol 40:1138–1148

Pointing SB, Chan Y, Lacap DC, Lau MCY, Jurgens JA, Farrell RL (2009) Highly specialized microbial diversity in hyper-arid polar desert. Proc Nat Acad Sci USA 106:19964–19969

Poli A, Esposito E, Orlando P, Lama L, Giordano A, de Appolonia F, Nicolaus B, Gambacorta A (2006) *Halomonas alkaliantarctica* sp. nov., isolated from saline lake Cape Russell in Antarctica, an alkalophilic moderately halophilic, exopolysaccharide-producing bacterium. Syst Appl Microbiol 30:31–38

Poglazova MN, Mitskevich IN, Abyzov SS, Ivanov MV (2001) Microbiological Characterization of the accreted ice of subglacial Lake Vostok, Antarctica. Microbiology 70:723–730

Pommier T, Canback B, Riemann L, Bostrom KH, Simu K, Lundberg P, Unlid A, Hagström Å (2007) Global patterns of diversity and community structure in marine bacterioplankton. Mol Ecol 16:867–880

Prescott GW (1978) How to know the freshwater algae, 3rd edn. Wm C Brown, Dubuque, IA, USA

Price PB (2006) Microbial life in glacial ice and implications for a cold origin of life. FEMS Microbiol Ecol 59:217–231

Price PB, Nagornov OV, Bay R, Chirkin D, He Y, Miocinovic P, Richards A, Woschnagg K, Koci B, Zagorodnov V (2002) Temperature profile for glacial ice at the South Pole: Implications for life in a nearby subglacial lake. Proc Natl Acad Sci U S A 99:7844–7847

Priscu JC, Adams EE, Lyons WB, Voytek MA, Mogk DW, Brown RL, McKay CP, Takacs CD, Welch KA, Wolf CF, Kirshtein JD, Avci R (1999) Geomicrobiology of subglacial ice above Lake Vostok, Antarctica. Science 286:2141–2144

Priscu JC, Kennicutt II, MC, Bell RE, Bulat SA, Ellis-Evans JC, Lukin VV, Petit J-R, Powell RD,Siegert MJ, Tabacco I (2005) Exploring subglacial Antarctic Lake environments. American Geophysical Union, EOS Transactions 86:193–200

Purves K, Macintyre L, Brennan D, Hreggviðsson GÓ, Kuttner E, Ásgeirsdóttir ME, Young LC, Green DH, Edrada-Ebel R, Duncan KR (2016) Using molecular networking for microbial secondary metabolite bioprospecting. Metabolites 6(1):2. doi:10.3390/metabo6010002

Rakusa-Suszczewski S (1980) The role of near-shore research in gaining and understanding of the functioning of the Antarctic ecosystem. Pol Arch Hydrobiol 27:229–233

Ravenschlag K, Sahm K, Pernthaler J, Amann R (1999) High bacterial diversity in permanently cold marine sediments. Appl Environ Microbiol 65:3982–3989

Reddy GS, Prakash JS, Vairamani M, Prabhakar S, Matsumoto GI, Shivaji S (2002) *Planococcus antarcticus* and *Planococcus psychrophilus* spp. nov. isolated from cyanobacterial mat samples collected from ponds in Antarctica. Extremophiles 6:253–261

Reddy GS, Raghavan PU, Sarita NB, Prakash JS, Nagesh N, Delille D, Shivaji S (2003) *Halomonas glaciei* sp. nov. isolated from fast ice of Adelie Land, Antarctica. Extremophiles 7:55–61

Reid IN, Sparks WB, Lubow S, McGrath M, Livio M, Valenti J, Sowers KR, Shukla HD, MacAuley S, Miller T; Suvanasuthi R, Belas R, Colman A, Robb FT, DasSarma P, Müller JA, Coker JA, Cavicchioli R, Chen F, DasSarma S (2006) Terrestrial models for extraterrestrial life: methanogens and halophiles at Martian temperatures. Int J Astrobiol 5:89–97

Riedel T, Held B, Nolan M, Lucas S, Lapidus A, Tice H, Del Rio TG, Cheng JF, Han C, Tapia R, Goodwin LA, Pitluck S, Liolios K, Mavromatis K, Pagani I, Ivanova N, Mikhailova N, Pati A, Chen A, Palaniappan K, Land M, Rohde M, Tindall BJ, Detter JC, Göker M, Bristow J, Eisen JA, Markowitz V, Hugenholtz P, Kyrpides NC, Klenk HP, Woyke T (2012) Genome sequence of the Antarctic rhodopsins-containing flavobacterium *Gillisia limnaea* type strain (R-8282(T)). Stand Genomic Sci 7(1):107–119. doi:10.4056/sigs.3216895

Robador A, Müller AL, Sawicka JE, Berry D, Hubert CR, Loy A, Jørgensen BB, Brüchert V (2016) Activity and community structures of sulfatereducing microorganisms in polar, temperate and tropical marine sediments. ISME J 10:796–780 doi: 10.1038/ismej.2015.157

Rogers AD, Tyler PA, Connelly DP, Copley JT, James R, Larter RD, Linse K, Mills RA, Garabato AN, Pancost RD, Pearce DA, Polunin NV, German CR, Shank T, Boersch-Supan PH, Alker BJ, Aquilina A, Bennett SA, Clarke A, Dinley RJ, Graham AG, Green DR, Hawkes JA, Hepburn L, Hilario A, Huvenne VA, Marsh L, Ramirez-Llodra E, Reid WD, Roterman CN, Sweeting CJ, Thatje S, Zwirglmaier K (2012) The discovery of new deep-sea hydrothermal vent communities in the southern ocean and implications for biogeography. PLoS Biol 10(1):e1001234. doi:10.1371/journal.pbio.1001234

Sattler B, Puxbaum H, Psenner R (2001) Bacterial growth in supercooled cloud droplets. Geophys Res Lett 28:239–242

Saunders NF, Thomas T, Curmi PM, Mattick JS, Kuczek E, Slade R, Davis J, Franzmann PD, Boone D, Rusterholtz K, Feldman R, Gates C, Bench S, Sowers K, Kadner K, Aerts A, Dehal P, Detter C, Glavina T, Lucas S, Richardson P, Larimer F, Hauser L, Land M, Cavicchioli R (2003) Mechanisms of thermal adaptation revealed from the genomes of the Antarctic Archaea *Methanogenium frigidum* and *Methanococcoides burtonii*. Genome Res 13: 1580–1588

Säwström C, Anesio MA, Granéli W, Laybourn-Parry J (2007) Seasonal viral loop dynamics in two large ultraoligotrophic Antarctic freshwater lakes. Microb Ecol 53:1–11

Schiraldi C, De Rosa M (2002) The production of biocatalysts and biomolecules from extremophiles. Trends Biotechnol 20:515–521

Sharma A, Scott JH, Cody GD, Fogel ML, Hazen RM, Hemley RJ, Huntress WT (2002) Microbial activity at gigapascal pressures. Science 295:1514–1516

Sharp M, Parks J, Cragg B, Fairchild I, Lamb H, Tranter M (1999) Widespread bacterial population at glacier beds and their relationship to rock weathering and carbon cycling. Geology 27:107–110

Sheridan PP, Loveland-Curtze J, Miteva VI, Brenchley JE (2003) *Rhodoglobus vestalii* gen. nov. sp. nov., a novel psychrophilic organism isolated from an Antarctic Dry Valley Lake. Int J Syst Evol Microbiol 53:985–994

Shivaji S, Reddy GSN, Raghavan PUM, Sarita NB, Delille D (2004) *Psychrobacter salsus* sp. nov. and *Psychrobacter adeliensis* sp. nov. isolated from fast ice from Adelie Land, Antarctica. Syst Appl Microbiol 27:628–635

Siddiqui KS, Cavicchioli R (2006) Cold adapted enzymes. Ann Rev Biochem 75:403–433

Siegert MJ, Tranter M, Ellis-Evans JC, Priscu JC, Berry Lyons W (2003) The hydrochemistry of Lake Vostok and the potential for life in Antarctic subglacial lakes. Hydrol Process 17:795–814

Sjöling S, Cowan DA (2003) High 16S rDNA bacteria diversity in glacial meltwater lake sediment, Bratina island, Antarctica. Extremophiles 7:275–282

Smith JJ, Ah Tow L, Stafford W, Cary C, Cowan DA (2006) Bacterial diversity in three different Antarctic cold desert mineral soils. Microb Ecol 51:413–421

Sonjak S, Frisvad JC, Gunde-Cimerman N (2006) *Penicillium* mycobiota in Arctic subglacial ice. Microb Ecol 52:207–216

Spring S, Merkhoffer B, Weiss N, Kroppenstedt RM, Hippe H, Stackebrandt E (2003) Characterization of novel psychrophilic clostridia from an Antarctic microbial mat: description of *Clostridium frigoris* sp. nov., *Clostridium lacusfryxellense* sp. nov., *Clostridium bowmanii* sp. nov. and *Clostridium psychrophilum* sp. nov. and reclassification of *Clostridium laramiense* as *Clostridium estertheticum* subsp. *laramiense* subsp. nov. Int J Syst Evol Microbiol 53:1019–1029

Staley JT, Gosink JJ (1999) Poles apart: biodiversity and biogeography of sea ice bacteria. Annu Rev Microbiol 53:189–215

Stingl U, Cho J-C, Foo W, Vergin KL, Lanoil B, Giovannoni SJ (2008) Dilution-to-extinction culturing of psychrotolerant planktonic bacteria from permanently ice-covered lakes in the McMurdo Dry Valleys, Antarctica. Microb Ecol 55:395–405

Taton A, Grubisic S, Balthasart P, Hodgson DA, Laybourn-Parry J, Wilmotte A (2006) Biogeographical distribution and ecological ranges of benthic cyanobacteria in East Antarctic lakes. FEMS Microbiol Ecol 57(2):272–289

Tatur A (2002) Ornithogenic ecosystems in the Maritime Antarctic – formation, development and disintegration. In: Beyer L, Bölter M (eds) Geoecology of Antarctic ice-free coastal landscapes. Series Ecological Studies, vol 154. Springer, Berlin, Heidelberg, pp. 161–184

Telling J, Anesio AM, Tranter M, Fountain AG, Nylen T, Hawkings J, Singh VB, Kaur P, Musilova M, Wadham JL (2014) Spring thaw ionic pulses boost nutrient availability and microbial growth in entombed Antarctic Dry Valley cryoconite holes. Front Microbiol 5:694. doi:10.3389/fmicb.2014.00694

Terauds A, Chown SL,Morgan F, Peat HJ, Watts DJ, Keys H, Convey P, Bergstrom DM (2012) Conservation biogeography of the Antarctic. Diversity Distrib 18(7). doi:10.1111/j.1472-4642.2012.00925.x

Thomas DN, Dieckmann GS (2002) Antarctic sea ice -a habitat for extremophiles. Science 295:641–644

Tindall B (2004) Prokaryotic diversity in the Antarctic: The tip of the iceberg. Microb Ecol 47:271–283

Ting L, Williams TJ, Cowley MJ, Lauro FM, Guilhaus M, Raftery MJ, Cavicchioli R (2010) Cold adaptation in the marine bacterium, *Sphingopyxis alaskensis* assessed using quantitative proteomics. Environ Microbiol 12:2658–2676

Turner J, King JC, Lachlan-Cope TA, Jones PD (2002) Recent temperature trends in the Antarctic. Nature 418:291–292

Tytgat B, Verleyen E, Obbels D, Peeters K, De Wever A, D'hondt S, De Meyer T, Van Criekinge W, Vyverman W, Willems A (2014) Bacterial diversity assessment in Antarctic terrestrial and aquatic microbial mats: a comparison between bidirectional pyrosequencing and cultivation. PLoS One 9(6):e97564. doi:10.1371/journal.pone.0097564 eCollection 2014

Tytgat B, Verleyen E, Sweetlove M, D'hondt S, Clercx P, Van Ranst E, Peeters K, Roberts S, Namsaraev Z, Wilmotte A, Vyverman W, Willems A (2016) Bacterial community composition in relation to bedrock type and macrobiota in soils from the Sør Rondane Mountains, East Antarctica.FEMS Microbiol Ecol 92(9).pii: fiw126. doi: 10.1093/femsec/fiw126. Epub 2016 Jul 10

Van Trappen S, Mergaert J, Eygen SV, Dawyndt P, Cnockaert MC, Swings J (2002) Diversity of 746 heterotrophic bacteria isolated from microbial mats from ten Antarctic lakes. Syst Appl Microbiol 25:603–610

Van Trappen S, Mergaert J, Swings J (2003) *Flavobacterium gelidilacus* sp. nov., isolated from microbial mats in Antarctic lakes. Int J Syst Evol Microbiol 53:1241–1245

Van Trappen S, Vandecandelaere I, Mergaert J, Swings J (2004a) *Algoriphagus antarcticus* sp. nov., a novel psychrophile from microbial mats in Antarctic lakes. Int J Syst Evol Microbiol 54:1969–1973

Van Trappen S, Mergaert J, Swings J (2004b) *Lokanella salsilacus* gen. nov., sp. nov., *Lokanella fryxellensis* sp. nov. and *Lokanella vestfoldensis* sp. nov., new members of the *Rhodobacter* group, isolated from microbial mats in Antarctic lakes. Int J Syst Evol Microbiol 54:1263–1269

Van Trappen S, Vandecandelaere I, Mergaert J, Swings J (2004c) *Flavobacterium degerlachei* sp. nov., *Flavobacterium frigoris* sp. nov. and *Flavobacterium micromati* sp. nov., novel psychrophilic bacteria isolated from microbial mats in Antarctic lakes. Int J Syst Evol Microbiol 54:85–92

Van Trappen S, Vandecandelaere I, Mergaert J, Swings J (2005) *Flavobacterium fryxellicola* sp. nov. and *Flavobacterium psychrolimnae* sp. nov., novel psychrophilic bacteria isolated from microbial mats in Antarctic lakes. Int J Syst Evol Microbiol 55:769–772

Vishnivetskaya TA, Siletzky R, Jefferies N, Tiedje JM, Kathariou S (2007) Effect of low temperature and culture media on the growth and freeze-thawing tolerance of *Exiguobacterium* strains. Cryobiology 54:234–240

Wilson WH, Lane D, Pearce DA, Ellis-Evans JC (2000) Transmission electron microscope analysis of virus-like particles in the freshwater lakes of Signy Island, Antarctica. Polar Biol 23:657–660

Willerslev E, Hansen A, Christensen B, Steffensen J, Arctander P (1999) Diversity of Holocene life forms in fossil glacier ice. Proc Natl Acad Sci U S A 96:8017–8021

Yergeau E, Schoondermark-Stolk SA, Brodie EL, Déjean S, DeSantis TZ, Gonçalves O, Piceno YM, Andersen GL, Kowalchuk GA (2008) Environmental microarray analyses of Antarctic soil microbial communities. ISME J 3:340–351

Yi H, Oh HM, Lee JH, Kim SJ, Chun J (2005a) *Flavobacterium antarcticum* sp. nov., a novel psychrotolerant bacterium isolated from the Antarctic. Int J Syst Evol Microbiol 55:637–641

Yi H, Yoon HI, Chun J (2005b) *Sejongia antarctica* gen. nov., sp. nov. and *Sejongia jeonii* sp. nov., isolated from the Antarctic. Int J Syst Evol Microbiol 55:409–416

Yoshimune K, Galkin A, Kulakova L, Yoshimura T, Esaki N (2005) Cold-active DnaK of an Antarctic psychrotroph *Shewanella* sp. Ac10 supporting the growth of dnaK-null mutant of *Escherichia coli* at cold temperatures. Extremophiles 9:145–150

Yu Y, Xin YH, Liu HC, Chen B, Sheng J, Chi ZM, Zhou PJ, Zhang DC (2008) *Sporosarcina antarctica* sp. nov., a psychrophilic bacterium isolated from the Antarctic. Int J Syst Evol Microbiol 58:2114–2117

Desert Cyanobacteria: Potential for Space and Earth Applications

6

Daniela Billi, Mickael Baqué, Cyprien Verseux, Lynn Rothschild, and Jean-Pierre de Vera

Contents

6.1 Introduction

In cold and hot deserts, where life is pushed to its physical limits due to extreme water deficit and/or extreme temperatures, microorganisms escape life-limiting conditions by colonizing rocks. They form hypolithic biofilms at the stone-soil interface or endolithic communities in the upper few millimeters to centimeters of rock (Golubic et al. 1981).

Research of photosynthesis-based communities in deserts was pioneered by E. Imre Friedmann and Roseli Ocampo-Friedmann. They first described the

D. Billi (✉) • C. Verseux
University of RomeTor Vergata, Department of Biology, Rome, Italy
e-mail: billi@uniroma2.it

M. Baqué • J.-P. de Vera
German Aerospace Center (DLR), Institute of Planetary Research, Berlin, Germany

L. Rothschild
Earth Sciences Division, NASA Ames Research Center, Moffett Field, California, USA

H. Stan-Lotter, S. Fendrihan (eds.), *Adaption of Microbial Life to Environmental Extremes*, DOI 10.1007/978-3-319-48327-6_6

Chroococcidiopsis-dominated communities in the Negev Desert, Israel (Friedmann et al. 1967). Later on, they described endolithic communities in the Dry Valleys (Friedmann and Ocampo1976), the largest ice-free region on the Antarctic continent, which was previously thought to be sterile (Horowitz et al. 1972). Their field campaigns continued in several deserts worldwide, leading to the establishment of the Culture Collection of Microorganisms from Extreme Environments. Today, this collection gathers about 250 desert strains from the genus *Chroococcidiopsis* and a few related genera. It is maintained at the University of Rome "Tor Vergata."

The Atacama Desert, the driest nonpolar desert in the world, is often referred to by astrobiologists as an analogue of Mars due to its environmental conditions. Studies based on this analogy suggest that if microhabitats exist (or have existed) on Mars, they will be difficult to detect as dispersed in virtually lifeless surroundings (Warren-Rhodes et al. 2006). Indeed, in the Atacama, hypolithic communities diminish along the aridity gradient. But even at the hyperarid core, rare *Chroococcidiopsis*-based communities exist, albeit in small spatially isolated islands amidst a microbially impoverished soil (Warren-Rhodes et al. 2006). They colonize halite deposits (Stivaletta et al. 2012; Wierzchos et al. 2006), where they take advantage of halite deliquescence (Davila et al. 2013): inactive most of the time, microbial cells recover metabolic activity when relative humidity is high enough for halite crystals to form a saturated brine droplet with absorbed atmospheric water.

The adaptability of *Chroococcidiopsis* spp. to extreme environments was further demonstrated in another desert used as a Mars analogue: the Mojave Desert. There, *Chroococcidiopsis* spp. isolated from different rock types (talc, marble, quartz, white carbonate, and red-coated carbonate) were found to have shifted the photosynthetic pigment emissions, presumably as an adaptation to the rocks' different light transmission properties (Smith et al. 2014).

The endurance of cyanobacteria isolated from desert lithic communities is currently tested in space and under simulated Mars conditions in low Earth orbit (LEO) outside the International Space Station, notably within the BOSS (Biofilm Organisms Surfing Space) and BIOMEX (Biology and Mars Experiment) experiments of the EXPOSE-R2 space mission (de Vera et al. 2012; Baqué et al. 2013b, c). This endeavor is of prime importance to search for life beyond Earth (Cottin et al. 2015). In parallel to those experiments, ground-based simulations of space and planetary environments are performed. Such experiments allow the astrobiology community to: (i) understand the limits of life and potential habitability of the solar system and beyond (Baglioni et al. 2007; Cockell et al. 2016); (ii) identify suitable biosignatures to search for past or extant life on Mars (de Vera et al. 2012); (iii) test the lithopanspermia theory, i.e., the possibility of interplanetary transport of life by means of material ejected by asteroid and meteorite impacts (Horneck et al. 2008; Nicholson 2009; Stöffler et al. 2007); (iv) improve procedures for planetary protection, to avoid contamination of bodies of interest in our solar system with terrestrial life via probes and rovers; and (v) design life-support systems for beyond-Earth settlements (Verseux et al. 2016a).

6.2 Cyanobacteria Under Dry Conditions on Earth

Water is essential for life as we know it, and its removal is lethal to most organisms due to damage induced, by various mechanisms, at every level of the cellular organization. For instance, the removal of the hydration shell from phospholipids of the membrane bilayers causes a transition to a gel phase at environmentally relevant temperatures (Crowe et al. 1992). Reactive oxygen species (ROS) are produced – especially in organisms performing oxygenic photosynthesis, such as cyanobacteria – and induce oxidative damage to proteins, lipids, and nucleic acids (França et al. 2007). The Haber-Weiss, Fenton, and Maillard reactions further damage proteins and nucleic acids (Potts 2001).

However, a small group of taxonomically diverse organisms can withstand air-drying by entering an ametabolic state referred to as anhydrobiosis or anabiosis (Crowe et al. 1992; Feofilova 2003). Such organisms are found among cyanobacteria, yeasts, lichens, algae, mosses, plants, nematodes, rotifers, and tardigrades; some anhydrobiotes are particularly resistant to desiccation when in dormant forms such as bacterial spores, cyanobacterial akinetes, plant seeds, and shrimp cysts (Alpert 2006; Crowe et al. 1992; Rebecchi et al. 2007).

How anhydrobiotic cyanobacteria manage to cope with desiccation remains largely unknown (Billi 2012; Billi and Potts 2000, 2002), but investigations made clear that the involved structural, physiological, and molecular mechanisms are both numerous and highly diverse. In hot and cold deserts, rock-inhabiting phototrophic communities are wetted for only a few hours per year and persist in an ametabolic dry and/or frozen state for the greater part of their life. They repair accumulated damage upon rewetting. Under laboratory desiccation, desert strains of *Chroococcidiopsis* were reported to survive more than 4 years of dry storage, by avoiding and/or limiting the genomic fragmentation and ROS generation (Billi 2009). Unlike *Deinococcus radiodurans*, in which desiccation induces DNA degradation (Mattimore and Battista 1996), dried cells of *Nostoc commune* and of *Chroococcidiopsis* sp. maintain DNA integrity (Shirkey et al. 2003; Billi 2009). This indicates that protection mechanisms can be critical to desiccation resistance, even more than an exceptional ability to repair DNA damage.

A crucial structural mechanism in the adaptation of *Chroococcidiopsis* to anhydrobiosis is the production of abundant polysaccharide-rich envelopes (Grilli Caiola et al. 1996). It has been proposed that extracellular polysaccharides (EPS) significantly contribute to cyanobacterial desiccation tolerance, by providing both a repository for water and a matrix which stabilizes desiccation-related enzymes and molecules (Wright et al. 2005). EPS production might act in synergy with an accumulation of the disaccharide trehalose in the cytoplasm, which occurs in most anhydrobiotes upon desiccation. By substituting water molecules, trehalose prevents the phase transition of cellular membranes and stabilizes dried proteins (Crowe 2007).

6.3 Cyanobacteria Under Ground-Based, Simulated Extraterrestrial Conditions

Various anhydrobiotes have demonstrated an outstanding capability to survive conditions not found on Earth, such as high doses of UV and ionizing radiations. This may be partially explained as a by-product of their desiccation tolerance mechanisms (Mattimore and Battista 1996; Pavlopoulou et al. 2016). For instance, as both desiccation and radiation induce ROS-mediated damage, mechanisms mitigating oxidative stress and evolved under desiccation stress may be relevant for coping with radiation.

In line with this, desert strains of *Chroococcidiopsis* can withstand high doses of UVC and ionizing radiation. Hydrated cells survived 13 kJ/m^2 of UVC radiation, likely due to the occurrence of multicellular aggregates which attenuate the dose reaching the inner cells (Baqué et al. 2013a). Dried monolayers of *Chroococcidiopsis* also survived 130 kJ/m^2 of a simulated Martian UV flux, thus displaying a greater tolerance than *Bacillus subtilis* spores (Cockell et al. 2005). Both hydrated and dried cells survived 2 kGy of heavy ions and 15 kGy of photon irradiation, with a higher resistance of dried cells (Billi et al. 2000; Verseux et al. 2017). Other studies have evidenced radiation resistance in other cyanobacteria. *Anabaena* sp. PCC7120 and *Arthrospira* sp. PCC 8005, for instance, can cope with 5 kGy and 6.4 kGy of gamma radiation, respectively (Badri et al. 2015; Singh et al. 2013). Further investigations to verify the correlation between the capability to withstand prolonged desiccation (years) and the level of radio resistance would be relevant.

When desert strains of *Chroococcidiopsis* were exposed in the dried state to ground-based space and Martian simulations carried out in the frame of the EXPOSE-E space mission (Rabbow et al. 2012), they survived a UV radiation dose (200–400 nm) corresponding to that expected from a 1.5-year exposure in LEO, if shielded under 3 mm of sandstone (Billi et al. 2011). Additional space and Martian simulations were carried out on dried cells of desert strains of *Chroococcidiopsis* in the frame of the BOSS and BIOMEX experiments of the EXPOSE-R2 space mission (Baqué et al. 2013b), as described below.

BOSS aims at investigating whether biofilm-forming bacteria are more resistant to extraterrestrial environmental stressors than their planktonic counterparts. Under laboratory simulations, dried biofilms of *Chroococcidiopsis* tolerated up to 5×10^5 kJ/m^2 of polychromatic UV radiation (attenuated with a 0.1 % neutral density filter), whereas dried planktonic samples underwent extensive DNA damage and lost their colony-forming ability at the same dose. This enhanced resistance has been attributed to the presence of EPS, with an abundant lipidic component, which will need further characterization to understand their involvement in resistance to environmental factors and their potential use as biosignatures for searching life on Mars or other planetary bodies (Baqué et al. 2013c).

In BIOMEX, a number of selected extremophiles and their constituents (pigments, cell wall components, etc.), mixed with Martian and lunar mineral analogues, are investigated in the context of the lithopanspermia theory, the habitability of Mars, and to establish a biosignature database for searching life on Mars (de Vera et al. 2012). The presence of lunar and Martian mineral analogues had a protective effect on the preservation of biomarkers like DNA and photosynthetic pigments upon UV irradiation (either monochromatic at 254 nm or polychromatic with the whole UV spectrum from 200 to 400 nm) (Baqué et al. 2014, 2015). β-Carotene was also investigated as a potential cyanobacterial biomarker by Raman spectroscopy (Böttger et al. 2012; Vitek et al. 2012). This technology, which was selected for upcoming search-for-life-missions to be operated on Mars, will rely on Raman spectrometers for nondestructively identifying both the mineral context and putative biosignatures such as photoprotective pigments like β-carotene (Beegle et al. 2014; Hutchinson et al. 2014). Even though UV irradiation proved to substantially damage the β-carotene signature in fully exposed samples, it was still detectable at 2–5 % of its original value and remained fully detectable in Martian-simulated atmosphere-exposed samples (Baqué et al. 2015).

6.4 Cyanobacteria Under Low Earth Orbit Conditions

Because the major damaging effect caused by space vacuum is desiccation (Horneck et al. 2010), organisms exposed to the space environment must be desiccation-tolerant and able to repair the accumulated damage upon recovery (Billi et al. 2013). Furthermore, concerning the harsh radiation environment found in space due to Solar and Galactic Cosmic Radiation (SCR and GCR), a strong correlation has been shown between desiccation and radiation tolerance (Musilova et al. 2015). Due to their high desiccation and radiation resistance, cyanobacteria thus appeared to have all the prerequisites to survive space exposure.

The first space experiment with cyanobacteria was performed in 1994 aboard ESA's BIOPAN-I facility: it carried the halophilic cyanobacterium *Synechococcus* (Nägeli), inhabiting gypsum-halite crystals from a marine intertidal area along the coast of Baja California in Mexico (Mancinelli et al. 1998). Later, during the BIOPAN-VI mission in 2007, epilithic and endolithic cyanobacteria from rock-dwelling microbial communities were exposed to the space environment, and after 10 days a single *Gloeocapsa*-like cyanobacterium was selected within a sample from coastal limestone (Olsson-Francis et al. 2010).

By taking advantage of the possibility to allocate samples at the exterior of the Foton spacecraft, *Chroococcidiopsis* cells were inoculated within rocks, at a depth at which light is available for photosynthesis, and tested for survival upon reentry into the Earth atmosphere in the context of the lithopanspermia theory (Cockell et al. 2007). Lithopanspermia is currently divided into three stages: (i) launch of microbe-bearing rocks from a donor planet into space; (ii) transit through space to a recipient planet; and (iii) entry into a recipient planet. Survival of shockwaves simulating pressures which are appearing in the spallation and ejection zone of an impact crater, based on values obtained by studies of Martian meteorites, was reported for spores of *B. subtilis* as well as photobiont and mycobiont partners of the lichen *Xanthoria elegans* which, embedded in a Martian analogue rock, encompassed shock pressures up to 40 GPa (Horneck et al. 2008). By contrast,

Chroococcidiopsis survived shock pressures ranging from 5 to 10 GPa. Therefore, given the low frequency of weakly shocked meteorites, the chances for interplanetary transport of cyanobacteria-type organisms seem reduced (Horneck et al. 2008). Since *Chroococcidiopsis* did not survive the reentry process, due to the heating well above the upper temperature limit for life, it was suggested that eventually non-photosynthetic organisms living deeper within rocks have a better chance of surviving the exit and entry processes (Cockell et al. 2007).

Results from the first EXPOSE-E mission showed survival to a longer permanence in space (548 days) of rock-inhabited communities than previously tested; within an epilithic community augmented with akinetes of *Anabaena cylindrica*, and vegetative cells of *Nostoc commune* and *Chroococcidiopsis*, only the latter survived after exposure to space vacuum and extraterrestrial ultraviolet spectrum (Cockell et al. 2011). Furthermore, during the following EXPOSE-R mission (2009–2011), *Chroococcidiopsis* cells inoculated into impact-shocked gneiss survived 22 months of space exposure (Bryce et al. 2015); also the halophilic cyanobacterium *Synechococcus* (Nägeli) showed survival when partially shielded from UVs (Mancinelli 2015).

The resilience of desert strains of *Chroococcidiopsis* was further tested in the EXPOSE-R2 space mission (2014–2016), during which they were exposed to space conditions, but also to a simulated Mars-like environment in space (CO_2 atmosphere and UV > 200 nm). The EXPOSE-R2 space mission successfully reached the International Space Station (ISS) on July 24, 2014, on board the space cargo Progress 56P. On August 18, it was installed outside the ISS on the Russian Zvezda module for a 16-month exposure until it was taken back inside the station on February 3, 2016.

6.5 Cyanobacteria and the Search for Life Beyond Earth

Mars is still considered the prime target for discovering life outside of Earth (Schulze-Makuch et al. 2008). Recent data from NASA's Mars Science Laboratory Curiosity rover have confirmed and consolidated models on Mars' potential past habitability: a lacustrine environment was, for example, discovered by Curiosity as a putative past habitable environment (Grotzinger et al. 2014), and sedimentary rocks were found to have been recently (78 ± 30 million years) exposed to surface conditions, thus providing protection to potential biological remnants or dormant organisms (Farley et al. 2014). If life exists on present-day Mars, it is likely underground and photosynthesis independent, given the harsh conditions at the surface where solar light is available (e.g., Cockell 2014; Westall et al. 2013). However, if environmental conditions on Mars were life permissive long enough for life to originate there (Cockell 2014; McKay 2010; Westall et al. 2013) and to evolve metabolic systems as complex as oxygenic photosynthesis, and/or if life was exchanged between the two planets (e.g., Nicholson 2009), remains from life-forms analogous to desiccation- and radiation-resistant cyanobacteria or other photosynthetic microorganisms might still exist and be detectable on Mars. Indeed, cyanobacteria are among the oldest life-forms on Earth, with earliest evidence in the fossil record between 3.5 and 2.8 billion years ago (Bosak et al. 2013; Schopf 2002; Schwartzman et al. 2008), at a time when environmental conditions (noteworthy water availability and UV irradiation) are thought to be similar on the Earth and on Mars (Cockell et al. 2000; McKay 1997).

Since Mars lacks an intrinsic magnetic field (possibly lost, at least 3.5 billion years ago), ionizing radiation of solar energetic protons (SEP), and galactic and solar cosmic rays (GCRs and SCRs) reach today's Martian surface and subsurface with high energies. Recent data from NASA's Mars Science Laboratory rover, Curiosity, reported absorbed dose rates at the Martian surface at Gale Crater of 76 mGy/year (Hassler et al. 2013), while previous models calculated values between 50 mGy/year (Pavlov et al. 2012) and 150 mGy/year (Dartnell et al. 2007). These GCRs/SCRs dose rates suggest that, although over geological time scales dormant microorganisms would be inactivated even at depths between 1 and 2 m below the Martian surface, transient habitable conditions would allow them to repair the accumulated damages and reset their "inactivation clock" (Dartnell et al. 2007; Hassler et al. 2013). The extraordinary radiation resistance of desert cyanobacteria is thus a seducing argument for imagining a potential Martian biosphere. For example, being able to resist up to 15 kGy of ionizing radiation in the active form, *Chroococcidiopsis* could potentially withstand a radiation environment equivalent to 200,000 years on the Martian surface. Its radiation resistance in the dormant desiccated state and at low temperatures could moreover be enhanced by the interplay of protection mechanisms involved in both desiccation and radiation resistance (Daly 2009, 2012; Musilova et al. 2015).

Furthermore, even if an active Martian biosphere in the subsurface is highly speculative today, biosignatures from early life-forms could have been preserved over geological time scales, based on the latest Curiosity measurements (Hassler et al. 2013).

Future space missions dedicated to the search for life on Mars are therefore foreseen by ESA/Roscosmos and NASA through the ExoMars (Barnes et al. 2006; Vago et al. 2006) and Mars2020 rovers (Fries et al. 2010). They will, to this end, carry new types of instrumentation to detect traces of life in extraterrestrial environments by, e.g., Raman spectrometry (Beegle et al. 2014; Hutchinson et al. 2014). Extensive field and laboratory investigations have been focused on the detection of carotenoids and photosynthetic pigments in Mars terrestrial analogues, via the development of Raman and fluorescence emission techniques (Edwards et al. 2005; Groemer et al. 2014; Jorge-Villar and Edwards 2013; Smith et al. 2012; Vítek et al. 2009) or their optimization in the presence of Martian analogue minerals (Böttger et al. 2012; Jehlička et al. 2014; Vandenabeele et al. 2012; Vítek et al. 2009). The assessment of potential biosignature degradation under Mars-like conditions is also of prime importance in the preparation of these future missions (Dartnell and Patel 2014; Dartnell et al. 2011, 2012; Stromberg et al. 2014). In this context, cyanobacteria have been extensively used as models as they contain photoprotective and antioxidant molecules (e.g., carotenoids), ultraviolet screening compounds (e.g., scytonemin), photosynthetic pigments (e.g., chlorophyll *a* and phycobiliproteins), and their degradation products (e.g., porphyrins), which have been classified as high priority targets for biomolecule detection (Parnell et al. 2007).

Moreover, the EPS composition of *Chroococcidiopsis* samples, assessed by CLSM imaging combined with fluorescent probes, revealed the presence of polysaccharides and abundant lipidic compounds in biofilm samples but also (to a lesser extent) in planktonic ones (Baqué et al. 2013c). Lipid hydrocarbons, much like DNA, are a ubiquitous characteristic of life as they compose cellular membranes. By selecting and producing only certain types of lipids for membrane composition, the biology can easily be distinguished from the chemistry thanks to the "LEGO" principle (Georgiou and Deamer 2014; McKay 2004). Besides, as biological lipids are among the most stable biomarkers over geological time scales and are easily analyzed, the characterization of *Chroococcidiopsis* EPS's lipidic component stability will give us a supplementary potential biosignature for guiding future search for life missions.

The recent results from ground-based and LEO exposure, presented in the above sections, showed that even a limited UV shielding, provided by a few micrometers of regolith powder or by multilayers of cells in the biofilm form, proved to be sufficient in protecting pigment autofluorescence and DNA from degradation. As hypolithic or endolithic photosynthetic communities are expected to be protected from UV irradiation as an adaptive strategy, this finding corroborates our need to search for putative extinct or extant life in the Martian subsurface.

6.6 Cyanobacterium-Based Life-Support Systems in Space and on Earth

Cyanobacterial adaptation to extreme environments, notably to terrestrial deserts, makes them promising components of future life-support systems for future manned Mars missions (Verseux et al. 2016a). Indeed, in order for a manned Mars mission to allow extensive human scientific activity and yield meaningful scientific data, scientists will need to spend a considerable period on site. For long-term missions to be financially sustainable, explorers need to learn how to live "off the land" there rather than sending supplies in amounts proportional to the mission's duration. Biological systems (mostly plants and microorganisms) could be extremely useful in this context, possibly even more so after genetic engineering (see for instance Verseux et al. 2016b). But most plants and microorganisms are unable to exploit Martian resources, and sending substrates from Earth to support their metabolism would strongly limit the cost-effectiveness and sustainability of their cultivation. However, resources needed to grow specific cyanobacteria are available on Mars due to their photosynthetic abilities, nitrogen-fixing activities, and lithotrophic lifestyles. They could be used directly for various applications, including the production of food, fuel, and oxygen, but also indirectly: products from their culture could support the growth of other organisms, opening the way to a wide range of life-support biological processes based on Martian resources (Verseux et al. 2016a). Work aimed at developing cyanobacterium-based life-support systems is ongoing.

In particular, the PowerCell project led by Lynn Rothschild – where an engineered strain of *Anabaena* sp. secretes the disaccharide sucrose, used to feed *Bacillus subtilis* without the need to disrupt cyanobacterial cells – will be tested in space, onboard the Euglena & Combined Regenerative Organic-Food Production in Space (Eu:CROPIS) mission satellite planned to be launched in March 2017.[1]

Most technologies developed in the space sector find applications on Earth, and cyanobacterium-based technologies developed for Martian outposts will likely be no exception. Their principles could be transferred to Earth for the creation of sustainable industrial processes based on carbon dioxide, solar energy, water, and minerals. Such systems could lead to tremendous economic and ecological benefits in developed countries. It could also be highly valuable in resource-poor areas and, in the most extreme case, deserts, where the abundant mineral resources, solar energy, and surface area could be exploited. There, some of the constraints – notably the harsh environmental conditions and the low availability of resources – are somewhat similar to constraints found on Mars. The environment is however more favorable to life-support technologies due to presence of 1-G gravity, more abundant and direct sunlight, higher atmospheric pressure, breathable air, lower consequences of accidental contamination, low levels of UV and ionizing radiation, and much reduced constraints on mass and volume. Finally, strategies for transferring to Earth (notably to deserts) technologies developed for space life-support systems have been proposed (Nelson et al. 2003; Polyakov et al. 2010); similarly to the way they could be used on Mars to feed life-support systems, cyanobacteria could be used on Earth to feed transferred technologies.

Out of sunlight, minerals, and water, the latter will generally be the most difficult element to provide – particularly so in deserts. Several solutions can be worked on to face it. First, technologies developed for water recycling during spaceflight can be applied. In space stations, recycling technologies have allowed reclamation rates of 90 % (Bobe et al. 2007). Even though targeting such efficiencies would be extremely ambitious at large scale and with limited resources, terrestrial processes can definitely take advantage of water reclamation technologies developed for space. Second, some water mining techniques developed for Mars exploration can be adapted to terrestrial desert, notably those relying on water extraction from the atmosphere (see, e.g., Meyer and McKay 1984) and from the soil and rocks (see, e.g., Ethridge and Kaulker 2012). Third, water use can be reduced by, for example (depending on species and targeted products), immobilizing cyanobacteria within polymeric matrices (Hertzberg and Jensen 1989) or growing cyanobacteria as biofilms in a semi-closed environment where adequate moisture is provided. Finally, for areas where seawater is available, marine cyanobacteria can be used. Alternatively, genetic engineering could be used to make freshwater cyanobacteria thrive in seawater. This has been demonstrated with a freshwater *Synechococcus* sp. which, engineered for increased salt resistance, could grow in seawater (Waditee et al. 2002).

Desert cyanobacteria studied in space could thus, eventually, lead to game-changing applications back where they were discovered.

[1] https://www.nasa.gov/centers/ames/engineering/projects/powercell. Accessed March 8.

Acknowledgments Researches were funded by the Italian Space Agency, the Italian Ministry of Foreign Affairs, the Directorate General for Development Cooperation, the German Aerospace Center (DLR), and the German Ministry of Economy and Energy. MB acknowledges support from the Alexander von Humboldt Foundation. JPdV acknowledges support of the Helmholtz Gemeinschaft where his work was partly supported through the Helmholtz Alliance "Planetary Evolution and Life."

References

Alpert P (2006) Constraints of tolerance: why are desiccation-tolerant organisms so small or rare? J Exp Biol 209:1575–1584

Badri H, Monsieurs P, Coninx I, Wattiez R, Leys N (2015) Molecular investigation of the radiation resistance of edible cyanobacterium *Arthrospira* sp. PCC 8005. MicrobiologyOpen 4:187–207

Baglioni P, Sabbatini M, Horneck G (2007) Astrobiology experiments in low earth orbit—facilities, instrumentation and results. In: Rettberg P, Horneck G (eds) Complete course in astrobiology. Wiley-VCH, Berlin, pp. 273–319

Baqué M, Viaggiu E, Scalzi G, Billi D (2013a) Endurance of the endolithic desert cyanobacterium *Chroococcidiopsis* under UVC radiation. Extremophiles 17:161–169

Baqué M, de Vera J-P, Rettberg P, Billi D (2013b) The BOSS and BIOMEX space experiments on the EXPOSE-R2 mission: endurance of the desert cyanobacterium *Chroococcidiopsis* under simulated space vacuum, Martian atmosphere, UVC radiation and temperature extremes. Acta Astronaut 91:180–186

Baqué M, Scalzi G, Rabbow E, Rettberg P, Billi D (2013c) Biofilm and planktonic lifestyles differently support the resistance of the desert cyanobacterium *Chroococcidiopsis* under space and Martian simulations. Orig Life Evol Biospheres 43:377–389

Baqué M, Verseux C, Rabbow E, de Vera J-P, Billi D (2014) Detection of macromolecules in desert cyanobacteria mixed with a Lunar mineral analogue after space simulations. Orig Life Evol Biospheres 44:209–221

Baqué M, Verseux C, Böttger U, Rabbow E, de Vera J-P, Billi D (2015) Preservation of biomarkers from cyanobacteria mixed with Marslike regolith under simulated Martian atmosphere and UV flux. Orig Life Evol Biospheres 46:289–310

Barnes D, Battistelli E, Bertrand R, Butera F, Chatila R, Del Biancio A, Draper C, Ellery A, Gelmi R, Ingrand F et al (2006) The ExoMars rover and Pasteur payload Phase A study: an approach to experimental astrobiology. Int J Astrobiol 5:221–241

Beegle LW, Bhartia R, DeFlores L, Darrach M, Kidd RD, Abbey W, Asher S, Burton A, Clegg S, Conrad PG et al (2014) SHERLOC: scanning habitable environments with Raman and luminescence for organics and chemicals, an investigation for 2020. LPI Contrib 1783:5101

Billi D (2009) Subcellular integrities in *Chroococcidiopsis* sp. CCMEE 029 survivors after prolonged desiccation revealed by molecular probes and genome stability assays. Extremophiles 13:49–57

Billi D (2012) Anhydrobiotic rock-inhabiting cyanobacteria: potential for astrobiology and biotechnology. In: Stan-Lotter H, Fendrihan S (eds) Adaption of microbial life to environmental extremes: novel research results and application. Springer Wien, New York, pp. 119–132

Billi D, Potts M (2000) Life without water: responses of prokaryotes to desiccation. In: Storey KB, Storey JM (eds) Cell and molecular response to stress: environmental stressors and gene responses. Elsevier Science, Amsterdam, pp. 181–192

Billi D, Potts M (2002) Life and death of dried prokaryotes. Res Microbiol 153:7–12

Billi D, Friedmann EI, Hofer KG, Caiola MG, Ocampo-Friedmann R (2000) Ionizing-radiation resistance in the desiccation-tolerant cyanobacterium *Chroococcidiopsis*. Appl Env Microbiol 66:1489–1492

Billi D, Viaggiu E, Cockell CS, Rabbow E, Horneck G, Onofri S (2011) Damage escape and repair in dried *Chroococcidiopsis* spp. from hot and cold deserts exposed to simulated space and martian conditions. Astrobiology 1:65–73

Billi D, Baqué M, Smith HD, McKay CP (2013) Cyanobacteria from extreme deserts to space. Adv Microbiol 3:80–86
Bobe L, Samsonov N, Gavrilov L, Novikov V, Tomashpolskiy M, Andreychuk P, Protasov N, Synjak Y, Skuratov V (2007) Regenerative water supply for an interplanetary space station: the experience gained on the space stations "Salut","Mir", ISS and development prospects. Acta Astronaut 61:8–15
Bosak T, Knoll AH, Petroff AP (2013) The meaning of stromatolites. Annu Rev Earth Planet Sci 41:21–44
Böttger U, de Vera J-P, Fritz J, Weber I, Hübers H-W, Schulze-Makuch D (2012) Optimizing the detection of carotene in cyanobacteria in a martian regolith analogue with a Raman spectrometer for the ExoMars mission. Planet Space Sci 60:356–362
Bryce CC, Horneck G, Rabbow E, Edwards HGM, Cockell CS (2015) Impact shocked rocks as protective habitats on an anoxic early Earth. Int J Astrobiol 14:115–122
Cockell CS (2014) Trajectories of martian habitability. Astrobiology 14:182–203
Cockell CS, Catling DC, Davis WL, Snook K, Kepner RL, Lee P, McKay CP (2000) The ultraviolet environment of Mars: biological implications past, present, and future. Icarus 146:343–359
Cockell CS, Schuerger AC, Billi D, Friedmann EI, Panitz C (2005) Effects of a simulated martian UV flux on the cyanobacterium, *Chroococcidiopsis* sp. 029. Astrobiology 5:127–140
Cockell CS, Brack A, Wynn-Williams DD, Baglioni P, Brandstätter F, Demets R, Edwards HGM, Gronstal AL, Kurat G, Lee P et al (2007) Interplanetary transfer of photosynthesis: an experimental demonstration of a selective dispersal filter in planetary island biogeography. Astrobiology 7:1–9
Cockell CS, Rettberg P, Rabbow E, Olsson-Francis K (2011) Exposure of phototrophs to 548 days in low Earth orbit: microbial selection pressures in outer space and on early earth. ISME J 5:1671–1682
Cockell CS, Bush T, Bryce C, Direito S, Fox-Powell M, Harrison JP, Lammer H, Landenmark H, Martin-Torres J, Nicholson N et al (2016) Habitability: a review. Astrobiology 16:89–117
Cottin H, Kotler JM, Bartik K, Ii HJC, Cockell CS, deVera J-P, Ehrenfreund P, Leuko S, Loes ten Kate I, Martins Z, et al. (2015) Astrobiology and the possibility of life on Earth and elsewhere… Space Sci Rev:1–42. doi: 10.1007/s11214-015-0196-1
Crowe JH (2007) Trehalose as a "chemical chaperone": fact and fantasy. Adv Exp Med Biol 594:143–158
Crowe JH, Hoekstra FA, Crowe LM (1992) Anhydrobiosis. Annu Rev Physiol 54:579–599
Daly MJ (2009) A new perspective on radiation resistance based on *Deinococcus radiodurans*. Nat Rev Microbiol 7:237–245
Daly MJ (2012) Death by protein damage in irradiated cells. DNA Repair 11:12–21
Dartnell LR, Patel MR (2014) Degradation of microbial fluorescence biosignatures by solar ultraviolet radiation on Mars. Int J Astrobiol 13:112–123
Dartnell LR, Desorgher L, Ward JM, Coates AJ (2007) Modelling the surface and subsurface martian radiation environment: implications for astrobiology. Geophys Res Lett 34:L02207
Dartnell LR, Storrie-Lombardi MC, Mullineaux CW, Ruban AV, Wright G, Griffiths AD, Muller J-P, Ward JM (2011) Degradation of cyanobacterialbiosignatures by ionizing radiation. Astrobiology 11:997–1016
Dartnell LR, Page K, Jorge-Villar SE, Wright G, Munshi T, Scowen IJ, Ward JM, Edwards HGM (2012) Destruction of Raman biosignatures by ionising radiation and the implications for life detection on Mars. Anal Bioanal Chem 403:131–144
Davila AF, Hawes I, Ascaso C, Wierzchos J (2013) Salt deliquescence drives photosynthesis in the hyperarid Atacama Desert. Environ Microbiol Rep 5:583–587
de Vera J-P, Boettger U, de laTorre NR, Sánchez FJ, Grunow D, Schmitz N, Lange C, Hübers H-W, Billi D, Baqué M et al (2012) Supporting Mars exploration: BIOMEX in low earth orbit and further astrobiological studies on the Moon using Raman and PanCam technology. Planet Space Sci 74:103–110
Edwards HGM, Moody CD, Jorge-Villar SE, Wynn-Williams DD (2005) Raman spectroscopic detection of key biomarkers of cyanobacteria and lichen symbiosis in extreme Antarctic habitats: evaluation for Mars Lander missions. Icarus 174:560–571

Ethridge EC, Kaulker WF (2012) Microwave extraction of volatiles for Mars science and ISRU. In: Concepts and approaches for Mars exploration, LPI Contribution No. 1679. Houston, p 4328

Farley KA, Malespin C, Mahaffy P, Grotzinger JP, Vasconcelos PM, Milliken RE, Malin M, Edgett KS, Pavlov AA, Hurowitz JA et al (2014) In situ radiometric and exposure age dating of the Martian surface. Science 343:247166

França MB, Panek AD, Eleutherio ECA (2007) Oxidative stress and its effects during dehydration. Comp Biochem Physiol A Mol Integr Physiol 146:621–631

Feofilova EP (2003) Deceleration of vital activity as a universal biochemical mechanism ensuring adaptation of microorganisms to stress factors: a review. Appl Biochem Microbiol 39:1–18

Friedmann EI, Ocampo R (1976) Endolithicblue-green algae in the dry valleys: primary producers in the antarctic desert ecosystem. Science 193:1247–1249

Friedmann I, Lipkin Y, Ocampo-Paus R (1967) Desert algae of the Negev (Israel). Phycologia 6:185–200

Fries MD, Bhartia R, Beegle LW, Gursel Y, Mungas GS (2010) Microscopic sample interrogation through multi-wavelength spectroscopy coupled with variable magnification imaging. LPI Contrib 1538:5214

Georgiou CD, Deamer DW (2014) Lipids as universal biomarkers of extraterrestrial life. Astrobiology 14:541–549

Golubic S, Friedmann EI, Schneider J (1981) The lithobiontic ecological niche, with special reference to microorganisms. J Sediment Res 51:475–478

Grilli Caiola M, Billi D, Friedmann EI (1996) Effect of desiccation on envelopes of the cyanobacterium *Chroococcidiopsis* sp. (Chroococcales). Eur J Phycol 31:97–105

Groemer G, Sattler B, Weisleitner K, Hunger L, Kohstall C, Frisch A, Józefowicz M, Meszyński S, Storrie-Lombardi M, Bothe C et al (2014) Field trial of a dual-wavelength fluorescent emission (L.I.F.E.) instrument and the magma white rover during the MARS 2013 Mars analog mission. Astrobiology 14:391–405

Grotzinger JP, Sumner DY, Kah LC, Stack K, Gupta S, Edgar L, Rubin D, Lewis K, Schieber J, Mangold N et al (2014) A habitable fluvio-lacustrine environment at Yellowknife Bay, Gale Crater, Mars. Science 343:1242777

Hassler DM, Zeitlin C, Wimmer-Schweingruber RF, Ehresmann B, Rafkin S, Eigenbrode JL, Brinza DE, Weigle G, Böttcher S, Böhm E et al (2013) Mars' surface radiation environment measured with the Mars Science Laboratory's Curiosity rover. Science 343:1244797

Hertzberg S, Jensen A (1989) Studies of alginate immobilized marine microalgae. Bot Mar 32:267–274

Horneck G, Stöffler D, Ott S, Hornemann U, Cockell CS, Moeller R, Meyer C, de Vera J-P, Fritz J, Schade S et al (2008) Microbial rock inhabitants survive hypervelocity impacts on Mars-like host planets: first phase of lithopanspermia experimentally tested. Astrobiology 8:17–44

Horneck G, Klaus DM, Mancinelli RL (2010) Space microbiology. Microbiol Mol Biol Rev 74:121–156

Horowitz NH, Cameron RE, Hubbard JS (1972) Microbiology of the dry valleys of antarctica. Science 176:242–245

Hutchinson IB, Edwards HGM, Ingley R, Harris L, McHugh M, Malherbe C, Jehlicka J, Marshall C, Parnell J (2014) Preparations for the launch of the EXOMARS Raman laser spectrometer–a review of recent studies which highlight the astrobiological and geological capabilities of portable Raman instrumentation. LPI Contrib 1783:5093

Jehlička J, Edwards HGM, Oren A (2014) Raman spectroscopy of microbial pigments. Appl Environ Microbiol 80:3286–3295

Jorge-Villar S, Edwards HG (2013) Microorganism response to stressed terrestrial environments: a Raman spectroscopic perspective of extremophilic life strategies. Life 3:276–294

Mancinelli RL (2015) The affect of the space environment on the survival of *Halorubrum chaoviator* and *Synechococcus* (Nägeli): data from the space experiment OSMO on EXPOSE-R. Int J Astrobiol 14:123–128

Mancinelli RL, White MR, Rothschild LJ (1998) Biopan-survival I: exposure of the osmophiles *Synechococcus* sp. (Nägeli) and *Haloarcula* sp. to the space environment. Adv Space Res 22:327–334

Mattimore V, Battista JR (1996) Radio resistance of *Deinococcus radiodurans*: functions necessary to survive ionizing radiation are also necessary to survive prolonged desiccation. J Bacteriol 178:633–637

McKay CP (1997) The search for life on Mars. Orig Life Evol Biosph 27:263–289

McKay CP (2004) What is life–and how do we search for it in other worlds? PLoS Biol 2:e302

McKay CP (2010) An origin of life on Mars. Cold Spring Harb Perspect Biol 2:a003509

Meyer TR, McKay CP (1984) The atmosphere of Mars-resources for the exploration and settlement of Mars. In: Boston PJ (ed) The case for Mars. Univelt, San Diego, CA: Univelt, pp. 209–232

Musilova M, Wright G, Ward JM, Dartnell LR (2015) Isolation of radiation-resistant bacteria from Mars analog Antarctic Dry Valleys by preselection, and the correlation between radiation and desiccation resistance. Astrobiology 15:1076–1090

Nelson M, Allen J, Alling A, Dempster WF, Silverstone S (2003) Earth applications of closed ecological systems: relevance to the development of sustainability in our global biosphere. Adv Space Res 31:1649–1655

Nicholson WL (2009) Ancient micronauts: interplanetary transport of microbes by cosmic impacts. Trends Microbiol 17:243–250

Olsson-Francis K, de la Torre R, Cockell CS (2010) Isolation of novel extreme-tolerant cyanobacteria from a rock-dwelling microbial community by using exposure to low Earth orbit. Appl Environ Microbiol 76:2115–2121

Parnell J, Cullen D, Sims MR, Bowden S, Cockell CS, Court R, Ehrenfreund P, Gaubert F, Grant W, Parro V et al (2007) Searching for life on Mars: selection of molecular targets for ESA's aurora ExoMars mission. Astrobiology 7:578–604

Pavlopoulou A, Savva GD, Louka M, Bagos PG, Vorgias CE, Michalopoulos I, Georgakilas AG (2016) Unraveling the mechanisms of extreme radio resistance in prokaryotes: lessons from nature. Mutat Res 767:92–107

Pavlov AA, Vasilyev G, Ostryakov VM, Pavlov AK, Mahaffy P (2012) Degradation of the organic molecules in the shallow subsurface of Mars due to irradiation by cosmic rays. Geophys Res Lett 39:L13202

Polyakov YS, Musaev I, Polyakov SV (2010) Closed bioregenerative life support systems: applicability to hot deserts. Adv Space Res 46:775–786

Potts M (2001) Desiccation tolerance: a simple process? Trends Microbiol 9:553–559

Rabbow E, Rettberg P, Barczyk S, Bohmeier M, Parpart A, Panitz C, Horneck G, von Heise-Rotenburg R, Hoppenbrouwers T, Willnecker R et al (2012) EXPOSE-E: an ESA astrobiology mission 1.5 years in space. Astrobiology 12:374–386

Rebecchi L, Altiero T, Guidetti R (2007) Anhydrobiosis: the extreme limit of desiccation tolerance. Invertebr Sur J 4:65–81

Schopf JW (2002) The fossil record: tracing the roots of the cyanobacterial lineage. In: Whitton BA, Potts M (eds) The ecology of cyanobacteria, Springer Netherlands, pp 13–35

Schulze-Makuch D, Fairén AG, Davila AF (2008) The case for life on Mars. Int J Astrobiol 7:117–141

Schwartzman D, Caldeira K, Pavlov A (2008) Cyanobacterial emergence at 2.8 Gya and greenhouse feedbacks. Astrobiology 8:187–203

Shirkey B, McMaster NJ, Smith SC, Wright DJ, Rodriguez H, Jaruga P, Birincioglu M, Helm RF, Potts M (2003) Genomic DNA of *Nostoc commune* (Cyanobacteria) becomes covalently modified during long-term (decades) desiccation but is protected from oxidative damage and degradation. Nucl Acids Res 31:2995–3005

Singh H, Anurag K, Apte SK (2013) High radiation and desiccation tolerance of nitrogen-fixing cultures of the cyanobacterium *Anabaena* sp. strain PCC 7120 emanates from genome/proteome repair capabilities. Photosynth Res 118:71–81

Smith HD, Duncan AG, Neary PL, Lloyd CR, Anderson AJ, Sims RC, McKay CP (2012) In situ microbial detection in Mojave Desert soil using native fluorescence. Astrobiology 12:247–257

Smith HD, Baqué M, Duncan AG, Lloyd CR, McKay CP, Billi D (2014) Comparative analysis of cyanobacteria inhabiting rocks with different light transmittance in the Mojave Desert: a Mars terrestrial analogue. Int J Astrobiol 13:271–277

Stivaletta N, Barbieri R, Billi D (2012) Microbial colonization of the salt deposits in the driest place of the Atacama Desert (Chile). Orig Life Evol Biospheres 42:187–200

Stöffler D, Horneck G, Ott S, Hornemann U, Cockell CS, Moeller R, Meyer C, de Vera J-P, Fritz J, Artemieva NA (2007) Experimental evidence for the potential impact ejection of viable microorganisms from Mars and Mars-like planets. Icarus 186:585–588

Stromberg JM, Applin DM, Cloutis EA, Rice M, Berard G, Mann P (2014) The persistence of a chlorophyll spectral biosignature from martian evaporite and spring analogues under Mars-like conditions. Int J Astrobiol 13:203–223

Vago J, Gardini B, Kminek G, Baglioni P, Gianfiglio G, Santovincenzo A, Bayón S, van Winnendael M (2006) ExoMars–searching for life on the Red Planet. ESA Bull 126:16–23

Vandenabeele P, Jehlička J, Vítek P, Edwards HGM (2012) On the definition of Raman spectroscopic detection limits for the analysis of biomarkers in solid matrices. Planet Space Sci 62:48–54

Verseux C, Baqué M, Lehto K, deVera J-P, Rothschild LJ, Billi D (2016a). Sustainable life support on Mars – the potential roles of cyanobacteria. Int J Astrobiol 15:65–92

Verseux C, Paulino-Lima I, Baqué M, Billi D, Rothschild L (2016b) Synthetic biology for space exploration: promises and societal implications. In: Hagen K, Engelhard M, Toepfer G (eds) Ambivalences of creating life. Societal and philosophical dimensions of synthetic biology. Springer-Verlag, Springer International Publishing pp 73–100

Verseux C, Baqué M, Cifariello R, Fagliarone C, Raguse M, Moeller R, Billi D (2017) Evaluation of the resistance of *Chroococcidiopsis* spp. to sparsely and densely ionizing irradiation. Astrobiol 17(2). DOI: 10.1089/ast.2015.1450

Vítek P, Jehlička J, Edwards HGM, Osterrothová K (2009) Identification of β-carotene in an evaporitic matrix—evaluation of Raman spectroscopic analysis for astrobiological research on Mars. Anal Bioanal Chem 393:1967–1975

Vítek P, Jehlička J, Edwards HGM, Hutchinson I, Ascaso C, Wierzchos J (2012) The miniaturized Raman system and detection of traces of life in halite from the Atacama Desert: some considerations for the search for life signatures on Mars. Astrobiology 12:1095–1099

Waditee R, Hibino T, Nakamura T, Incharoensakdi A, Takabe T (2002) Overexpression of a Na+/H+ antiporter confers salt tolerance on a freshwater cyanobacterium, making it capable of growth in sea water. Proc Natl Acad Sci 99:4109–4114

Warren-Rhodes KA, Rhodes KL, Pointing SB, Ewing SA, Lacap DC, Gómez-Silva B, Amundson R, Friedmann EI, McKay CP (2006) Hypolithic cyanobacteria, dry limit of photosynthesis, and microbial ecology in the hyperarid Atacama Desert. Microb Ecol 52:389–398

Westall F, Loizeau D, Foucher F, Bost N, Betrand M, Vago J, Kminek G (2013) Habitability on Mars from a microbial point of view. Astrobiology 13:887–897

Wierzchos J, Ascaso C, McKay CP (2006) Endolithic cyanobacteria in halite rocks from the hyperarid core of the Atacama Desert. Astrobiology 6:415–422

Wright DJ, Smith SC, Joardar V, Scherer S, Jervis J, Warren A, Helm RF, Potts M (2005) UV irradiation and desiccation modulate the three-dimensional extracellular matrix of *Nostoc commune* (Cyanobacteria). J Biol Chem 280:40271–40281

Psychrophilic Microorganisms as Important Source for Biotechnological Processes

7

Sergiu Fendrihan and Teodor Gh Negoiță†

Contents

7.1 Introduction

The major parts of Earth's environments are cold and have temperatures below 5 °C (Gounot 1999; Russell and Cowan 2005). About 70 % of the freshwater is ice, and about 14 % from the Earth's biosphere is represented by terrestrial and aquatic polar areas (Priscu and Christner 2004). The depth of the oceans, the poles, and high mountains are the most important cold regions on Earth (Russell and Cowan 2005).

†Author was deceased at the time of publication.

S. Fendrihan (✉)
University "Vasile Goldiș", Arad, Romania

Romanian Bioresource Center, Bucharest, Romania

Research-Development Institute for Plant Protection, Bucharest, Romania
e-mail: sergiu.fendrihan@rbcar.ro; ecologos23@yahoo.com

T.G. Negoiță
Romanian Institute for Antarctic Polar Research, Bucharest, Romania

H. Stan-Lotter, S. Fendrihan (eds.), *Adaption of Microbial Life to Environmental Extremes*, DOI 10.1007/978-3-319-48327-6_7

Global ice, for example, covers 6.5 million km^2 which increases to 14.4 million km^2 in winter time (Perovich et al. 2002). Here we can meet representatives from all domains of the living world. Two categories of microorganisms were discovered in such cold environments: first, the psychrophiles with an optimum growth temperature of about 15 °C or even less, which cannot grow above 20 °C (Moyer and Morita 2007), and second, the psychrotolerants with an optimum growth temperature of 20–30 °C, which are able to grow and exhibit activity at temperatures close to the freezing point of water (Madigan and Jung 2003). The lowest temperature for life's activities is −20 °C under certain defined conditions (Rivkina et al. 2000; Gilichinsky 2002; D'Amico et al. 2006); others consider the temperature limits for reproduction as −12 °C and for metabolism as −20 °C (Bakermans 2008). *Colwellia psychrerythraea* strain 34H is motile at −10 °C, as observed by transmitted light microscopy (Junge et al. 2003). The organisms living in cold areas are practically psychrophiles, which have optimum growth temperatures below 15 °C, and psychrotolerants, which can survive at temperatures below 0 °C but grow optimally at 20–25 °C or even higher temperatures (De Maayer et al. 2014). Psychrophilic microorganisms are dominant in permanently cold environments such as Antarctic waters and have important roles in the biogeochemical cycles in the polar zones (Helmke and Weyland 2004). Not only are prokaryotes adapted to the cold but so are many eukaryotes such as algae (Takeuchi and Kohshima 2004) and macroorganisms from crustaceans to fishes. The present work will focus on prokaryotes and some microscopic eukaryotes of biotechnological importance.

7.2 Diversity of Cold-Adapted Microorganisms

The psychrophilic and psychrotolerant microorganisms belong to all three of life's principal domains, Archaea, Bacteria, and Eukarya. It is interesting to note that viruses are omnipresent and even so in those inhospitable places.

Viruses Viruses from the families Podoviridae, Siphoviridae, and Myoviridae were identified in cold environments (Wells 2008). Bacteriophages were identified in inner polar waters and in ice (Säwström et al. 2007) infecting psychrophilic microorganisms; for example, phage 9 A of *Colwellia psychrerythraea* strain 34 is capable of forming plaques at low temperatures but not at 13 °C (Wells 2008).

Archaea Archaea found in cold environments are methanogens, for example, from the genera *Methanogenium*, *Methanococcoides*, and *Methanosarcina*, but halophilic (*Halorubrum*) and other strains can also occur (Cavicchioli 2006). The strain *Methanolobus psychrophilus* R15 was isolated from a Tibetan plateau wetland and grows at temperature between 0 and 25 °C. The proteomic study of the cultures grown at 30°, 18°, and 4 °C showed differentially expressed proteins and some chaperones which increased at 4 °C (Chen et al. 2015). *Methanococcoides burtonii*, a methanogen isolated from Ace Lake in Antarctica, was genetically analyzed in order to understand genes related to cold adaptation (Dong and Chen 2012).

Archaea exist in the active layer and in permafrost in acidic Arctic wetland, the second being an archaean community, too. At the same time, they are accompanied by a bacterial community. It looks like wet cold-temperature soil, as habitat, even in different geographical locations, could host the same archaeal and bacterial strains (Wilhelm et al. 2011).

Bacteria The majority of isolates from polar areas belong to the groups of *Beta-*, *Gamma-*, and *Delta proteobacteria*, *Actinobacteria*, *Acidobacteria*, the *Cytophaga–Flexibacter–Bacteroides* group, and green nonsulfur bacteria. Many strains of Bacteria as well as Archaea and Eukarya were revealed by 16S rRNA and 18S rRNA gene clone libraries (Tian et al. 2009). Soils of the McMurdo Dry Valleys host species of *Pseudonocardia*, *Nocardioides*, *Geodermatophilus*, *Modestobacter*, *Sporichthya*, and *Streptomyces* (Babalola et al. 2008). Cyanobacteria as photoautotrophs were retrieved from ice, soils, rocks, lakes, ponds, marine ecosystems, and alpine areas (Zakhia et al. 2008). *Chamaesiphon* sp., *Chroococcidiopsis* (from sandstone), and *Synechococcus* sp. (from lakes, marine water, and others) are examples of cyanobacterial genera with cold-adapted strains. Some bacteria, belonging to 22 phylotypes (*Flavobacterium*, *Alcanivorax*, and many others) with a concentration of 10^4–10^6 cells/gram of cyanobacteria biomass (Prasad et al. 2013), were present in a mass of Arctic cyanobacteria.

Algae Species of *Chlamydomonas* were retrieved from water derived from melting glacier ice and from some layer species of *Rhodomonas* and *Chromulina*. Species of *Tribonemataceae* were found in Antarctic terrestrial environments (Rybalka et al. 2008). Several microalgae can be found in all known cold environments as in snow (*Chlamydomonas* and *Chloromonas*), seawater (diatoms), in sea ice (diatoms and dinoflagellates), on rocks as endoliths (*Hemichloris antarctica*), in ice-covered lakes (*Chloromonas* sp., *Chlamydomonas intermedia*, and *Chlamydomonas raudensis*), and at high altitudes (reviewed by Mock and Thomas 2008). Samples from the Tyndall Glacier in Patagonia, Chile, contained algal species of the genera *Mesotaenium*, *Cylindrocystis*, *Ancylonema*, *Closterium*, and *Chloromonas* and some cyanobacteria (Takeuchi and Kohshima 2004).

Yeasts Yeast strains such as *Sporobolomyces*, *Cryptococcus*, and *Rhodotorula* sp. were isolated from Lake Vanda (Goto et al. 1969) and from other Antarctic and alpine environments, including psychrophilic yeasts such as the novel species *Mrakia robertii*, *M. blollopis*, and *M. niccombsi* (Thomas-Hall et al. 2010). Several yeasts, which are producers of lipases and proteases, were isolated from cold marine water and freshwater (Rashidah et al. 2007), such as *Cryptococcus antarcticus* and *Cryptococcus albidosimilis*, *Basidioblastomycetes* (Vishniac and Kurtzman 1992), *Cryptococcus nyarrowii* (Thomas-Hall and Watson 2002), *Cryptococcus watticus* (Guffogg et al. 2004), and *Leucosporidium antarcticum* – the latter from Antarctic waters (Turkiewicz et al. 2005) – and Mrakia strains (Thomas-Hall et al. 2010). From puddles of the Midre Lovenbreen Glacier, about 132 of yeast strains were isolated and were identified as *Cryptococcus gastricus*, *C.terricolus*, *Rhodotorula* *glacialis*, *R. muscorum*, *Mrakia psychrophila*, and *M. gelida*, which can produce different enzymes (Pathan et al. 2010). It was observed that the amount of unsaturated fatty acids increased at a growth temperature of 8 °C, compared with 22 °C. Yeasts were isolated from many glacial habitats, not only from the Antarctic, and were found to be producers of cold-active enzymes (Buzzini et al. 2012).

Fungi Fungi were isolated from numerous cold environments. For example, *Penicillium*, *Aspergillus*, *Paecilomyces*, *Cladosporium*, *Mortierella*, *Candida*, and *Rhodotorula* were isolated from soils of Terra Nova Bay and Edmonson Point, Antarctica (Gesheva 2009). *Rhodotorula svalbardensis*, a novel strain of yeast, was isolated from cryoconite holes from Svalbard (Singh et al. 2014). Some authors described isolates from soils, such as *Chrysosporium* sp., *Phoma exigua*, *Heterocephalum aurantiacum*, *Aureobasidium pullulans*, *Fusarium oxysporum*, *Trichoderma viride*, and *Penicillium antarcticum* (Negoiță et al. 2001a). From the soils of Schirmacher Oasis, Antarctica, fungi such as *Acremonium*, *Aspergillus*, and *Penicillium* were isolated, the majority surviving as spores in those harsh environments, and some species possess unique features of their mycelia (Singh et al. 2006). Frisvad (2008b) reviewed the fungi from cold ecosystems and indicated their isolation from soils and permafrost, caves, rocks, mosses and lichens, glacier ice, freshwater, as well as from frozen foods. The fungi belong to the Ascomycetes (*Acremonium antarcticum*, *A. psychrophilum*, and *Penicillium antarcticum*), Zygomycetes (*Mortierella alpina* and *Absidia psychrophila*), and basidiomycetous yeasts, which are very rare in cold areas. Endolithic fungi resistant to low temperature and low water activity were isolated by Onofri et al. (2007).

7.3 Ecology and Biology

Some of the microorganisms are polyextremophiles, for example, halo-psychrophiles, or piezo-psychrophiles, which tolerate high pressure (Nogi 2008) and cannot grow at atmospheric pressure and at temperatures above 20 °C, such as strains of *Shewanella*, *Colwellia*, *Moritella*, and *Psychromonas*. In these categories all the physiological and metabolic types can be found – anaerobes and aerobes, methanogens, methanotrophs, chemolithotrophs, sulfate reducers, and organotrophs. Anaerobic cold-adapted *Clostridium* sp. (e.g., *C. frigoris*, *C. bowmannii*, and *C. psychrophilum*) were isolated from Antarctic microbial mats (Spring et al. 2003) or some psychrotolerants, such as *C. frigidicarnis* and *C. algidixylanolyticum*, from frozen products (Finster 2008). Sulfate-reducing psychrophiles *Desulfotalea*, *Desulfofaba*, *and Desulfofrigus* (Knoblauch et al. 1999); sulfur-oxidizing bacteria (SOB), occurring in such organic carbon-depleted environments as subglacial waters (Sattley and Madigan 2006); as well as denitrifying microorganisms in sea ice (Rysgaard et al. 2008) were found. Ammonia oxidizers were identified by genetic methods in all of the samples taken from lakes Fryxell, Bonney, Hoare, Joyce, and Vanda in Antarctica, belonging to the *Proteobacteria*

(Voytek et al. 1999). Acetogenic bacterial sequences originating from *Acetobacterium tundrae* and others related to *Acetobacterium bakii* (Sattley and Madigan 2007) were isolated from sediments of Lake Fryxell. From the same lake, different phototrophic purple bacteria were identified with molecular methods (Karr et al. 2003) as well as methanogenic and other Archaea (Karr et al. 2006). Biological methane oxidation and sulfate reduction by Archaea occur in alpine lakes (such as Lake Lugano deeps) in anoxic zones (Blees et al. 2010). Methanogens were detected in soils, water sediments, sea, and lake waters from cold environments (Cavicchioli 2006). Methanotrophy was detected indirectly in Lake Untersee (Antarctica) by identification of hopanoids and two steroids (4-methyl steroid and 4,4-dimethyl steroid), one hopanoid (diplopterol) having a specific low isotopic ^{13}C content and originating from the aerobic methylotroph *Methylococcus* sp. (Niemann et al. 2010). Some *Shewanella* and *Pseudomonas* strains from Antarctic lakes are able to mediate redox reactions of manganese under stimulation by Co and Ni (Krishnan et al. 2009). In any case, the microbiomes from cold environments differ from habitat to habitat. For example, variation due to local condition, nutrients, cations, dissolved organic carbon in the streams, and lithology caused differences in bacterial community structure from stream habitat, sediments, and epilithon (on rocks) biofilms in the Noatak National Preserve, Alaska (Larouche et al. 2012). This is also true for Antarctic soil communities which have the capacity to contribute to biogeochemical cycles of important elements. They can be found in different edaphic niches being crypto-endolithic-hypoliths and endoliths, or living in microbial mats and permafrost in different communities which are able to interfere in carbon cycling, polymer degradation, and nitrogen fixation (Cowan et al. 2014).

7.4 Cold Environments

The diversity of cold environment is remarkable. In the Antarctic continent alone, there are lots of habitats where psychrophiles and psychrotolerants can live, for example, hyperarid soils, rocky soils, cryptic hypolithic habitats, ornithogenic soils, thermally heated soils, mixed lichen communities, wet soils, permafrost, and others (Cowan et al. 2014).

Cold Deserts There are cold deserts in Antarctica where the precipitation is very low, the temperatures range between −55 and 10 °C, UV radiation is high, and water activity is low; these are some of the most extreme environments on Earth. Many microorganisms can be found in endolithic communities composed of cyanobacteria such as *Acaryochloris marina* and *Gloeocapsa* species (de los Rios et al. 2007). Endolithic bacteria, fungi, archaea, green algae, yeasts, and lichens were found in McMurdo Dry Valley (Gounot 1999), analyzed by staining with the BacLight LIVE/DEAD kit and observed with confocal laser scanning microscopy to demonstrate their survival (Wierzchos et al. 2004). In the same valley, many bacteria were isolated, with a predominance of *Acidobacteria* (Bakermans et al. 2014).

Soils Covered with Snow From Arctic wetland soil, methanotrophic bacteria were retrieved such as *Methylocystis rosea* (Wartiainen et al. 2006). In soils of Lapland, microbial communities were discovered, similarly as in soils from alpine zones, where the temperatures can reach −25 °C in winter time. In addition, soils from Spitsbergen contained many fungi such as *Mucor*, *Mortierella*, *Alternaria*, *Fusarium*, and *Zygorhynchus* (Negoiţă et al. 2001b), genera which are very probably psychrotolerants.

Permafrost Permafrost soils in the geological sense stay below 0 °C for two consecutive years or more and are specific for arctic areas covering about 26 % of the surface of the Northern Hemisphere. The average temperature is −16 °C; in Siberia −11 °C and in Antarctica −18 to −27 °C were measured (Vorobyova et al. 1997). From those soils over 100 bacterial strains were isolated, also some methanogenic archaea from the families *Methanomicrobiaceae*, *Methanosarcinaceae*, and *Methanosetaceae* (Ganzert et al. 2007), methane-oxidizing bacteria (Liebner and Wagner 2007), sulfate-reducing bacteria, aerobic and anaerobic heterotrophs (Gilichinsky 2002), denitrifiers, and iron and sulfate reducers (Rivkina et al. 1998). The majority of strains included species of *Micrococcus, Bacillus, Paenibacillus, Rhodococcus*, *Arthrobacter*, *Haloarcula*, and *Halobaculum* (Steven et al. 2007), which were isolated in quantities of 10^7–10^9 cells per gram of dry soil. From layers of permafrost which were demonstrated to be about 3–5 million years old, viable cells of bacteria were isolated (Rodrigues-Diaz et al. 2008), which have to face low temperatures and natural irradiation by radionuclides (Gilichinsky et al. 2008). From a layer of an arctic permafrost ice wedge from Canada (temperature −17.5 °C, pH 6.5, salt concentration 14.6 g/l, age about 25,000 years), bacteria were isolated (Katayama et al. 2007) belonging to the classes *Actinobacteria* and *Gammaproteobacteria*. Molecular methods revealed in permafrost samples microorganisms able of active metabolism decomposing organics by methanogenesis and using acetate, methanol, and methylamine after thawing (Coolen and Orsi 2015). The authors found microorganism from *Bacteroidetes*, *Firmicutes*, ascomycete fungi, and methanogenic archaea. In thawing permafrost (wet fen), high activity of microbial methanogenesis and other activities were detected. Gittel et al. (2014) found that *Actinobacteria* and others are responsible for production of oxidative enzymes and degradation of phenolics.

Snow, Ice, and Glaciers The ice glaciers in Antarctica contain approximately 90 % of the ice of our planet according to the National Snow and Ice Data Centre of the USA (http://nsidc.org/, cited by Christner et al. 2008). Some aspects of the soils covered partially with ice on the shore of the Antarctic sea are shown in Figs. 7.1 and 7.2. Sea and lake ice glaciers are hosting considerable quantities of biological material, consisting of microorganisms, bacteria, spores, and pollen grains, the majority being transported there by air. The microbiota can survive in crevices and capillary tunnels containing concentrated ionic solutions with a lower freezing point (Price 2006). In ice from the Collins Glacier, psychrophilic microorganisms were found, of which the *Betaproteobacteria* were the most frequent, about 35 %;

Fig. 7.1 Larsemann Hills Coast, Law-Racovita Base area, East Antarctica, 69°23′0″S; 76°23′0″E (Photo T.G. Negoiță)

Gammaproteobacteria (18.5 %), *Alphaproteobacteria* (16.6 %), and *Cytophaga–Flavobacterium–Bacteroides* (13 %) were also present (Garcia-Echauri et al. 2011). Some psychrophilic and psychrotolerant strains of yeast were obtained from Patagonian glaciers, such as *Agaricomycotina (Tremellales)* and from the order *Cystofilobasidialles* (de Garcia et al. 2012). Other authors (Maccario et al. 2014) found by metagenomic approaches in snow samples bacteria belonging to *Bacteroidetes*, *Proteobacteria*, and fungi. In Icelandic glaciers and ice caps and especially in snow communities of microorganisms containing algae (*Chloromonas polyptera, Raphidonema sempervirens,* some *Chlamydomonadace*ae), bacteria (most of them belonging to *Proteobacteria* and *Bacteroidetes*) and some archaea (mainly belonging to the *Nitrososphaerales*) were found (Lutz et al. 2015). A psychrophilic bacterium, *Cryobacterium levicorallium*, was isolated from a glacier in Northwest China (Liu et al. 2013).

The number of viable microorganisms decreases with the depth of the ice layers; there is a supraglacial community (bacteria, viruses, diatoms, tardigrades, rotifers), a subglacial community (aerobic and anaerobic) and an endoglacial community (Hodson et al. 2008). Hollibaugh et al. (2007) studied the sea ice microbial community (SIMCO) formed of bacteria, algae, and fungi of various metabolic types: sulfate reducers, chemolithotrophs, methanogens, anaerobic nitrate reducers (Skidmore et al. 2000), and viruses (Deming 2007). In ice there is an incredible

Fig. 7.2 Antarctic ice cap in the Schirmacher Oasis, 70°46′ S; 11°50′ E, 100 km inside the continent (Photo T.G. Negoiţă)

diversity of *Proteobacteria*, of the phylum *Cytophaga–Flavobacterium–Bacteroides*, high-GC Gram positives and low-GC Gram positives (Miteva 2008), and a large metabolic and physiological diversity. Some of them were entrapped for very long periods of time, such as the strain *Herminiimonas glaciei*, a Gram-negative ultramicrobacterium, which was isolated from a 3042 m deep drilling core from a Greenland glacier of about 120,000-year-old ice (Miteva and Brenchley 2005; Loveland-Curtze et al. 2009), or *Chryseobacterium greenlandense* (Loveland-Curtze et al. 2010) and sequences from *Pseudomonas* and *Acinetobacter*, which stem from 750,000-year-old ice from the Qinghan-Tibetan plateau in Western China (Christner et al. 2003a). Many prokaryotes isolated from snow melt water belong to the *Betaproteobacteria* (21.3 %), *Sphingobacteria* (16.4 %), *Flavobacteria* (9.0 %), *Acidobacteria* (7.7 %), *Alphaproteobacteria* (6.5 %), and other groups (Larose et al. 2010). The cryoconite holes form another microhabitat containing various forms of life, such as diatoms, algae, prokaryotes, fungi, rotifers, and tardigrades (Wharton et al. 1985; Christner et al. 2003b). The snow packs from the Arctic harbor a very diverse microbiota fungi; bacteria as *Proteobacteria* and *Bacteroidetes* were the most frequent in the samples but *Firmicutes, Actinobacteria, Cyanobacteria, Acidobacteria*, and Archaea, too (Maccario et al. 2014).

A method for quantifying the abundance of microbial cells in surface ice from the Greenland ice sheet (GrIS) was used by Stibal et al. (2015). This included several physical and molecular methods like fluorescence microscopy, flow cytometry,

Q-PCR, and, for calculations, multivariate analysis determinations, which yielded 2 × 10^3 to ~2 × 10^6 cells. Some particles of an organic dust were detected in the samples.

Cold Caves Caves represent a constant temperature environment with low organic content, sometimes only 1 mg of organic matter per liter. Some strains are chemolithotrophs such as *Gallionella*. In many cases there are more psychrotolerants than psychrotrophs. Some stenothermic bacterial strains were isolated which can grow at 10–20 °C and only few which grow at 2 °C or 28 °C (Gounot 1999). The strain *Arthrobacter psychrophenolicum* was isolated from an Austrian ice cave (Margesin et al. 2004). Generally, caves are oligotrophic habitats; the microorganism must be chemolithotrophs adapted to dark and low nutrients. The ice caves from Mt. Erebus (Antarctica) offer conditions with atmospheric carbon and volcanic emissions. The microbial communities there contained bacteria and fungi, but no archaea were found (Tebo et al. 2015).

Cold Lakes The cold lakes in the polar and alpine zones can be covered with an ice layer (Antarctic lakes), which practically isolates the lake from the rest of the environment, and their content of organic carbon and oxygen is rather low. Christner et al. (2008) pointed out in their comprehensive review that there are 141 subglacial Antarctic lakes having a total volume of about 10,000 km^3. One of the largest lakes is Lake Vostok of 14,000 km^2, covered by a 4000 m thick ice sheet and being 400–800 m deep. The bottom is covered with a thick sediment layer. The temperature of the ice layer is about −55 °C, but the lake has a constant temperature of −2.65 °C (Di Prisco 2007). The ice layer is about 15 million years old. Alpine lakes are only temporarily covered by ice, and the microbiota there are subject to seasonal fluctuations (Pernthaler et al. 1998). The cold Antarctic lake environment is chemically driven, with reactions such as sulfide and iron oxidation (Christner et al. 2008), and contains methanogens such as *Methanosarcina*, *Methanoculleus*, and anoxic methanotrophs (Karr et al. 2006). The saline lakes host euryhalophiles related to *Halomonas and Marinobacter* (Naganuma et al. 2005). From accretion ice from Vostok Lake by metagenomic techniques, many sequences were isolated, belonging to Bacteria (94 %), Eukarya (6 %), and two of them to Archaea, as well as some fungi (Rogers et al. 2013), showing the diversity of this oligotrophic habitat.

From Mc Murdo Dry Valley lakes, bacteria from the groups *Alphaproteobacteria* (Priscu et al. 1998), *Gammaproteobacteria* (Voytek et al. 1999), and purple nonsulfur bacteria (0.29 %) were isolated and identified (Karr et al. 2003). Other studies of the subglacial Lake Whillans (SLW) in Antarctica examined the existing genes involved in sulfur oxidation and reduction, which revealed a community using the energy issued from these processes containing *Thiobacillus* sp., *Desulfobacteraceae*, and *Desulfotomaculum* (Purcell et al. 2014).

Cold Marine Waters From marine waters of Ushuaia, a subantarctic town in Argentina, many sequences were identified belonging to the *Alpha*- and *Gammaproteobacteria*, *Cytophaga–Flavobacterium–Bacteroidetes* group, and the genera *Marinomonas, Colwellia, Cytophaga, Glaciecola, Cellulophaga, Roseobacter, Staleya, Sulfitobacter, Psychrobacter, Polaribacter, Ulvibacter, Tenacibacter, Arcobacter, and Formosa* (Prabagaran et al. 2007). In the depth of the ocean, the temperature is about 3 °C, and a considerable pressure exists (the pressure increases by 1 atm per each 10 m of depth). Here a very diverse bacterial community can be retrieved, for example, from the sediments of the Japanese Trench (Hamamoto 1993). In the sediment microbiota belonging to all domains of life can be obtained, as, for example, previous data by Foreman et al. (2004) showed, and later, eukaryotes like *Fungi, Cercozoa, Alveolata, and Metazoa* were identified by molecular methods (Tian et al. 2009). The archaeal sequences can reach 17 % of total microbiota in marine coastal waters (Murray et al. 1998). Sulfate-reducing bacteria form a large community in sediments of the Arctic Ocean, being active at about 2.6 °C with a sulfate reduction rate similar to that under mesophilic conditions (Knoblauch et al. 1999). Psychrophilic and psychrotolerant bacteria were isolated from Argentina's Beagle Channel (Cristóbal et al. 2015) and identified by molecular methods (*Shewanella, Colwellia, Serratia, Pseudoalteromonas, Aeromonas, Glaciecola, Pseudomonas*). Some strains were producers of cellulases, β-galactosidases, xylanases, and proteases. Interestingly, the same authors identified in the intestine of bottom invertebrates (benthos) more than 200 bacterial strains, most of them psychrotolerants, and some psychrophilic able to produce protease at 4–15 °C (Cristóbal et al. 2011).

In the Bransfield Strait (northwest of the Antarctic Peninsula), samples from different layers were taken and analyzed for abiotic factors and microbial communities. The main components of the communities below depths of 100 m were *Thaumarchaeota*, *Euryarchaeota*, and *Proteobacteria* (*Gamma-*, *Delta-*, *Beta-*, and *Alphaproteobacteria*) (Signori et al. 2014). In the layers from above 100 m, *Bacteroidetes* and *Proteobacteria* were predominant. In East Antarctica, 33 aerobic bacteria were isolated belonging to four phyla, some psychrotolerant and some psychrophilic, producing esterases, β-glucosidase, and proteases. Some were producing lipases, amylase, and chitinase. It was suggested that Antarctic habitats must be further prospected for useful bacteria (Yu et al. 2011).

Strains of alginate lyase producers associated with the algae *Laminaria*, such as *Psychrobacter*, *Psychromonas*, and *Polaribacter* were isolated from Arctic marine waters (Dong et al. 2012). In some sediments the sulfate reduction activity was demonstrated, and the dominant strain was *Desulfofrigus* sp. (Kraft et al. 2013).

Anthropic Cold Environments Artificial cooling and freg systems can be visualized as man-made environments. *Pseudomonas fluorescens* is one of the lipolytic food-spoiling bacteria which is active in the cold, and its hydrolytic activity at low temperatures was studied as a function of water activity (Andersson et al. 1979). The lower temperatures and lower water activity did not affect the enzymatic activity since the substrates were hydrophobic. In cooling devices the bacterium *Pseudomonas fragi* is frequently found, which is supported by temperatures between 2 and 35 °C; it possesses some cold-shock proteins (Csps) and degrades frozen foods. Another bacterium from water-cooling systems is *Chryseobacterium aqui-*

frigidense (Park et al. 2008). The psychrophilic strain *Lactobacillus algidus* was isolated from refrigerated, packed beef meat (Kato et al. 2000).

Air From the Antarctic continent air, several psychrotolerant microorganisms were isolated, such as *Sphingomonas aurantiaca, Sphingomonas aerolata*, and *Sphingomonas faeni* sp. nov. (Busse et al. 2003).

7.5 Adaptation to Cold Environments

Growth and Activity The temperature has a direct influence on microbial growth and the relationship between growth and temperature generally conforms to the Arrhenius law (Gounot 1999). Christner (2002) reported the incorporation of DNA and protein precursors by *Arthrobacter* and *Psychrobacter* at −15 °C. *Polaromonas hydrogenivorans* has a lower temperature limit of 0 °C for growth (Sizova and Panikov 2007), and psychrophilic methanotrophs can grow at about 2 °C (Liebner and Wagner 2007). The psychrophilic strain *Psychromonas ingrahamii* showed growth at −12 °C with a slow rate of 10 days of generation time (Breezee et al. 2004). The activity of microorganisms was proven by measurement of ATP as a result of biomass activity in soils and permafrost (Cowan and Casanueva 2007); truly psychrophilic microorganisms showed an increase of the ATP content at lower temperatures, which is the opposite reaction of mesophiles (Napolitano and Shain 2004). Other information can be obtained by determination of the Indicator of Enzymatic Soil Activity Potential, the Indicator of Vital Activity Potential, and Biologic Synthetic Indicator (Negoiţă et al. 2001b). These indicators were introduced by Stefanic (1994) in order to obtain comprehensive information about the biological activity of soils and to compare them for agricultural uses. Energy metabolism was studied (Amato and Christner 2009) by growing the bacterium *Psychrobacter cryohalolentis* at different temperatures (22 °C, 15 °C, 5 °C, 0 °C, −5 °C, −15 °C, −20 °C, and −80 °C) without shaking, and then the content of ATP/ADP was measured by luminescence. The stability of some enzymes in low-temperature conditions was demonstrated, for example, cytochrome c552 was characterized by UV–VIS absorption spectroscopy and determination of melting temperature and van't Hoff enthalpy (Oswald et al. 2014).

Membrane Lipids There are differences regarding the composition of membrane lipids, and there are clear contributions to cold adaptation, depending also on bacterial taxonomy. The cytoplasmic membrane contains lipids with fatty acids of lengths ranging mainly between C14 and C18. Gram-negative bacteria possess in addition an outer membrane containing lipopolysaccharides. Archaea contain ether-linked glycerol alkyl lipids instead of fatty acids, and eukaryotes contain sterols (Russell 2008). Membrane fluidity depends on the degree of saturation of the polar lipids; the membranes from psychrophiles contain a higher amount of unsaturated and/or polyunsaturated and branched fatty acids, with methyl groups and a larger percentage of double bonds of the cis type (Chintalapati et al. 2004). The changes in amount and type of methyl-branched fatty acids of Gram-positive bacteria are a possibility for increasing membrane fluidity at low temperatures. The amount of unsaturated fatty acids contributes to the flexibility of the membrane structure in cold-adapted microorganisms, including eukaryotic photobionts such as diatoms and algae (Morgan-Kiss et al. 2006). The presence of polyunsaturated fatty acids (PUFAs) does not completely explain the adaptation to cold environments, because there are many marine strains without them (Russell and Nichols 1999). In yeast the main adaptation mechanism to low temperatures was the unsaturation of fatty acids in facultative psychrophiles and the increased amount of alpha linoleic acid in obligate psychrophiles (Rossi et al. 2009). Archaeal adaptation to the cold showed a similar increase in desaturation of their isoprenoids containing lipids; *Methanococcoides burtonii*, for example, generates unsaturated lipids during growth at low temperatures by selective saturation and not by using a desaturase such as bacteria (Cavicchioli 2006). Lipid A, the membrane anchor of lipopolysaccharide, presented also tiny differences such as single-methylene variation in the acyl moiety, showing psychrotolerants and psychrophiles are capable of metabolic changes to the fluidity of lipid A (Sweet et al. 2015).

The Proteome Cold-adapted bacterial proteins have a reduced amount of arginine, glutamic acids, and proline (salt bridge-forming residues) and reduced amounts of hydrophobic clusters (Grzymski et al. 2006). A comparison of the contents of amino acids of psychrophilic enzymes was made by Gianese et al. (2001); they found that generally Arg and Glu residues in the exposed sites of alpha helices were replaced by Lys and Ala in psychrophiles. Studying the crystal structure of the β-lactamase from several psychrophilic strains (*P. fluorescens* and others), some authors found that the enzymes from psychrophiles have a lower content of arginine in comparison with lysine and a lower proline content than mesophilic enzymes (Michaux et al. 2008). The peptidyl prolyl cis-trans isomerases (PPIases) play a role in cold adaption. An example is a cold-shock protein from *Shewanella*, which possesses foldase and chaperone activities, helping in cold adaptation (Budiman et al. 2011).

The lysine residues are of great importance for the cold adaptation mechanism in enzymes, for example, in α-amylase from *Pseudoalteromonas haloplanktis* (Siddiqui et al. 2006). A similar replacement is observed in Archaea having a higher content of non-charged amino acids (as glutamine and threonine) and lower contents of hydrophobic amino acids such as leucine (Cavicchioli 2006). At the same time, the number of hydrogen bonds (Michaux et al. 2008) and the number of disulfide bridges are reduced (Sælensminde et al. 2009). The cellulase Cel5G from *P. haloplanktis* possesses a catalytic domain and a carbohydrate-binding domain which are joined by a long-linker region containing three loops closed by disulfide bridges. By experimental shortening of this linker region, the enzyme became less flexible approaching the activity of its mesophilic counterpart, which suggested that a long-linker region is an appropriate adaptation of this enzyme to low temperatures (Sonan et al. 2007). Studying the thermal adaptations of psychrophilic, mesophilic, and thermophilic DNA ligases, it was concluded that "the active site of the cold-adapted enzyme is destabilized by an excess of hydrophobic surfaces and contains a decreased number of charged residues compared with its thermophilic counterpart" (Georlette et al. 2003). The proteins must keep a balance between their stability and flexibility, especially enzymes, which are to be active at lower temperatures than their mesophilic counterparts (Georlette et al. 2003). An intensive study of the proteomics of psychrophilic microorganisms (Kurihara and Esaki 2008) showed that there are various proteins involved in transcription, folding of RNA and proteins, modulation of gene expression, and others, which are inducibly produced at low temperatures. The enzymes from psychrophilic microorganisms have flexible structures which allow "a reduction in activation enthalpy and a more negative activation entropy" compared with other microorganisms. The consequences are that a decrease of activity reaction rates is more slowly than for the enzymes from thermophiles. They have high catalytic activity even at low temperature, supported by low morphological stability, necessary for flexibility at those temperatures (Cavicchioli et al. 2011). Genomic and proteomic studies showed they are able to reduce the free energy barrier of the transition state and reduce the number of weak bonds in their structure that means a reduced stability (Feller 2013). The application of cold-adapted enzymes can be very efficient for catalysis in lower temperature, in order to protect temperature sensitive products. The same author commented that in fact, thermophiles and psychrophiles improve the activity of their enzymes in high and low temperatures, playing with the flexibility and stability of the structure but in opposite sense (Feller 2010). Another finding was the dependence of the presence of some ions in the cell, for example, activity and temperature dependency of *Shewanella* pyrophosphatase is influenced by the presence of Mg ions (Ginting et al. 2014).

The following three main types of proteins are of interest for mechanisms of adaptation:

Cold-Shock Proteins (Csps) Cold-shock proteins, which are induced by exposure to low temperatures, are involved in several cellular processes (fluidity of membranes, transcription, translation). Proteins from psychrophiles (Caps, cold acclimation proteins) are similar to the Csps. The regulation of the CspA protein takes place at the transcriptional level at the level of stabilization of mRNA (CspA mRNA) and at the level of translation (Phadtare and Inouye 2008). Numerous Csps and proteins helping in the adaptation to low temperatures were isolated and characterized (Russell 2008). They play a role in the cold adaptation during stress response and also act as RNA chaperones. Similar proteins can be found in Archaea, such as *Methanogenium frigidum*, which are bound to nucleic acids (Giaquinto et al. 2007). Cold-shock proteins as enzymes are useful for a lot of applications. Also, a cold-shock inducible system for expression of psychrophilic proteins was described (Bjerga and Williamson 2016). Studying the thermodynamics of growth rates and temperature of thermophilic, psychrophilic, and mesophilic bacteria and some archaea and eukaryotes, several scientists tried to predict characteristics of enzymes, implementing the Arrhenius Equation and Bayesian prediction method to link these two facts, in order to set up a credible model of temperature dependence of growth rate which describes the protein reactions, too. A strong relation between temperature dependence and biochemical levels of both unicellular and multicellular life forms was found (Corkrey et al. 2014). The proteins from psychrophiles appear to have more soft structures at the surface (Isaksen et al. 2014). A RNA helicase protein CsdA from *Psychrobacter arcticus* 273–4, which grows at −10 °C, was discussed (Kuhn 2012). Even at those temperatures, the genes for energy metabolism, maintenance of the membrane, cell walls, and nucleic acids motions are upregulated and also the expression of cold-shock-induced RNA helicase protein A (CsdA), which, if found to be functional, might extend the known minimum growth temperature. The membrane proteins of cold-adapted *Vibrionaceae* showed only changes in amino acid composition and a lower hydrophobicity in psychrophilic membrane proteins (Kahlke and Thorvalsen 2012).

Antifreeze Proteins (AFPs) AFPs and antifreeze glycoproteins (AFGPs) can lower the temperature of the freezing point of water (D'Amico et al. 2006; Kawahara 2008). They can inhibit the formation of ice crystals and prevent the penetration of ice into cells (Zachariassen and Lundheim 1999). One of the examples is the protein Hsc25 produced by the bacterium *Pantoea ananatis* KUIN 3, which helps to refold denatured proteins in the cold (Kawahara 2008). The antifreeze proteins inhibit the formation of ice crystals, and ice nucleation proteins induce the formation of ice crystals. They have been found to use the same strategy of binding ice, even though they belong to different protein classes (Lorv et al. 2014).

Antinucleating Proteins (ANPs) ANPs and other compounds inhibit ice nucleation and formation of intracellular ice crystals, avoiding thereby the damage of cells. *Acinetobacter aceticus* can release such antinucleating proteins with a mass of 550 kDa. The proteins can be used in the preservation of livers (in a concentration of 20 mg/ml) at subzero temperatures without freezing, with addition of an antioxidant such as ascorbic acid (Kawahara 2008). An ice-binding protein of a mass of 54 kDa, isolated from a bacterial strain from an ice core of over 3000 m depth, was able to inhibit the recrystallization of ice (Raymond et al. 2007).

Enzymes The rate of a chemical reaction is temperature dependent, according to the Arrhenius equation $K = A \exp. (Ea/RT)$, where K is the reaction rate, Ea is the activation energy, R is the gas constant, T is absolute temperature in Kelvin, and A is a constant. It is well known that biological reactions showed a 16- to 80-fold reduction when the temperature is reduced from 37 to 0 °C (Collins et al. 2008).

While psychrophiles exhibit a high metabolic rate at cold temperatures, they are usually inactivated at mesophilic temperatures because of the flexibility and lower stability as a consequence of the plasticity of catalytic zones of their molecules, due to a reduction of hydrophobic and hydrogen bonds and a lower content of arginine and proline (D'Amico et al. 2006). Shifting to optimum activation energy allows them to keep normal reaction rates at low temperatures (Siddiqui and Cavicchioli 2006). At the same time, the 3D structure is also important (Tkaczuk et al. 2005), such as intramolecular bond modifications (Feller and Gerday 1997; Bae and Phillips 2004), modification of amino acids in or near catalytic domains of enzymes (Papaleo et al. 2008), and a lower content of hydrogen bonds (Michaux et al. 2008). An intensive search for cold-adapted enzymes was performed by Morita et al. (1997) who isolated more than 130 bacterial strains and tested the properties of amylases, proteases, and lipases, showing that they were easily inactivated at above optimum temperatures. Proteomic studies can reveal the mechanism of cold adaptation. For example, *P. haloplanktis* was growing fast at the moderate temperature of 16 °C, showing high levels of cold acclimation and oxidative stress proteins, an abundance of transporters, and generally very efficient protein synthesis (Wilmes et al. 2011).

Other Substances Other substances which can play a role in cold adaptation and cryoprotection are carotenoids, which contribute to the stability of cellular membranes (Russell 2008); extracellular polymeric substances (EPSs), some of them of high molecular weight or heteropolysaccharides (with additions of proteins), which are released by some microorganism into the neighboring environment and form a kind of gel with cryoprotective effects (Krembs and Deming 2008); polyhydroxyalkanoates (PHAs), which can reduce oxidative stress in the cold, maintaining the redox state (Ayub et al. 2009); and trehalose, which is able to protect cells under conditions of shock exposure to high and low temperatures and osmotic stress (Phadtare and Inouye 2008) by stabilizing the cell membrane and removing free radicals, thus preventing denaturation of proteins.

Generally speaking, when comparing with thermophiles and mesophiles, it appears that cold-adapted microorganisms adopted the strategy of more entropy by molecular mechanisms, which are allowing an enhanced flexibility for maintaining dynamics and functions of the molecules (Feller 2007). Polysaccharides released by *Colwellia psychrerythraea* 34 H, isolated from marine sediments, form a capsule which have an effect similar to antifreeze proteins (Carillo et al. 2015; see below). Some osmolytes from psychrophiles can increase thermal stability and specific activity at higher temperature (De Santi et al. 2012).

Genetic Features as Adaptation Mechanisms So far, the following genomes from psychrophiles and psychrotolerants have been sequenced: *Methanococcoides burtonii* (Allen et al. 2009), *Methanogenium frigidum* (Saunders et al. 2003), *Colwellia psychrerythraea* 34H (Methé et al. 2005), *Desulphotalea psychrophila* (Rabus et al. 2004), *Idiomarina loihiensis* L2TR (Hou et al. 2004), *Pseudoalteromonas haloplanktis* TAC125 (Médigue et al. 2005), *Shewanella frigidimarina* (Copeland et al. 2006), *Psychrobacter arcticus* 253–4 (Ayala-del-Rio et al. 2010), *Psychromonas ingrahamii* (Riley et al. 2008), 14 *Shewanella* strains (Hau and Gralnick 2007), *Shewanella violacea* (Aono et al. 2010), and several others, which are partially sequenced. The analysis of the genes showed some principal features of the mechanisms for cold adaptation (Bowman 2008): a lower content in arginine and proline, which influences the flexibility of proteins, was observed, especially in sequences related to growth and development (Ayala-del-Rio et al. 2010). Nucleic acids of psychrophiles showed a different proportion of uracil in 16S rRNA sequences, such as an inverse proportional relation to their optimum growth temperature (Khachane et al. 2005). The proteome of about 40 families of enzymes from thermophiles, mesophiles, and psychrophiles were studied with bioinformatic tools in order to predict by Bayesian inference of the genome sequences the adaptation and activity at different ranges of temperatures based on optimal temperature (Jensen et al. 2012). Single-stranded DNA-binding proteins with a role in recombination and repair of nucleic acids, obtained from some psychrophilic bacteria, possess high stability (Nowak et al. 2014). Activity at the level of genetic material under conditions of low temperatures in permafrost was demonstrated (Tuorto et al. 2014) by studies of a microcosm detecting labeled ^{13}C DNA synthesized at subzero temperatures.

Usually, for practical purposes, psychrophilic enzymes are produced using mesophilic expression systems (Bjerga et al. 2016). The authors tried to develop more efficient low-temperature expression systems. The plasmids of cold-adapted bacterial strains were analyzed and found to harbor genes of enzymes involved in adaptation processes, removal of reactive oxygen species, energy conversion for metabolism, resistance to heavy metals, and the use of normally toxic compounds (Dziewit and Bartosik 2014).

Comparisons of DEAD-box proteins from *E. coli* and the psychrophilic bacteria *Pseudoalteromonas haloplanktis* and *Colwellia psychrerythraea* showed that rather low activation energy is necessary for psychrophilic enzymes and that this could facilitate RNA metabolism in the cold (Cartier et al. 2010). In general the expression of genes in *E. coli* is used on a large scale, but sometimes thermolabile proteins cannot be expressed. As alternative, an expression system of mesophilic genes in the strain *Pseudoalteromonas* sp. SM20429 that could not use the *E. coli* system was developed (Yu et al. 2015).

7.6 Applications of Psychrophilic Microorganisms

Bioprospecting and bioscreening of psychrophilic microbial resources (Nichols et al. 2002) have become real challenges and opportunities for biotechnology. Cold-adapted bioactive substances provide advantages in different areas, such as activity at low temperatures, the possibility of challenging reactions with a sufficiently high reaction rate, energy savings, efficient production with lower processing costs, thermal protection of the products, and better quality of products. Presently the market for bioactive products and industrial enzymes is growing. Archaea, Bacteria, and Eukarya can be sources of valuable products. Huston (2008) reviewed the enzymes from cold-adapted microorganisms, identifying compounds and enzymes for the food and cosmetic industry, pharmaceuticals, biofuels, substances for molecular biology studies, and even for nanobiotechnology. An extensive compilation of applications of psychrophilic and psychrotolerant microorganisms is presented in Table 7.1.

Table 7.1 Applications of psychrophilic and psychrotolerant microorganisms isolated from cold environments

Microorganism	Enzymes and other metabolites	Applications	References
Bacteria			
Acinetobacter sp.	Lipases	Lipid hydrolysis, detergent additives	Ramteke et al. (2005); Joseph et al. 2007
Achromobacter sp.	Lipases	Lipid hydrolysis	Ramteke et al. (2005) ; Joseph et al. (2007)
Aeromonas sp.	Lipases	Lipid hydrolysis	Lee et al. (2003)
Arthrobacter strains	Antibiotics	Pharma industry	Lo Giudice et al. (2007); Benešova et al. (2005)
Arthrobacter sp.	Alkaline phosphatases	Alkaline phosphatase (removal of 5' phosphate groups from DNA and RNA)	De Prada et al. (1996)
Arthrobacter strain C2-2	α-Glucosidaseβ-Glucosidase	Cleavage of maltose at β1-4 bonds; pharma industry, medicine	Benešova et al. (2005)
Arthrobacter strain 20B	β-Galactosidases	Lactose hydrolysis	Białkowska et al. (2009)
Arthrobacter psychrolactophilus strain F2	β-Galactosidase	Produces trisaccharides from lactose; food industry	Nakagawa et al. (2007)
Arthrobacter psychrophenolicus		Degradation of phenol and phenolic compounds	Margesin et al. (2004)
Bacillus subtilis strain MIUG 6150	α-, β-Amylases	Starch hydrolysis; food industry	Bahrim and Negoiţă (2004)
"	Proteases	Protein hydrolysis	Bahrim and Negoiţă (2004)
Brevibacterium antarcticum		Bioremediation; resistant to metals in soils (Cu, Cr, Hg, and others)	Tashyrev (2009)
Colwellia demingiae	Protease (azocasein)		Nichols et al. (1999)
"	Protease (azoalbumin)		Nichols et al. (1999)
"	Trypsin-like protease	Protein hydrolysis	Nichols et al. (1999)
"	Phosphatase		Nichols et al. (1999)
Colwellia demingiae	Synthesizes docosahexaenoic acid	PUFAs as precursors for prostaglandins, thromboxanes, leukotrienes; medicine, pharma industry	Bowman et al. (1998); Lees (1990)
Colwellia-like strain	Trypsin-like enzyme		Nichols et al. (1999)

(continued)

Table 7.1 (continued)

Microorganism	Enzymes and other metabolites	Applications	References
"	Phosphatase		Nichols et al. (1999)
"	β-Galactosidase	Lactose hydrolysis	Nichols et al. (1999)
"	Protease (azocasein)		Nichols et al. (1999)
"	Protease (azoalbumin)	Protein hydrolysis	Nichols et al. (1999)
"	Trypsin		Nichols et al. (1999)
"	β-Galactosidase	Removal of lactose	Nichols et al. (1999)
"	α-Amylase	Starch hydrolysis	Nichols et al. (1999)
"	Alkaline phosphatase		Nichols et al. (1999)
Colwellia hornerae	Synthesizes docosahexaenoic acid	Pharma industry	Bowman et al. (1998)
Colwellia maris	Malate synthaseisocitrate lyase	Bioethanol and biomethane production; waste water treatment, bioremediation	Brenchley (1996); Cavicchioli (2002)
Colwellia rossensis	Synthesizes docosahexaenoic acid	Pharma industry	Bowman et al. (1998)
Colwellia psychrotropica	Synthesizes docosahexaenoic acid	Pharma industry	Bowman et al. (1998)
Dactylosporangium roseum	Antibiotics	Pharma industry, medicine	Raja et al. (2010)
Erythrobacter litoralis strain HTCC2594	Epoxide hydrolase	Enantio-selective hydrolysis of styrene oxide	Woo et al. (2007)
Fibrobacter succinogenes	Cellulase	Animal food industry, detergents, textile industry	Cavicchioli et al. (2002)
Flavobacterium sp.	β-Mannanase	Decreases viscosity in food products	Zakaria et al. (1998)
Flavobacterium frigidarium	Xylanolytic and laminarinolytic	Xylane degradation	Humphry et al. (2001)
Flavobacterium frigidimaris	Malate dehydrogenase		Oikawa et al. (2005)
Flavobacterium hibernum	β–Galactosidase	Lactose degradation	McCammon et al. (1998)
Flavobacterium limicola		Organic polymer degradation	Tamaki et al. (2003)
Glaciecola chathamensis	Polysaccharide-producing strain	Exopolysaccharides, industrial applications	Matsuyama et al. (2006)
Glaciecola chathamensis	Exopolysaccharides	Food processing industry; medical and industrial uses	Matsuyama et al. (2006)
Intrasporangium sp.	Antibiotics	Pharma industry	Raja et al. (2010)
Janibacter sp.	Antibiotics	Pharma industry, medicine	Lo Giudice et al. (2007)

Kordiimonas gwangyangensis	Cold-adapted enzymes	Capable of degrading polycyclic aromatic hydrocarbons (PAHs)	Kwon et al. (2005)
Micromonospora sp.	Antibiotics	Pharma industry	Raja et al. (2010)
Moraxella sp.	Lipases	Pharma industry, medicine, food additives	Ramteke et al. (2005); Joseph et al. (2007)
Oceanibulbus indolifex	Indole and several indole derivatives	Cosmetic industry, pharma industry, cancer prevention	Wagner-Döbler et al. (2004); Auborn et al. (2003)
"	Cyclic dipeptides, cyclo-(Leu-Pro), cyclo-(Phe-Pro),and cyclo-(Tyr-Pro)	Compounds with antiviral, antibiotic, and antitumor activity	Wagner-Döbler et al. (2004); Milne et al. (1998)
"	Tryptanthrin	Activity against some Gram-positive bacteria and fungi	Wagner-Döbler et al. (2004)
Oleispira antarctica	Cold-adapted enzymes	Hydrocarbonoclastic; for bioremediation	Yakimov et al. (2003)
Photobacterium frigidiphilum	Lipases	Lipid hydrolysis	Seo et al. (2005)
Planococcus sp.	β-Galactosidase	Lactose hydrolysis	Sheridan and Brenchley (2000)
Polaromonas naphthalenivorans	Enzymes	Degradation of naphthalene	Jeon et al. (2004)
Polaromonas sp.strain JS666	Enzymes	*cis*-1,2-Dichloroethene (cDCE) as carbon source, for bioremediation	Mattes et al. (2008)
Polaromonas vacuolata		Metal- resistant strain, recovery of mercury compounds	Barkay and Poulain (2007)
Pseudoalteromonas sp.	Protease (azocasein)		Nichols et al. (1999)
"	Trypsin-like enzyme		Nichols et al. (1999)
Pseudoalteromonas sp.	Antibiotics	Pharma industry, medicine	Lo Giudice et al. (2007)
Pseudoalteromonas sp.	Lipases		Ramteke et al. (2005)
Pseudoalteromonas haloplanktis strain TAE 47	β-Galactosidase	Lactose hydrolysis	Hoyoux et al. (2001)
Pseudoalteromonas sp. strain SM9913	Subtilase		Yan et al. (2009)
Pseudomonas sp. strain B11-1	Lipases, esterases		Suzuki et al. (2001)
Psychrobacter okhotskensis	Lipase-producing strain		Yumoto et al. (2003)

(continued)

Table 7.1 (continued)

Microorganism	Enzymes and other metabolites	Applications	References
Psychrobacter sp.	Lipase		Joseph et al. (2008)
Rhodococcus sp.	Antibiotics	Pharma industry, medicine	Lo Giudice et al. (2007)
Rhodococcus sp.		Biodegradation of phenol and phenolic compounds	Margesin and Shinner (1999)
Rhodococcus strain N774	Nitrile hydratase	Acrylamide production	Kobayashi et al. (1992)
Rhodoccocus sp.strain Q15		Degrades short-and long-chain aliphatic alkanes from diesel fuel	Whyte et al. (1998)
Rhodococcus ruber		Product "Rhoder" for bioremediation of oil-polluted environments	Muryghina et al. (2000)
Rhodococcus erythrococcus		Product "Rhoder" for bioremediation of oil-polluted environments	Muryghina et al. (2000)
Serratia proteamaculans	Trypsin-like protease		Mikhailova et al. (2006)
Shewanella sp.	Produces omega-3 fatty acids	Essential fatty acid for humans	Hau and Gralnick (2007)
"		Waste removal of radionuclides (uranium, technetium)	Hau and Gralnick (2007)
"		Reduction of organic chlorine compounds	Hau and Gralnick (2007)
Shewanella donghaensis	High levels of polyunsaturated fatty acids	Medicine, food supplements	Yang et al. (2007)
Shewanella gelidimarina	ß-Galactosidase	Lactose hydrolysis	Nichols et al. (1999)
Shewanella frigidimarina	Eicosapentaenoic acid (20:w503)	Food additives	Bowman et al. (1997); Bozal et al. (2002)
Shewanella pacifica	Produces polyunsaturated fatty acid	Food additives, pharma industry	Ivanova et al. (2004)
Serratia proteamaculans	Trypsin-like protease	Protein hydrolysis	Mikhailova et al. (2006)
Serratia sp.	Lipases	Lipid hydrolysis	Ramteke et al. (2005)
Sphingomonas paucimobilis	Proteases	Meat industry, detergent industry; molecular biology	Cavicchioli et al. (2002)
Sphingopyxis alaskensis		Metal-resistant bacteria implied in mercury biogeochemistry, bioremediation	Barkay and Poulain (2007)

Streptomyces sp.	Amylases, proteases, cellulases, lipases, antibiotics, other bioactive compounds	Detergent additives, starch industry, bread industry, antibiotics, immunosuppressants, anticancer agents, extracellular hydrolytic enzymes, degradation oflignocellulosic materials	Galante and Formantici (2003); Morita et al. (1997)
Streptomyces anulatus	Dextranase	Dextrane hydrolysis, sugar industry	Doaa Mahmoud and Wafaa Helmy (2009)
Streptoverticillium sp.	Antibiotics	Pharma industry	Raja et al. (2010)
Fungi			
Candida antarctica	Lipases		Joseph et al. (2007)
"		Conversion of n-alkanes into glycolipid; biosurfactants	Kitamoto et al. (2001)
Cryptococcus albidus	Xylanase	Hydrolyzing xylane for improvement of food; food industry, waste treatments	Amoresano et al. (2000)
Cryptococcus laurentii	Phytase	Animal feeding	Pavlova et al. (2008)
"	ß-Galactosidases	Dairy industry	Law and Goodenough (1995)
Cryptococcus cylindricus	Pectinases	Clarification of fruit juices; improving filterability and extractability of juices	Nagakawa et al. (2004)
Cystofilobasidium capitatum	Pectinases	Clarification of fruit juices; improving filterability and extractability of juices	Nagakawa et al. (2004)
Mrakia frigida	Pectinases	Clarification of fruit juices; improving filterability and extractability of juices	Nagakawa et al. (2004)
Pichia lynferdii strain Y-7723	Lipase		Kim et al. (2010)
Rhodotorula psychrophenolica		Degradation of phenolic compounds	Margesin et al. (2007)
Algae			
Porphyridium cruentum	Eicosapentaenoic acid, arachidonic acid	Pharma industry	Cohen (1990)

Bioscreening for valuable products is generally not made any more in the classical way by isolation and cultivation of microorganisms. Now high-throughput culturing technologies enable the isolation of a major proportion of the microbiota in environmental samples; combined with metagenomics and gene expression studies, genome data mining permits an efficient search for bioproducts (see Huston 2008).

The benefits of the application of cold-adapted enzymes are that they can be used in reactions that must be carried on at low temperatures, and at the same time, high activity is desired. They can also been selectively inactivated in a reaction mixture. For example, using cold-adapted proteases for meat tenderizing should avoid the formation of undesirable by-products. They are required in detergent, textiles, food industry, and molecular biology and possibly others (Sarmiento et al. 2015).

Psychrophilic proteins, for example, can have some interesting applications, and their production can be achieved directly or expressed in an adequate host such as *Escherichia coli,* which was used for the α-amylase from *P. haloplanktis* (Feller et al. 1998). It can be difficult to obtain a stable production, due to autolytic deterioration. A possible solution is overexpression using a plasmid vector from *P. haloplanktis* pMTBL and the plasmid of *E. coli* pJB3 (Tutino et al. 2001). This type of expression technology promises a wide application for the problem of efficient gene expression systems and rapid purification steps. Recombinant proteins can be obtained by expressing them in prokaryotic cells of cold-adapted (*P. haloplanktis* TAC125) and eukaryotic cells (*Saccharomyces cerevisiae*; Parrilli et al. 2008). From *Pseudomonas haloplanktis*, a two-component regulatory system, which regulates the synthesis of an outer membrane porin and is under control of L-malate in the culture medium, was used to construct an inducible expression vector for production of recombinant enzymes in this psychrophilic microorganism (Rippa et al. 2012). A study of cold-adapted enzymes in comparison with mesophilic and thermophilic homologues (Siglioccolo et al. 2010) showed segments of local flexibility or rigidity which help enzyme activity at different temperatures. The relative backbone flexibility of the enzymes was suggested to be related to cold adaptation. A lower number of salt bridges and cation-pi interactions were demonstrated in *Shewanella* sp. acetate kinase, compared with *E. coli* acetate kinase, which caused weakening of intramolecular electrostatic interactions and more flexible structures, permitting activity at low temperatures (Tang et al. 2014). A study of acylaminoacyl peptidase from *Sporosarcina psychrophila* (SpAAP) showed it possesses lysine to arginine substitutions, a high number of glycine residues, and a lack of disulfide bridges, compared with enzymes from mesophiles and thermophiles (Papaleo et al. 2014).

Enzymes from psychrophiles have different structures and composition, despite their similarity with homologous mesophilic counterparts. For example, there is a difference in salt bridge residues, especially on the surface of proteins, which often form not very stable ion pairs. Mutations in these residues from a psychrophilic homologue of aqualysin I were more flexible, and, with one exception, no distinct involvement in enhanced thermal stability was deduced (Jónsdóttir et al. 2014). Homologous α-amylases from psychrophilic bacteria, thermophilic actinomycetes, and mesophilic multicellular organisms, like fruit fly and pig, were compared, and their thermal stability was studied by intrinsic fluorescence, circular dichroism, and differential scanning calorimetry (Cipolla et al. 2012). Inactivation of the enzymes occurred by melting down of the structure, but in contrast to meso- and thermophilic enzymes, the psychrophilic enzyme was heat inactivated before its structure was destroyed. This confirmed previous assumptions that a very flexible and therefore heat-labile active site is required for activity at low temperatures (Cipolla et al. 2012). Some researchers demonstrated that osmoprotectants like trimethylamine N-oxide can enhance thermal stability and increase of specific activity (De Santi et al. 2012). Manipulation of AMP genes and expression in *E. coli* showed that as in psychrophiles the concentration of nucleotides increased as temperature declined and that modified *E. coli* was capable of growing up to ca. 70 % faster at low temperatures and became up to ca. Tenfold more cold tolerant relative to wild type (Parry and Shain 2011). Studies of the structure of active sites in wild type and mutant lipases showed the latter remained rather rigid at low temperatures, keeping their geometry and showing even higher enzymatic activity (Kamal et al. 2012). Glutathione synthetase, a nearly ubiquitous enzyme whose catalytic mechanism and adaption to extreme environments is not understood, was isolated, expressed and purified from Antarctic *Pseudoalteromonas haloplanktis* (PhGshB),and crystallized, for the first structural characterization of a psychrophilic glutathione synthetase (Merlino et al. 2011).

The enzymes from psychrophilic bacteria have high catalytic activity and low thermal stability, which differentiates them from their mesophilic and thermophilic counterparts. Improvements of their adaptive structural features have significantly benefitted the enzyme industry (Maiangwa et al. 2015).

Antibiotics The isolates from the Antarctic Ocean, Ross Bay, were shown to have antibiotic activities which were tested with the terrestrial bacteria *E. coli, Pseudomonas aeruginosa, Staphylococcus aureus, Micrococcus luteus, Bacillus subtilis, Proteus mirabilis, Salmonella enterica*, and the yeast *Candida albicans*, following incubation at 37 °C on nutrient agar (Lo Giudice et al. 2007). The isolates were identified by 16S rRNA and characterized by biochemical tests. From 580 isolated strains belonging to *Arthrobacter, Rhodococcus*, *Pseudoalteromonas*, and *Janibacter*, some were able to inhibit the test strains. The actinomycetes *Intrasporangium* sp., *Micromonospora* sp., *Streptoverticillium* sp., *Streptomyces* sp., and *Dactylosporangium roseum*, isolated from the soils in India at over 4000 m altitude, showed different antibiotic activities against strains of *Streptococcus* isolated from dental plaque (Raja et al. 2010). From cold environments, strains close to *Serratia* and *Pseudomonas* were isolated; both are producers of antimicrobials, probably class II microcins acting in the cold (Sanchez et al. 2009).

Enzymes Many psychrophilic and psychrotolerant bacteria possess the capacity to produce extracellular enzymes such as lipases, proteases, amylases, cellulases, chitinases, and β-galactosidases, when induced by the presence of specific substrates. Producers were strains of sea ice microorganisms, for example, for lipases *Colwellia psychrerythraea, Shewanella living stonensis*, and *Marinomonas* *prymoriensis*; for hydrolysis of polysaccharides *Colwellia, Marinomonas, Pseudoalteromonas, Pseudomonas, and Shewanella;* and for hydrolysis of chitin *Pseudoalteromonas tetraodonis, Pseudoalteromonas elyakovii, Bacillus firmus,* and *Janibacter melonis,* which degrade the organic matter from phyto- and zooplankton (Yu et al. 2009). A simple screening of 137 cold-water isolates belonging to *Moraxella*, *Pseudomonas*, *Aeromonas*, *Chromobacterium*, *Vibrio*, and others showed that about 62 % can produce gelatinase, 71 % proteases, 31 % lipases, 47 % amylases, 36 % chitinases, 36 % β-galactosidases, 47 % cellulases, and 25 % alginate lyases (Ramaiah 1994).

Groudieva et al. (2004) found that from 116 strains isolated from the Spitsbergen sea ice, 40 % possessed the ability to degrade skim milk, casein, lipids, starch, and proteins. Enzymes such as dehydrogenase from cold-adapted microorganisms can also be used as biosensors or in biotransformations (Gomes and Steiner 2004). Protease-producing strains from the genera *Pseudoalteromonas, Shewanella, Colwellia*, and *Planococcus* were isolated from the subantarctic marine sediments of Isla de Los Estados (Olivera et al. 2007).

The enzymes from psychrophilic and psychrotolerant strains can be divided into three categories: (1) heat sensitive but similar to mesophilic enzymes, (2) heat sensitive and more active at low temperatures than mesophilic enzymes, and (3) heat sensitive exactly as mesophilic enzymes but more active at lower temperatures (Ohgiya et al. 1999). Another example is a complex of enzymes generated by an Antarctic isolate, *B. subtilis* strain MIUG 6150, which produces α-and β-amylases and proteases (Bahrim and Negoiță 2004). The productivity of the microorganisms showed a strong dependency on the culture media used for growth and other conditions (Bahrim et al. 2007). In fact some enzymes keep its thermal properties through structural changes for increasing flexibility adapted to temperature of living environments, and some regions of the enzyme preserve their flexibility in order to keep its function (Kovacic et al. 2016).

Sometimes in order to make the production efficient, genetic and chemical modifications are applied, in order to respond to the industry's demands (Siddiqui et al. 2015). An example of chemical modification is the treatment of the oligopeptidase B (produced by *Serratia proteamaculans*) with chymotrypsin, and the truncated enzyme remains active on low-molecular-weight substances (Mikhailova et al. 2015). The enzymes can retain about 80 % activity in low-temperature condition, like the lipase from *Colwellia psychrerythraea* 34H (Do et al. 2013). A folding funnel model was proposed to describe the structure of an α-amylase from psychrophiles – in fact that enzyme has a population of conformations, assuring more flexibility (Mereghettti et al. 2010). The cold-adapted strains of *Streptomyces* can be a source of valuable enzymes (Cotârleț et al. 2008). The strategy for production of psychrophilic enzymes uses two lines– one is an expression in mesophilic hosts adapted to the cold or development of new psychrophilic hosts, which means genetic engineering in order to obtain a good expression system (Bjerga et al. 2016) and assure enzyme production.

A biocatalyst reaction system was used to study the psychrophilic enzymes and achieve the conversion of aspartic acid from fumaric acid without formation of L malic acid. The catalyst system was immobilized in alginate and can increase the yield of target compounds (Tajima et al. 2015). The same method was used to obtain a 3-hydroxypropionaldehyde from glycerol (Tajima et al. 2013). The activation of the cold-active enzymes is achieved by reduction of number and strength of the interactions and bonds in the enzyme and by improvement of the active site, and therefore it was suggested that directed evolution appears to be the most suitable methodology to engineer cold activity in biological catalysts (Struvay and Feller 2012). Another method for controlling and improving the catalytic activity is the use of copper oxide nano particles (corresponding to copper laccase) entrapped in nanotubes, which causes the enzyme to retain its activity even after repeated episodes of freezing and thawing (Mukhopadhayay et al. 2015).

Hydrolases are a generic name for a class of hydrolytic enzymes containing amylases, cellulases, peptidases, and lipases, for example. The enzymes from psychrophiles can be used in extreme conditions in industrial processes. They are employed in the pharmaceutical, food, and chemical industries and also in biofuel production (Dalmaso et al. 2015). The need for hydrolases is growing in different industries, and now over 500 enzymes from microorganisms are employed.

Proteases Strains producing proteases belong to the genera *Bacillus* and *Pseudomonas* and were isolated from Antarctic cyanobacterial mats. They showed good production at a temperature of about 20 °C, with glucose and maltose as carbon sources (*Pseudomonas* sp.) and soybean meal and peptone as nitrogen sources (Singh and Ramana Venkata 1998). A trypsin-like protease was identified and characterized which is produced by *Serratia proteamaculans* (Mikhailova et al. 2006). These proteases are used in the dairy industry to enhance the flavor development in cheese. In the chemical industry, the enzymes from psychrophilic strains are used in the detergent and food industry and also in leather manufacturing (Cavicchioli et al. 2002). Strains producing proteases were isolated from Antarctic soils and identified as *Sporosarcina aquimarina* and *Algoriphagus antarcticus* (Santos et al. 2015). Numerous other psychrophilic and psychrotolerant strains (*Colwellia*, *Clostridium*, *Curtobacterium*, *Flavobacterium*, *Leucosporidium*, *Pseudomonas*, *Pseudoalteromonas*, *Serratia*, *Shewanella*, *Stenotrophomonas*, and others) produce different proteases, with wide uses in biotechnological applications (Kuddus and Ramteke 2012; Joshi and Satyanarayana 2013).

Alkaline Phosphatase Alkaline phosphatase was isolated and purified from the strain *Shewanella* sp. (Ishida et al. 1998). Interestingly, the enzyme showed a maximum activity at 40 °C, 39 % of that activity at 0 °C, and a tendency to loose activity at 20 °C. Two different extracellular alkaline phosphatases were identified from an *Arthrobacter* sp. strain (De Prada et al. 1996). Purified recombinant serine alkaline protease (in *E. coli*) from another *Shewanella* strain showed activity between 5 and 15 °C (Kulakova et al. 1999). A new alkaline protease was purified from a cold-adapted strain of *Pseudomonas aeruginosa* (HY1215) having a high activity in the range of 15–35 °C with a maximum at 25 °C (Hao and Sun 2015). A protease was isolated from *Pseudomonas lundensis* from the deep marine waters and was found to be active even at 4 °C (Yang et al. 2010). The enzymes like the microorganisms themselves face sometimes diverse polyextreme conditions, e.g., the cold-active protease MCP-03 from *Pseudoalteromonas* sp. SM9913 was active and stable also in a high-salt environment of about 3 M NaCl/KCl (Yan et al. 2009). Some bacteria producing proteases were isolated from a glacier from Tian Shan Mountains; they were growing at 15–24 °C and belonged to the genera *Pseudomonas*, *Polaromonas*, *Brevundimonas*, *Arthrobacter*, and *Kocuria* (Ni et al. 2013).

β-Galactosidases One possible application of this enzyme is obtaining an ice cream with reduced lactose content for lactose-intolerant peoples (Phadtare and Inouye 2008). The removal of lactose from milk is very important for persons with lactose intolerance. The cold-active β-galactosidases (EC3.2.1.23) isolated from psychrophilic yeasts (e.g., *Cryptococcus laurentii*; Law and Goodenough 1995) and fungi can be used to hydrolyze lactose, and therefore a new method of supplementation of milk with dormant cultures was proposed (Somkutl and Holsinger 1997). The utilization of cold-active β-galactosidase (optimum activity at 10 °C) from *Arthrobacter psychrolactophilus* strain F2, which was overexpressed in E. coli, for the production of trisaccharides from lactose was also tested for applications in the food industry (Nakagawa et al. 2007). Some strains contain isoenzymes (C2–2-1 and C2–2-2) such as an *Arthrobacter* strain (Karasova et al. 2002), which was found – as a first example – being able to catalyze transglycosylation reactions in the cold. The cold-active β-galactosidase from *Thalassospira* sp. 3SC-21 has a specific activity between 4 and 20 °C and is completely inactivated at 45 °C (Ghosh et al. 2012).

Aminotransferases Aminotransferases (EC 2.6.1.57) are used to obtain natural and nonnatural amino acids by transamination reaction. Some researchers consider the cold-active aminotransferases as more advantageous. The study of aminotransferase from *Psychrobacter* B6, an Antarctic isolate, together with some computational studies showed they can use both aspartate and aromatic amino acids as substrate (Bujacz et al. 2015). It appears that aspartate aminotransferase is involved in cold adaptation, since its expression was enhanced at 4 °C compared with 22 °C (Sundareswaran et al. 2010).

Serine Hydroxymethyltransferase Serine hydroxymethyltransferase (SHMTEC 2.1.2.1) was obtained from *Psychromonas ingrahamii* and analyzed as recombinant enzyme expressed in the *E coli* system, catalyzing the conversion of L-serine to glycine and glutamic compounds. Many β-hydroxy-α-amino acids are SHMT substrates and are important for the pharma industry, as food additives and for agriculture (Angelaccio et al. 2012).

Lipases Cold-active lipases (triacylglycerol acylhydrolases, EC 3.1.1.3) were isolated from many psychrophilic strains and can have many industrial applications such as additives for detergents and additives in food products, in bioremediation, and in molecular biology (Joseph et al. 2007, 2008). Strains of *Acinetobacter*,

Achromobacter, *Moraxella*, *Psychrobacter*, *Pseudoalteromonas*, *Serratia*, and others are lipase producers. An *Aeromonas* strain produces a cold-active lipase (Lee et al. 2003), and fungal lipase producers such as *Candida antarctica*, *Geotrichum*, and *Aspergillus* sp. were described. From 137 anaerobic strains isolated from soils in Schirmacher Oasis, Antarctica, 49 isolates showed lipolytic activity on Tween-agar medium (Ramteke et al. 2005). *Psychrobacter okhotskensis* was isolated from Okhotsk seawater (Yumoto et al. 2003) and is a producer of cold-active lipases. These cold-active enzymes have a larger K coefficient and high efficiency down to temperatures of zero degree; they are inactivated by raising the temperature. A cold-active lipase gene was isolated from a new *Pseudomonas* sp. strain AMS8 and expressed in *E coli* system, the lipase being very active at 20 °C but retains about 50 % of its activity at 10 C (Ali et al. 2013). They are used in the food industry and chemical industry; the latter utilizes them for catalyzing reactions with compounds which are unstable at higher temperatures (Suzuki et al. 2001), for example, lipases and esterases from *Pseudomonas* sp. strain B11–1. Many detergents contain a mixture of proteases, lipases, and amylases (Ohgiya et al. 1999). Some lipases can be used to remove fatty stains from various textiles (Araújo et al. 2008). Lipases can also be used for the synthesis of biopolymers, biodiesel, pharmaceuticals, and certain aromatic products (Joseph et al. 2007). The authors reported also that lipases from psychrophilic and psychrotolerant strains can be used in cold environments for the bioremediation areas contaminated by oil and grease, as well as in the detergent industry. Production of cold-active lipase from *Moritella* sp. strain in large culture was demonstrated (Wang et al. 2013). From deep-sea sediments of the Pacific ocean, *Photobacterium frigidiphilum*, a lipolytic psychrophilic bacterium, was isolated (Seo et al. 2005). Some yeasts such as *Pichia lynferdii* strain Y-7723 produce a cold-adapted lipase (Kim et al. 2010). Some strains of *Bacillus* like *B. subtilis* LipJ were adapted to cold environments by substituting Ile to Met residues in the catalytic domain, which caused optimal activity at 20 °C, whereas the parent enzyme was most active at 37 °C (Goomber et al. 2016). Many strains isolated from glacier and firn ice from Svalbard were reported (*Bacillus barbaricus*, *Pseudomonas orientalis*, *P. syncyanea*, *P. fluorescens*, *P. oryzihabitans*, *Sphingomonas dokdonensis*, *S. phyllosphaerae*), some of the strains exhibiting lipase, protease, cellulase, and amylase activity (Singh et al. 2015). *Rhodococcus cercidiphylli* BZ22, which produces a cold-active lipase (Yu and Margesin 2014), was isolated from alpine soil contaminated with hydrocarbons. Maximum growth was in the range of 1–10 °C, and the maximum of enzyme production was in the same temperature range. A strain of *Bacillus pumilus* ArcL5 from Artic Ocean produced a lipase, showing an optimum at pH 9 and 20 °C, but it retained 85 % of its activity at 5 °C (Wi et al. 2014). From Arctic sediments, researchers isolated lipase-producing strains (Rasol et al. 2014). Lipases have a wide range of applications as reviewed by Joseph et al. (2007, 2008), such as synthesis of fine chemicals, production of fatty acids, interesterification of fats, synthesis of detergent additives, synthesis of biodiesel, and removal of hydrocarbons, oils, and lipidic pollutants. Other uses are in the food industry and concern the improvement of food structure and gelling of fish meat (Cavicchioli and Siddiqui 2004). Nielsen et al. (1999) isolated two rather thermotolerant lipases A and B, with uses in the textile industry for the removal of waxes and lipids from fibers. The lipases obtained from microorganisms, which are used in different detergent formulations, are covered by many patents issued for industrial companies (Hasan et al. 2010). Lipase B from *C. antarctica* was immobilized onto epoxy-activated macroporous poly(methyl methacrylate) Amberzyme beads and on nanoparticles, in order to improve contact with the substrate and the reaction activity for polycondensation (Chen et al. 2008a). The enzyme was quickly adsorbed on the polystyrene porous particles (Chen et al. 2008b). Strains able to degrade olive oil and others having cold-active lipase activity were isolated from a glacier in the TianShan Mountains by molecular methods (Xu et al. 2011).

Superoxide Dismutases Superoxide dismutases from psychrophiles were purified and characterized. For example, the Mn superoxide dismutase from *Exiguobacterium* sp. strain OS-77 is growing at 20 °C (Nonaka et al. 2014). The distribution of hydration waters around psychrophilic, mesophilic, and thermophilic iron superoxide dismutases was investigated, and it was interesting to find that the hydrogen bond density in thermophilic Fe-SOD was higher than that in mesophilic Fe-SOD(Mou et al. 2014).

Pectinases Several cold-adapted yeast strains were isolated from the soil of Hokkaido Island (Japan), which were taxonomically affiliated with *Cryptococcus cylindricus*, *Mrakia frigida*, and *Cystofilobasidium capitatum*. The strains showed pectinolytic activity at temperatures less than 5 °C and can be used for the production of pectinolytic enzymes (pectin methylesterase EC3.1.1.11, endopolygalacturonase EC3.2.1.15) for the clarification of fruit juices at low temperatures (Nakagawa et al. 2004), improving at the same time the filterability and extractability of the juice. Polygalacturonase is used in the food processing industry, and the enzymes from psychrophiles are in some cases required and more efficient (Ramya and Pulicherla 2015). Analyzing the structure of polygalacturonase from *Pseudoalteromonas haloplanktis* detected motifs and domains and conserved sequences; compared with enzymes from mesophilic and thermophilic strain, a reduced content of arginine content was observed. The use of cold-adapted pectinases from psychrophilic microorganisms is a recent trend in the food industry whose advantages are outlined by Adapa et al. (2014).

Malate Dehydrogenases (EC 1.1.1.37) A malate dehydrogenase (EC 1.1.1.37) was purified from *Flavobacterium frigidimaris* KUC1 and characterized by Oikawa et al. (2005). It contains lower amounts of proline and arginine residues compared to other malate dehydrogenases and is dependent on NAD(P). The enzyme loses its activity at 55 °C within 30 min of incubation. The enzyme can be used for producing malate at low temperatures.

Glucose-6-Phosphate Dehydrogenase Glucose-6-phosphate dehydrogenase (EC1.1.1.49) was analyzed from the psychrophilic green alga *Koliella antarctica* (Ferrara et al. 2013).

Dextranases (EC 3.2.1.11) An important problem in the sugar industry is the removal of dextran(EC 3.2.1.11), a high-molecular-weight polymer of D-glucose, which can lower the recovery of sugar, interfere with material processing, and lead to a poor quality of the final product. Bacterial cold-active dextranases can resolve this problem at low temperatures of about 4 °C (Doaa Mahmoud and Wafaa Helmy 2009), such as the dextranase from psychrophilic strain *Streptomyces anulatus*.

β-Amylases (EC 3.2.1.1) The need for cold-active amylases (EC 3.2.1.1) and related starch-hydrolyzing enzymes, especially for obtaining sweeteners such as palatinose, a disaccharide of glucose and fructose, and cyclodextrin, was reported (Rendleman 1996).

Phytases (EC 3.1.3.8) A cold-active phytase (EC 3.1.3.8) is produced very efficiently by the Antarctic strain *Cryptococcus laurentii* AL 27 (Pavlova et al. 2008). Phytase is an enzyme which catalyzes the conversion of undigestible phytate to phosphorylated myoinositol derivatives and inorganic phosphate, which are digestible. Its applications are in the fields of animal food additives and the pharmaceutical industry.

Xylanases (EC 3.2.1.8) Xylanase (EC 3.2.1.8) from the yeast *Cryptococcus albidus*, isolated from Antarctica, is a glycoprotein; its structure was investigated by mass spectroscopy (Amoresano et al. 2000). The xylanases hydrolyze the heteropolysaccharide xylane (a hemicellulose containing a backbone chain of β-1, 4-linked xylanopyranoside residues) and have found wide applications, e.g., improvement of maceration processes, clarification of juices, improvement of filtration efficiency, maceration of grape skins in wine technology, reducing viscosity of coffee extracts, improvement of drying and lyophilization processes, and improvement of the elasticity of dough and bread textures. Xylanases can also be used to degrade xylane from agricultural wastes in order to obtain energy from biomass. Hydrolyzing xylane from the cell walls of plants at low temperatures will allow energy savings and the production of more accessible feedstock (Lee et al. 2006). Furthermore, xylanases are used for the pulping process in the paper industry and for biobleaching (Beg et al. 2001), thereby reducing the use of alkali. They also improve energy consumption in the textile industry, being used in the microbiological retting of textile materials, which replaces chemical retting. They are useful for obtaining fermentation products, bioethanol, and other chemicals as well as improving the separation of starch and gluten in the starch industry.

The glycoside hydrolase family 8 xylanases can be used in baking processes inorder to improve the flexibility of dough and product quality (Collins et al. 2006). The crystal structure of a xylanase from the glycoside hydrolase family was recently described showing higher flexibility of substrate-binding residues, a different ratio of amino acid residues than the normal xylanases, and the existence of more flexible loops in their structure, which explains the activity at low temperatures (Zhang et al. 2011).

Glutathione Synthase Glutathione synthase (EC 6.3.2.3) is an enzyme in the formation of glutathione from condensation of glutamyl cysteine and glycine (Albino et al. 2012). This was obtained from the cold-adapted bacteria *Pseudoalteromonas haloplanktis* sp.

β-Glucanase A novel glucanase (laminarinase) was obtained from the yeast *Glaciozyma antarctica* PI12 (Parvizpour et al. 2015) and was analyzed with the program MODELER 9v12 studying the 3D comparative structure of mesophilic, thermophilic, and psychrophilic enzymes. The cold-adapted enzymes showed amino acid substitutions on the surface, an increase of the number of aromatic residues, a reduction of hydrogen bonds and salt bridges, and other structural changes, assuring more flexibility and efficiency of catalysis.

β-Mannanase β-Mannanase (EC3.2.1.78) was isolated from *Flavobacterium* sp. and showed good activity at 4 °C. This enzyme can be used to decrease the viscosity in food products (Zakaria et al. 1998).

Chitinase Chitinase (EC 3.2.1.14) is a glycoside hydrolase produced by the strain *Pseudoalteromonas* DL-6 isolated from marine sediments. The gene Chi A was expressed in *E coli* with good activity even at 4 °C, with an optimum of 20 °C (Wang et al. 2014). The strain *Sanguibacter antarcticus* (KCTC 13143) was used to obtain a recombinant chitinase expressed in *Pichia pastoris*, which showed 63 % activity of the optimum at 10 °C and 44 % activity at 0 °C (Lee et al. 2010). The structure of a chitinase from the psychrophilic strain *Glaciozyma antarctica* PII was analyzed together with chitinases from mesophilic and thermophilic strains (Ramli et al. 2011) showing longer loops and an increased number of aromatic residues compared with homologous enzymes from meso- and thermophiles. The chitinase 60, obtained from *Moritella marina* (Mm Chi 60), has, like other chitinases, a structure formed by four domains having hinge regions in between, which makes it difficult to crystallize. The protein can adapt to other conformations in solution (Malecki et al. 2014).

Cellulases There are some glycoside hydrolases used in bleaching and biostoning of textile material (Gomes and Steiner 2004), and alkaline cellulases used in detergents are active toward amorphous cellulose (Ito et al. 1989).

Pullulanase (EC 3.2.1.41) was obtained from the arctic sea ice bacterium *Shewanella arctica* 40–3 with an optimal temperature of 45 °C but with a range of temperature activity from 10 to 50 °C (Qoura et al. 2014). The enzyme is a type-I pullulanase and belongs to rarely characterized pullulan-degrading enzymes from psychrophiles.

Carbonic anhydrases (EC 4.2.1.1) exist in at least six genetic families. Luca et al. (2015) report the cloning, purification, and physicochemical properties of NcoCA, a γ-CA isolated from the Antarctic cyanobacterium *Nostoc commune*. This enzyme showed a higher catalytic efficiency at lower temperatures compared to mesophilic counterparts belonging to α-, β-, and γ-classes.

Nitrile Hydratase Companies such as NitroChemical developed many years ago the production of acrylamide with the help of *Rhodococcus* sp. strain N774 (Kobayashi et al. 1992), which produces the enzyme nitrile hydratase (EC 4.2.1.84).

Trehalose In agriculture, trehalose-producing systems can be used for reducing crop losses due to the lower temperatures with the help of genetic engineering (Phadtare and Inouye 2008).

EPSs Extracellular polymeric substances are released by microorganisms, which promote the formation of biofilms and have presumably protective roles. They can be used in the chemical industry to produce biodegradable plastic materials. Several strains with potential for this type of production were investigated as well as the conditions for production, such as temperature, pressure, and pH (Marx et al. 2009). The authors reviewed useful microorganisms isolated from cold Antarctic environments, e.g., *Moraxella*, *Psychrobacter*, and *Aeromonas* from polar waters, and psychrotolerants such as *Pseudomonas* and *Photobacterium*. The strains showed a good production of EPS at −4 to −10 °C and resistance under high-pressure conditions between 1 and 200 atm. *Colwellia psychrerythraea* 34H, isolated from Arctic marine sediments, is able to secrete a capsular polysaccharide consisting of monomers of tetrasaccharide units (two amino sugars and two uronic acid molecules), which has a similar activity as antifreeze proteins (Carillo et al. 2015). Marine psychrophilic bacteria such as *Pseudoalteromonas* sp. are producing EPS consisting of heteropolysaccharides, for their protection in low-temperature and saline environments. They can be used in pharmaceutical industries, in foodprocessing industries, and as biosurfactants for petrochemical oil-polluted waters (Poli et al. 2010).

Medicinal Uses Besides improving the quality of foods, antifreeze proteins can also improve the preservability of human organs for transplants, for example, livers (Kawahara 2008). Frisvad (2008a) reviewed bioactive products from cold-adapted fungi, such as griseofulvin and cycloaspeptide A from *Penicillium soppi* and *P. lanosum* from cold soils; the latter compound can be used as an antimalarial product. Cycloaspeptides were found so far only in cold-adapted fungi. Irwin (2010) commented on the importance of extremophiles in veterinary medicine. A main category of extremophilic animal pathogens are psychrophilic and psychrotrophic microorganisms that cause fish diseases, e.g., *Flavobacterium psychrophilum*, *Moritella viscosa*, *Aliivibrio wodanis*, and *Aliivibrio salmonicida*. Duplantis et al. (2010) inserted essential genes from arctic psychrophiles into human pathogenic bacteria such as *Francisella tularensis*, *Salmonella enterica*, and *Mycobacterium smegmatis*, resulting in strains dying at body temperature and even less. These could be used for obtaining live vaccines and be injected in cool body sites. Proteases from cold-adapted microorganism can be used in cosmetics as antacid products and in medicine as boil disruptors, in wound management, and in removing dental plaque and the reduction the infectivity of viruses (Fornbacke and Clarsund 2013).

PUFAs PUFAs are produced by many different organisms, for instance, by strains such as *Shewanella frigidimarina* (Bowman et al. 1997). The strain *Psychroflexus torquis* was compared with the non-psychrophilic closely related *P. gondwanensis*, noting the evidence for horizontal gene transfer. *P. torquis* is an EPS and PUFA producer, and proteins putatively associated with icebinding and lightsensing showed the adaption to the sea ice habitat (Feng et al. 2014). PUFAs can be used as food supplements and medicinal products. Russell and Nichols (1999) showed that the bacterial PUFA-producing strain cannot compete with the fungal PUFA-producing strain but can be an alternative for feedstock in the food chains used in aquaculture. Some yeasts from cold environments produce PUFAs which, due to the decrease of saturation, lead to maintaining fluidity of membranes in conditions of low temperature and the production of compounds inhibiting ice crystallization (Alcaíno et al. 2015). Long-chain PUFAs are used successfully in food industry and as an alternative to fish oils for human consumption (Alcaíno et al. 2015).

Wine Industry Psychrophilic and psychrotolerant yeasts can grow at temperatures of 6–12 °C, secreting pectinolytic enzymes. A psychrophilic yeast *Cystofilobasidium capitatum* SPY11 and a psychrotolerant yeast *Rhodotorula mucilaginosa PT1* were isolated from soil, and their pectin methyl esterase (PME), endo-polygalacturonase (endo-PG), and exopolygalacturonase (exo-PG) were characterized. They are able to act at low temperatures and relatively low pH(3.5) and should be useful for the wine making industry (Sahay et al. 2013).

Biomining The biomining industry is developing processes at low temperatures in three-phase systems: the solid phase, which is represented by the mineral ore; a liquid phase, which contains the microorganisms and nutrients; and the gaseous phase (Rossi 1999). Such a process can be performed in stirred tank reactors or in airlift reactors. The process was used for the release and recovery of copper from sulfide minerals, of uranium, and for the pretreatment of gold ores (Ovalle 1987).

Bioremediation Psychrophilic strains can be used to degrade the organic pollutants from soils and waters at low temperatures. Many strains possessing biodegrading properties were isolated from polar and alpine areas, from soils and waters, but more research of some aspects is required, such as the stability of the bacterial community, the accessibility of the pollutant for microorganisms, and the low removal rate (Margesin and Schinner 2001). Petroleum spills can produce catastrophic damages, and their cleanup is an important goal. Petroleum is a complex mixture of water-soluble and water-insoluble compounds (linear cyclic alkanes, aromatic hydrocarbons, paraffin, asphalt, and waxy oils; Brakstad 2008), being very hard to degrade. About 200 bacterial, cyanobacterial, fungal, and algal genera possess the capacity to do it (Prince 2005). The main bacterial genera able to degrade petroleum are *Acinetobacter*, *Arthrobacter*, *Colwellia*, *Cytophaga*, *Halomonas*, *Marinobacter*, *Marinomonas*, *Pseudoalteromonas*, *Oleispira*, *Rhodococcus*, and *Shewanella* (Brakstad 2008). Alkane hydroxylases from psychrophiles can play a role in the degradation of alkanes and many other components of crude at low temperature, converting them to alcohols. There are an entire set of genes in different psychrophiles, which have been investigated in 19 genomes (Bowman and Deming 2014). Their products have the necessary characteristics for acting in low-temperature conditions. Both anaerobic and aerobic degradation are possible, and bioremediation can occur by stimulation of the local hydrocarbon degraders (using dispersants and nutrients) and less by bioaugmentation (inoculation of cultures of hydrocarbonoclastic bacteria (Margesin and Schinner 1999). *Dietzia psychralkaliphila* can grow on defined culture media containing n-alkenes as sole carbon source (Yumoto et al. 2002). The strain *Rhodococcus* sp. Q15 is able to degrade short- and long-chain aliphatic alkenes from diesel fuel at low temperatures of about 5C (Whyte et al. 1998). Two strains of *Rhodococcus, R. ruber* and *R. erythrococcus*, were used by Russian researchers for the product "Rhoder," which is applied for the removal of oil pollution (Murygina et al. 2000). Some strains isolated from alpine soils are able to degrade phenol and phenolic compounds (bacteria such as *Rhodococcus* spp., *Arthrobacter psychrophenolicus*, and *Pseudomonas*; yeasts such as *Rhodotorula psychrophenolica*, *Trichosporon dulcitum*, and *Leucosporidium watsoni*) and hydrocarbons from oil at low temperatures (Margesin 2007; Margesin et al. 2007), even though a complete biodegradation cannot be obtained. Polychlorophenols are toxic and persistent pollutants which are used as biocidal wood preservatives (Langwaldt et al. 2008). The authors list several genera such as *Ralstonia*, *Burkholderia*, *Arthrobacter*, and *Rhodococcus*, mycobacteria, and anaerobes such as *Desulfomonile* and *Desulfitobacterium*, which are able to degrade polychlorophenol compounds at low temperatures. *Polaromonas* sp. strain JS666 is an isolate which can grow on cis-1, 2-dichloroethene as carbon source; the investigation of its genome showed genes for the metabolism of aromatic compounds, alkanes, alcohols, and others (Mattes et al. 2008). Another strain, *Polaromonas naphthalenivorans*, was isolated from a contaminated freshwater environment and is capable of degrading naphthalene (Jeon et al. 2004). The isolation of the genes for naphthalene dehydrogenase from cold environments was reported (Flocco et al. 2009). For bioremediation, the genus *Shewanella*, which can use a wide range of electron acceptors, is important, since many members of this group show capabilities for degrading several pollutants. The genus showed the possibility to be used in bioremediation of radionuclide and reduction of elements such as Co, Hg, Cr, and As, as well as for the removal of organics such as halogenated compounds, e.g., tetrachloromethane or nitramine (an explosive contaminant), as reported in the review by Hau and Gralnick (2007). Many strains such as *Brevibacterium antarcticum* have demonstrated a polyresistance to heavy metals in high concentrations, resisting concentrations of Cu^{2+}, Hg^{2+}, and CrO_4, up to 6000 ppm (Tashyrev 2009). Several psychrophilic microorganisms are able to degrade natural organic polymers (starch, agar, and gelatin) such as *Flavobacterium limicola*, which was isolated from freshwater sediments (Tamaki et al. 2003).

Wastes and Wastewater Treatments The anaerobic treatment of wastewaters in treatment plants, using expanded granular sludge bed reactors at temperatures of 5–10 °C, looks very promising (Lettinga et al. 2001). A mixture of microorganisms such as *Methanobrevibacter* sp., *Methanosarcina* sp., and *Methanosaeta* sp. has been explored (Lettinga et al. 1999). The psychrophilic treatment of landfill leachates using anoxic/oxic biofilters appears to be a good solution to prevent water and soil pollution with such leachates containing organic matter and also heavy metals (Kalyuzhnyi et al. 2004). *Oleispira antarctica* is able to degrade hydrocarbons in cold marine waters (Yakimov et al. 2003). Aerobic treatment of wastewater in cold lagoons has been performed in Canada's cold areas with success (Smith and Emde 1999). The cold-adapted xylanases can be used for the hydrolysis of agricultural and food industry wastes. A selection of cold-active degrading microorganism for wastewater treatment was performed (Gratia et al. 2009), where *A. psychrolactophilus* Sp 31.3 was isolated, which had the desired characteristics and was used further. The recombinant *Pseudoalteromonas haloplanktis* TAC125 expressed the enzyme toluene xylene monooxygenase which is able to convert aromatic compounds into catechols, making possible the use as remediation of environments polluted with aromatic compounds (Parilli et al. 2010).

Acid mine drainage is the result of oxidation of certain sulfide minerals by exposure to environmental conditions and the activity of microorganisms. For example, ores containing pyrite and chalcopyrites are oxidized in the presence of water and oxygen and form highly acidic, sulfate-rich drainage. Ferrous iron (Fe^{2}) develops in the process, which can be re-oxidized by acidophilic bacteria and archaea to ferric iron (Fe^{3}), and the sulfur is oxidized to sulfate. These oxidations and the concomitant dissolution of sulfide minerals can take place in cold conditions, too. Sulfate reduction at low temperatures occurs with bacteria such as *Desulfofrigus*, *Desulfofaba*, *Desulfotalea*, and *Desulfovibrio* (Kaksonen et al. 2008). *Acidithiobacillus ferrooxidans* is also able to oxidize iron and sulfur compounds at low temperatures (Kaksonen et al. 2008).

Astrobiological Models A special theoretical application concerns astrobiology, since some scientists are considering the Antarctic a model of planet Mars or other planets, with respect to the low temperatures and water activity (Abyzov et al. 1998), and also a model of cold environments where certain microorganisms could possibly live (Deming 2007). At the same time, the protocols for sampling of ice and permafrost and their analysis can be used for the exploration of Mars and could be relevant and helpful in the isolation of potential Martian microbiota and for the development of future protocols for the decontamination of extraterrestrial samples (Christner et al. 2005).

Conclusions

1. The psychrophilic and psychrotolerant microorganisms can be retrieved from very diverse environments – ocean and fresh waters, hypersaline cold water, sediments, soils, permafrost, ice, glaciers, cold deserts, alpine soils, lakes and snow, cold man-made environments, and some micro ecosystems. The microorganisms in ice layers constitute not real ecosystems, even if some activity at subzero temperature was proven; instead, most of them are only opportunistic assemblages of mixtures of microorganisms brought together by air and water

currents from other environments. The diversity of so-called cold environments is much greater than was thought initially, and many microenvironments can be distinguished. In addition, the cold-adapted members of microbiota can have different other adaptations to extreme conditions – resistance to high radiation, oligotrophy, adaptation to high pressures, and perhaps others.

2. Psychrophiles are found in all the three domains of life and have a very diverse taxonomic origin. The most frequent taxonomic groups in cold environments are *Alpha-*, *Beta-*, *Delta-*, and *Gammaproteobacteria*, the phylum *Cytophaga–Flavobacterium–Bacteroidetes*, and *Actinobacteria.* Together with the prokaryotes (Archaea and Bacteria), numerous eukaryotes are present such as algae, yeasts, and fungi.
3. Special adaptations allowing the life in the cold include membrane lipids with branched unsaturated fatty acids, proteins, and enzymes with a more flexible 3D structure due to the reduction of the number of weak intramolecular bonds, reduction of salt bridges, reduction of aromatic interactions, density of charged surface residues, increased surface hydrophobicity, and increased clustering of glycine residues. Special proteins are cold-shock proteins (Csps), ice nucleation proteins, and antifreeze proteins, which protect the structure of the cells from cold and from formation of ice crystals.
4. The molecular biology of psychrophiles showed that the enzymes have low activation energy requirements due to their structure discussed in the text and to the flexibility of near active sites domains and that they are easily inactivated by higher temperatures.
5. The different possibilities of adaptation of the microorganisms, either psychrophiles or psychrotolerants, showed complicated mechanisms, combining adaptation features and environmental opportunities. Their adaptation mechanisms are thus much more flexible than we have thought, providing possibilities and strategies of survival in extreme conditions. There is evidence for survival of psychrophiles in such conditions for very long periods of time which suggests the possibility of survival of similar microorganisms on other planets.
6. Psychrophiles produce bioactive and useful compounds, especially enzymes, pharmaceuticals, biodegradable plastics, substances for medical care, agriculture, biomining, and bioremediation of wastes, all being usable in low-temperature conditions, which entails important energy savings.
7. More laboratory and field bioprospecting should be envisaged for the isolation and identification of appropriate microorganisms for psychrophilic biotechnology.

Acknowledgments We thank Dr. D.S. Nichols and his coworkers for the kind permission to integrate their data on enzymes of psychrophilic microorganisms in the table of this chapter. This work benefitted from the research funded by the Executive Unit for Funding of High Level Education and of Universities Scientific Research, Romania (U.E.F.I.S.C.S.U.) in the frame of contract nr.1, Europolar 2009–2010.

References

Abyzov SS, Mitskevich IN, Poglazova MN, Barkov NI, Lipenkov VY, Bobin NE, Koudryashov BB, Pashkevich VM (1998) Antarctic ice sheet as a model in search of life on other planets. Adv Space Res 22:363–368

Adapa V, Ramya LN, Pulicherla KK, Rao KR (2014) Cold active pectinases: advancing the food industry to the next generation. Appl Biochem Biotechnol 172:2324–2337

Albino A, Marco S, Di Maro A, Chambery A, Masullo M, De Vendittis E (2012) Characterization of a cold-adapted glutathione synthetase from the psychrophile *Pseudoalteromonas haloplanktis*. Mol Biosyst. 8:2405–2414

Alcaíno J, Cifuentes V, Baeza M (2015) Physiological adaptations of yeasts living in cold environments and their potential applications. World J Microbiol Biotechnol 31:1467–1473

Ali MS, Ganasen M, Rahman RN, Chor AL, Salleh AB, Basri M (2013) Cold-adapted RTX lipase from antarctic *Pseudomonas* sp. strain AMS8: isolation, molecular modeling and heterologous expression. Protein J 32:317–325

Allen MA, Lauro FM, Williams TJ, Burg D, Siddiqui KS, De Francisci D, Chong KW, Pilak O, Chew HH, De Maere MZ, Ting L, Katrib M, Ng C, Sowers KR, Galperin MY, Anderson IJ, Ivanova N, Dalin E, Martinez M, Lapidus A, Hauser L, Land M, Thomas T, Cavicchioli R (2009) The genome sequence of the psychrophilic archaeon, *Methanococcoides burtonii*: the role of genome evolution in cold adaptation. ISME J 3:1012–1035

Amato P, Christner BC (2009) Energy metabolism response to low-temperature and frozen conditions in *Psychrobacter cryohalolentis*. Appl Environ Microbiol 75:711–718

Amoresano A, Andolfo A, Corsaro MM, Zocchi I, Petrescu I, Gerday I, Marino G (2000) Structural characterization of a xylanase from psychrophilic yeast by mass spectrometry. Glycobiology 10:451–458

Andersson RE, Hedlund CB, Jonsson U (1979) Thermal inactivation of a heat-resistant lipase produced by the psychrotrophic bacterium *Pseudomonas fluorescens*. J Dairy Sci 62:361–367

Angelaccio S, Florio R, Consalvi V, Festa G, Pascarella S (2012) Serine hydroxymethyltransferase from the cold adapted microorganism *Psychromonas ingrahamii*: a low temperature active enzyme with broad substrate specificity. Int J Mol Sci 13:1314–1326

Aono E, Baba T, Ara T, Nishi T, Nakamichi T, Inamoto E, Toyonaga H, Hasegawa M, Takai Y, Okumura Y, Baba M, Tomita M, Kato C, Oshima T, Nakasone K, Mori H (2010) Complete genome sequence and comparative analysis of *Shewanella violacea*, a psychrophilic and piezophilic bacterium from deep sea floor sediments. Mol Biosyst. 6:1216–1226

Araújo R, Casal M, Cavaco-Paulo A (2008) Application of enzymes for textile fibres processing. Biocatal Biotransform 26:332–349

Auborn KJ, Fan S, Rosen EM, Goodwin L, Chandraskaren A, Williams DE, Chen D, Carter TH (2003) Indole-3-carbinol is a negative regulator of estrogen. J Nutr 133(Suppl):2470–2475

Ayala-del-Rio HL, Chain PS, Grzymski JJ, Ponder MA, Ivanova N, Bergholz PW, Di Bartolo G, Hauser L, Land M, Bakermans C, Rodrigues D, Klappenbach J, Zarka D, Larimer F, Richardson P, Murray A, Thomashow M, Tiedje JM (2010) The genome sequence of *Psychrobacter arcticus* 273-4, a psychroactive Siberian permafrost bacterium, reveals mechanisms for adaptation to lowtemperature growth. Appl Environ Microbiol 76:2304–2312

Ayub ND, Tribelli PM, Lopez N (2009) Polyhydroxyalkanoates are essential for maintenance of redox state in the Antarctic bacterium *Pseudomonas* sp. 14-3 during low temperature adaptation. Extremophiles 13:59–66

Babalola OO, Kirby BM, Le Roes-Hill M, Cook AE, Craig CS, Burton SG, Cowan DA (2008) Phylogenetic analysis of actinobacterial populations associated with Antarctic Dry Valley mineral soils. Environ Microbiol 11:566–576

Bae E, Phillips GN Jr (2004) Structures and analysis of highly homologous psychrophilic, mesophilic, and thermophilic adenylate kinases. J Biol Chem 279:28202–28208

Bahrim G, Negoiţă TG (2004) Effects of inorganic nitrogen and phosphorous sources on hydrolasecomplex production by the selected *Bacillus subtilis* Antarctic strain. Rom Biotechnol Lett 9:1925–1932

Bahrim GE, Scântee M, Negoiţă TG (2007) Biotechnological conditions of amylase and proteasecomplex production and utilization involving filamentous bacteria. In: Annals Univers "Dunărea de Jos" Galaţi, (University Press, Galati, Romania) Fasc VI Food Technol, University Press, Galati, Romania. pp 76–81

Bakermans C (2008) Limits for microbial life at subzero temperatures. In: Margesin R, Schinner R, Marx JC, Gerday C (eds) Psychrophiles: from biodiversity to biotechnology. Springer, Berlin/Heidelberg, pp. 17–28

Bakermans C, Skidmore ML, Douglas S, McKay CP (2014) Molecular characterization of bacteria from permafrost of the Taylor Valley, Antarctica. FEMS Microbiol Ecol 89:331–346

Barkay T, Poulain AJ (2007) Mercury (micro)biogeochemistry in polar environments. FEMS Microbiol Ecol 59:232–241

Beg QK, Kapoor M, Mhajan L, Hoondal GS (2001) Microbial xylanases and their industrial applications – a review. Appl Microbiol Biotechnol 56:326–338

Benešova E, Markova M, Kralova B (2005) Alpha glucosidase and beta glucosidase from psychrophilic strain *Arthrobacter* sp. C2-2. Czech J Food Sci 23:116–120

Białkowska AM, Cieslinski H, Niowakowska KM, Kur J, Turkievich M (2009) A new beta-galactosidase with a low temperature optimum isolated from the Antarctic *Arthrobacter* sp. 20B: gene cloning, purification, characterization. Arch Microbiol 191:825–835

Bjerga GEK, Lale R, Williamson A (2016) Engineering low-temperature expression systems for heterologous production of cold-adapted enzymes. Bioengineered 7:33–38. doi:10.1080/21655979.2015.1128589

Bjerga GEK, Williamson AK (2015) Cold shock induction of recombinant Arctic environmental genes. BMC Biotechnology 15:78. doi:10.1186/s12896-015-0185-1

Blees J, Wenk C, Niemann C, Schubert C, Zopfi J, Veronesi M, Simona M, Lehmann M (2010) The isotopic signature of methane oxidation in a deep south-alpine lake. Geophysical Research Abstracts 12:EGU2010–EGU2788

Bowman JS, Deming JW (2014) Alkane hydroxylases gene in psychrophilic genome and the potential for cold active catalysis. BMC Genomics 15:1120. doi:10.1186/1471-2164-15-1120

Bowman JP (2008) Genomic analysis of psychrophilic prokaryotes. In: Margesin R, Schinner F, Marx JC, Gerday C (eds) Psychrophiles: from biodiversity to biotechnology. Springer, Berlin/Heidelberg, pp. 265–284

Bowman JP, McCammon SA, Nichols DS, Skerratt JH, Rea SM, Nichols PD, McMeekin TA (1997) *Shewanella gelidimarina* sp. nov. and*Shewanella frigidimarina* sp. nov., novel Antarctic species with the ability to produce eicosapentaenoic acid (20:5 omega 3) and grow anaerobically by dissimilatory Fe(III) reduction. Int J Syst Bacteriol 47:1040–1047

Bowman JP, Gosink JJ, McCammon SA, Lewis TE, Nichols DS, Nichols PD, Skeratt JH, Staley JT, McMeekin TA (1998) *Colwellia demingiae* sp. nov., *Colwellia hornerae* sp. nov., *Colwellia rossensis* sp. nov. and*Colwellia psychrotropica* sp. nov.: psychrophilic Antarctic species with the ability to synthesize docosahexaenoic acid (22:6v3). Int J Syst Bacteriol 48:1171–1180

Bozal N, Montes MJ, Tudela E, Jimenez F, Guinea J (2002) *Shewanella frigidimarina* and *Shewanella livingstonensis* sp. nov. isolated from Antarctic coastal areas. Int J Syst Evol Microbiol 52:195–205

Brakstad OG (2008) Natural and stimulated biodegradation of petroleum in cold marine environments. In: Margesin R, Schinner F, Marx JC, Gerday C (eds) Psychrophiles: from biodiversity to biotechnology. Springer, Berlin/Heidelberg, pp. 389–428

Breezee J, Cady N, Staley JT (2004) Subfreezing growth of the sea ice bacterium *Psychromonas ingrahamii*. Microbial Ecol 47:300–304

Brenchley JE (1996) Psychrophilic microorganisms and their cold-active enzymes. J Ind Microbiol Biotechnol 17:432–437

Budiman C, Koga Y, Takano K, Kanaya S (2011) FK506-Binding protein 22 from a psychrophilic bacterium,a cold shock-inducible peptidyl prolyl isomerase with the ability to assist in protein folding. Int J Mol Sci 12:5261–5284. doi:10.3390/ijms1208526

Bujacz A, Rutkiewicz-Krotewicz M, Nowakowska-Sapota K, Turkiewicz M (2015) Crystal structure and enzymatic properties of a broad substrate-specificity psychrophilic aminotransferase from the Antarctic soil bacterium *Psychrobacter* sp. B6. Acta CrystallogD Biol Crystallogr 71:632–645. doi:10.1107/S1399004714028016

Busse H-J, Denner EBM, Buczolits S, Salkinoja-Salonen M, Bennasar A, Kämpfer P (2003) *Sphingomonas aurantiaca* sp. nov., *Sphingomonas aerolata* sp. nov. and *Sphingomonas faeni* sp. nov., air- and dustborne and Antarctic, orange pigmented, psychrotolerant bacteria, and emended description of the genus *Sphingomonas*. Int J Syst Evol Microbiol 53:1253–1260

Buzzini P, Branda E, Goretti M, Turchetti B (2012) Psychrophilic yeasts from worldwide glacial habitats: diversity, adaptation strategies and biotechnological potential. FEMS Microbiol Ecol 82:217–241

Carillo S, Casillo A, Pieretti G, Parrilli E, Sannino F, Bayer-Giraldi M, Cosconati S, Novellino E, Ewert M, Deming JW, Lanzetta R, Marino G, Parrilli M, Randazzo A, Tutino ML, Corsaro MM (2015) A unique capsular polysaccharide structure from the psychrophilic marine bacterium *Colwellia psychrerythraea* 34H that mimics antifreeze (glyco) proteins. J Am Chem Soc. 137:179–189

Cartier G, Lorieux F, Allemand F, Dreyfus M, Bizebard T (2010) Cold adaptation in DEAD-box proteins. Biochemistry 49:2636–2646

Cavicchioli R (2006) Cold adapted Archaea. Nat Rev Microbiol 4:331–343

Cavicchioli R, Siddiqui KS (2004) Cold-adapted enzymes. In: Paney A, Webb C, Socol CR, Larroche C (eds) Enzyme technology. Asia Tech Publishers, New Delhi, pp. 615–638

Cavicchioli R, Siddiqui KS, Andrews D, Sowers KR (2002) Low temperature extremophiles and their applications. Curr Opin Biotechnol 13:253–261

Cavicchioli R, Charlton T, Ertan H, Mohd Omar S, Siddiqui KS, Williams TJ (2011) Biotechnological uses of enzymes from psychrophiles. Microb Biotechnol 4:449–460

Chen Z, Feng D, Zhang B, Wang Q, Luo Y, Dong X (2015) Proteomic insights into the temperature responses of a cold-adaptive archaeon *Methanolobus psychrophilus* R15. Extremophiles 19:249–259

Chen B, Hu J, Miller EM, Xie W, Cai M, Gross RA (2008a) *Candida antarctica* lipase B chemically immobilized on epoxy-activated micro- and nanobeads: catalysts for polyester synthesis. Biomacromolecules 9:463–470

Chen B, Miller ME, Gross RA (2008b) Immobilization of *Candida antarctica* lipase B on porous polystyrene resins: protein distribution and activity. In: Polymer biocatalysis and biomaterials II, ACS symposium series, vol 999. American Chemical Society, Washington, DC, pp. 165–177

Chintalapati S, Kiran MD, Shivaji S (2004) Role of membrane lipid fatty acids in cold adaptation. Cell Mol Biol 50:631–642

Christner BC (2002) Incorporation of DNA and protein precursors into macromolecules by bacteria at -15 °C. Appl Environ Microbiol 68:6435–6438

Christner BC, Mosley-Thompson E, Thompson LG, Reeve JN (2003a) Bacterial recovery from ancient glacial ice. Environ Microbiol 5:433–436

Christner BC, Kvitko BH, Reeve JN (2003b) Molecular identification of bacteria and eukarya inhabiting an Antarctic cryoconite hole. Extremophiles 7:177–183

Christner BC, Mikucki JA, Foreman CM, Denson J, Priscu JC (2005) Glacial ice cores: a model system for developing extraterrestrial decontamination protocols. Icarus 174:572–584

Christner BC, Skidmore ML, Priscu JC, Tranter M, Foreman C (2008) Bacteria in subglacial environments. In: Margesin R, Schinner F, Marx JC, Gerday C (eds) Psychrophiles: from biodiversity to biotechnology. Springer, Berlin/Heidelberg, pp. 51–71

Cipolla A, Delbrassine F, Da Lage JL, Feller G (2012) Temperature adaptations in psychrophilic, mesophilic and thermophilic chloride-dependent alpha-amylases. Biochimie 94:1943–1950

Cohen Z (1990) The production potential of eicosapentaenoic acid and arachidonic acid of the red algae *Porphyridium cruentum*. J Am Oil Chem Soc 67:916–920

Collins T, Hoyoux A, Dutron A, Georis J, Genot B, Dauvrin T, Arnaut F, Gerday C, Feller G (2006) Use of glycoside hydrolase family 8 xylanases in baking. J Cereal Sci 43:79–84

Collins T, Roulling F, Piette F, Marx JC, Feller G, Gerday C, D'Amico S (2008) Fundamentals of cold adapted enzymes. In: Margesin R, Schinner F, Marx JC, Gerday C (eds) Psychrophiles: from biodiversity to biotechnology. Springer, Berlin/Heidelberg, pp. 211–228

Coolen MJL, Orsi WD (2015) The transcriptional response of microbial communities in thawing Alaskan permafrost soils. Front. Microbiol. 6:197. doi:10.3389/fmicb.2015.00197

Copeland A, Lucas S, Lapidus A, Barry K, Detter JC, Glavina del Rio T, Hammon N, Israni S, Dalin E, Tice H, Pitluck S, Fredrickson JK, Kolker E, McCuel LA, DiChristina T, Nealson KH, Newman D, Tiedje JM, Zhou J, Romine MF, Culley DE, Serres M, Chertkov O, Brettin T, Bruce D,Han CTapia R, Gilna P, Schmutz J, Larimer F, Land M, Hauser L, Kyrpides N, Mikailova N, Richardson P (2006) Complete sequence of *Shewanella frigidimarina* NCIMB 400. Submitted (Aug. Sept. 2006). Released 09/14/2006 by the DOE Joint Genome Institute.

Corkrey R, McMeekin TA, Bowman JP, Ratkowsky DA, Olley J et al (2014) Protein thermodynamics can be predicted directly from biological growth rates. PLoS ONE 9(5):e96100. doi:10.1371/journal.pone.0096100

Cotârleț M, Negoiță TG, Bahrim G, Stougaard P (2008) Screening of polar streptomycetes able to produce cold-active hydrolytic enzymes using common and chromogenic substrates. Rom Biotechnol Lett 13:69–80. [special issue, edited for Int Conf Ind Microbiol Appl Biotechnol]

Cowan DA, Casanueva A (2007) Stability of ATP in Antarctic mineral soils. Polar Biol 30:1599–1603

Cowan DA, Makhalanyane TP, Dennis PG, Hopkins DW (2014) Microbial ecology and biogeochemistry of continental Antarctic soils. Front Microbiol 5:154. doi:10.3389/fmicb.2014.00154

Cristóbal HA, Benito J, Lovrich GA, Abate CM (2015) Phylogenentic and enzymatic characterization of psychrophilic and psychrotolerant marine bacteria belong to γ-Proteobacteria group isolated from the sub-Antarctic Beagle Channel. Argentina.Folia Microbiol (Praha). 60:183–198

Cristóbal HA, López MA, Kothe E, Abate CM (2011) Diversity of protease-producing marine bacteria from sub-antarctic environments. J Basic Microbiol. 51:590–600

Dalmaso GZL, Davis Ferreira, Alane Beatriz Vermelho Marine (2015) Extremophiles: a source of hydrolases for biotechnological applications. Mar Drugs 13:1925–1965

D'Amico S, Collins T, Marx JC, Feller G, Gerday C (2006) Psychrophilic microorganisms: challenges for life. EMBO Rep 7:385–389

de Garcia V, Brizzio S, Broock MR (2012) Yeasts from glacial ice of Patagonian Andes, Argentina. FEMS Microbiol Ecol 82:540–550

de los Rios A, Grube M, LG S, Ascaso C (2007) Ultrastructural and genetic characteristics of endolithic cyanobacterial biofilms colonizing Antarctic granite rocks. FEMS Microbiol Ecol 59:386–395

De Luca V, Del Prete S, Vullo D, Carginale V, Di Fonzo P, Osman SM, AlOthman Z, Supuran CT, Capasso C (2015) Expression and characterization of a recombinant psychrophilic γ-carbonic anhydrase (NcoCA) identified in the genome of the Antarctic cyanobacteria belonging to the genus *Nostoc*. J Enzyme Inhib Med Chem 30:1–8

De Maayer P, Anderson D, Cary C, Cowan DA (2014) Some like it cold: understanding the survival strategies of psychrophiles. EMBO Reports 15:508–517

Deming JW (2007) Life in ice formations at very cold temperatures. In: Gerday C, Glansdorff N (eds) Physiology and biochemistry of extremophiles. ASM Press, Washington, DC, pp. 133–145

De Prada P, Loveland-Curtze J, Brenchley JE (1996) Production of two extracellular alkaline phosphatases by a psychrophilic *Arthrobacter* strain. Appl Environ Microbiol 62:3732–3738

De Santi C, Durante L, Vecchio PD, Tutino ML, Parrilli E, de Pascale D (2012) Thermal stabilization of psychrophilic enzymes: a case study of the cold-active hormone-sensitive lipase from *Psychrobacter* sp. TA144. Biotechnol Prog 28:946–952

Di Prisco G (2007) Lake Vostok and subglacial lakes of Antarctica: do they host life? In: Gerday C, Glansdorff F (eds) Physiology and biochemistry of extremophiles. ASM Press, Washington, DC, pp. 145–154

Do H, Lee JH, Kwon MH, Song HE, An JY, Eom SH, Lee SG, Kim HJ (2013) Purification, characterization and preliminary X-ray diffraction analysis of a cold-active lipase (CpsLip) from the psychrophilic bacterium *Colwellia psychrerythraea* 34H. Acta Crystallogr Sect F Struct Biol Cryst Commun 69(Pt 8):920–924

Doaa Mahmoud AR, Wafaa Helmy A (2009) Application of cold-active dextranase in dextran degradation and isomaltotriose synthesis by micro-reaction technology. Aust J Basic Appl Sci 3:3808–3817

Dong S, Yang J, Zhang XY, Shi M, Song XY, Chen XL, Zhang YZ (2012) Cultivable alginate lyase-excreting bacteria associated with the Arctic brown alga *Laminaria*. Mar Drugs 10:2481–2491

Dong X, Chen Z (2012) Psychrotolerant methanogenic archaea: diversity and cold adaptation mechanisms. Sci China Life Sci. 55(5):415–421. doi:10.1007/s11427-012-4320-0

Duplantis BN, Osusky M, Schmerk CL, Ross DR, Bosio CM, Nano FE (2010) Essential genes from Arctic bacteria used to construct stable, temperature-sensitive bacterial vaccines. Proc Natl Acad Sci USA 107:13456–13460

Dziewit L, Bartosik D (2014) Plasmids of psychrophilic and psychrotolerant bacteria and their role in adaptation to cold environments. Front Microbiol 5:596. doi:10.3389/fmicb.2014.00596

Feller G (2013) Psychrophilic enzymes: from folding to function and biotechnology.Scientifica (Cairo). ArticleID512840, http://dx.doi.org/10.1155/2013/512840

Feller G (2010) Protein stability and enzyme activity at extreme biological temperatures. J Phys Condens Matter 22:323101. doi:10.1088/0953-8984/22/32/323101

Feller G (2007) Life at low temperatures: is disorder the driving force? Extremophiles 11:211–216

Feller G, Gerday C (1997) Psychrophilic enzymes: molecular basis of cold-adaptation. Cell Mol Life Sci 53:830–841

Feller G, Le Bussy O, Gerday C (1998) Expression of psychrophilic genes in mesophilic hosts: assessment of the folding state of a recombinant a-amylase. Appl Environ Microbiol 64:1163–1165

Feng S, Powell SM, Wilson R, Bowman JP (2014) Extensive gene acquisition in the extremely psychrophilic bacterial species *Psychroflexus torquis* and the link to sea-ice ecosystem specialism. Genome Biol Evol 6:133–148

Ferrara M, Guerriero G, Cardi M, Esposito S (2013) Purification and biochemical characterisation of a glucose-6-phosphate dehydrogenase from the psychrophilic green alga *Koliella antarctica*. Extremophiles 17:53–62

Finster K (2008) Anaerobic bacteria and archaea in cold ecosystems. In: Margesin R, Schinner F, Marx JC, Gerday C (eds) Psychrophiles: from biodiversity to biotechnology. Springer, Berlin/Heidelberg, pp. 103–119

Flocco CG, Newton C, Gomes M, MacCormack W, Smalla K (2009) Occurrence and diversity of naphthalene dioxygenase genes in soil microbial communities from the maritime Antarctic. Environ Microbiol 11:700–714

Foreman CM, Wolf CF, Priscu JC (2004) Impact of episodic warming events on the physical, chemical and biological relationships of lakes in the McMurdo Dry Valley, Antarctica. Aq Geochem 10:239–268

Fornbacke M, Clarsund M (2013) Cold-adapted proteases as an emerging class of therapeutics. Infect Dis Ther 2:15–26

Frisvad JC (2008a) Cold adapted fungi as source of valuable metabolites. In: Margesin R, Schinner F, Marx JC, Gerday C (eds) Psychrophiles: from biodiversity to biotechnology. Springer, Berlin/Heidelberg, pp. 381–388

Frisvad JC (2008b) Fungi in cold environment. In: Margesin R, Schinner F, Marx JC, Gerday C (eds) Psychrophiles: from biodiversity to biotechnology. Springer, Berlin/Heidelberg, pp. 137–156

Galante YM, Formantici C (2003) Enzyme applications in detergency and in manufacturing industries. Curr Org Chem 7:1399–1422

Ganzert L, Jurgens G, Münster U, Wagner D (2007) Methanogenic communities in permafrost affected soils of the Laptev Sea Coast, Siberian Arctic, characterized by 16S rRNA gene fingerprints. FEMS Microbiol Ecol 59:476–488

García-Echauri SA, Gidekel M, Gutiérrez-Moraga A, Santos L, De León-Rodríguez A (2011) Isolation and phylogenetic classification of culturable psychrophilic prokaryotes from the Collins glacier in the Antarctica. Folia Microbiol (Praha). 56:209–214

Georlette D, Damien B, Blaise V, Depiereux E, Uversky VN, Gerday C, Feller G (2003) Structural and functional adaptations to extreme temperatures in psychrophilic, mesophilic and thermophilic DNA ligases. J Biol Chem 278:37015–37023

Gesheva V (2009) Distribution of psychrophilic microorganisms in soils of Terra Nova Bay and Edmonson Point, Victoria Land and their biosynthetic capabilities. Polar Biol 32:1287–1291

Ghosh M, Pulicherla KK, Rekha VP, Raja PK, Sambasiva Rao KR (2012) Cold active β-galactosidase from *Thalassospira* sp. 3SC-21 to use in milk lactose hydrolysis: a novel source from deep waters of Bay-of-Bengal. World J Microbiol Biotechnol 28:2859–2869

Gianese G, Argos P, Pascarella S (2001) Structural adaptation of enzymes to low temperatures. Prot Eng 14:141–148

Giaquinto L, Curmi PMG, Siddiqui KS, Poljak A, DeLong E, DasSarma S, Cavicchioli R (2007) Structure and function of cold shock proteins in archaea. J Bacteriol 189:5738–5748

Gilichinsky DA (2002) Permafrost. In: Bitton G (ed) Encyclopedia of environmental microbiology. John Wiley & Sons, New York, pp. 2367–2385

Gilichinsky D, Vishnivetskaya T, Petrova M, Spirina E, Mamkyn V, Rivkina E (2008) Bacteria in Permafrost. In: Margesin R, Schinner F, Marx JC, Gerday C (eds) Psychrophiles: from biodiversity to biotechnology. Springer, Berlin/Heidelberg, pp. 83–102

Ginting EL, Iwasaki S, Maeganeku C, Motoshima H, Watanabe K (2014) Expression, purification, and characterization of cold-adapted inorganic pyrophosphatase from psychrophilic *Shewanella* sp. AS-11. Prep Biochem Biotechnol 44:480–492. doi:10.1080/10826068.2013.833114

Gittel A, Bárta J, Kohoutová I, Schnecker J, Wild B, Čapek P, Kaiser C, Torsvik VL, Richter A, Schleper C, Urich T (2014) Site- and horizon-specific patterns of microbial community structure and enzyme activities in permafrost-affected soils of Greenland. Front Microbiol 5:541. doi:10.3389/fmicb.2014.00541

Gomes J, Steiner W (2004) The biocatalytic potential of extremophiles and extremozymes. Food Technol Biotechnol 42:223–235

Goomber S, Kumar A, Singh R, Kaur J (2016) Point mutation Ile137-Met near surface conferred psychrophilic behaviour and improved catalytic efficiency to *Bacillus* lipase of 1.4 subfamily. Appl Biochem Biotechnol 178:753–765

Goto S, Sugiyama J, Iizuka HA (1969) A taxonomical study of Antarctic yeasts. Mycologia 61:748–774

Gounot AM (1999) Microbial life in permanently cold soils. In: Margesin R, Schinner F (eds) Coldadapted organisms. Springer, Berlin/Heidelberg/New York, pp. 3–15

Gratia E, Weekers F, Margesin R, D'Amico S, Thonart P, Feller G (2009) Selection of a coldadapted bacterium for bioremediation of wastewater at low temperatures. Extremophiles 13:763–768

Groudieva T, Kambourova M, Yusef H, Royter M, Grote R, Trinks H, Antranikian G (2004) Diversity and cold-active hydrolytic enzymes of culturable bacteria associated with Arctic sea ice, Spitzbergen. Extremophiles 8:475–488

Grzymski JJ, Carter BJ, DeLong EF, Feldman RA (2006) Comparative genomics of DNA fragments from six Antarctic marine planktonic bacteria. Appl Environ Microbiol 72:1532–1541

Guffogg SP, Thomas-Hall S, Holloway P, Watson K (2004) A novel psychrotolerant member of the hymenomycetous yeasts from Antarctica: *Cryptococcus watticus* sp. nov. Int J Syst Evol Microbiol 54:275–277

Hamamoto T (1993) Psychrophilic microorganisms from deep sea environments. Riken Rev 3:9–10

Hasan F, Shah AA, Javed S, Hameed A (2010) Enzymes used in detergents: lipases. African J Biotechnol 9:4836–4844

Hau HH, Gralnick JA (2007) Ecology and biotechnology of the genus *Shewanella*. Ann Rev Microbiol 61:237–258

Hao JH, Sun M (2015) Purification and characterization of a cold alkaline protease from a psychrophilic *Pseudomonas aeruginosa* HY1215. Appl Biochem Biotechnol 175:715–722

Helmke E, Weyland H (2004) Psychrophilic versus psychrotolerant bacteria – occurrence and significance in polar and temperate marine habitats. Cell Mol Biol 50:553–561

Hodson A, Anesio AM, Tranter M, Fountain A, Osborn M, Priscu J, Laybourn-Parry J, Sattler B (2008) Glacial ecosystems. Ecol Monogr 78:41–67

Hollibaugh JT, Lovejoy C, Murray AE (2007) Microbiology in polar ocean. Oceanography 20:140–147

Hou S, Saw JH, Lee KS, Freitas TA, Belisle C, Kawarabayasi Y, Donachie SP, Pikina A, Galperin MY, Koonin EV, Makarova KS, Omelchenko MV, Sorokin A, Wolf YI, Li QX, Keum YS, Campbell S, Denery J, Aizawa S, Shibata S, Malahoff A, Alam M (2004) Genome sequence of the deep-sea gamma-proteobacterium *Idiomarina loihiensis* reveals amino acid fermentation as a source of carbon and energy. Proc Natl Acad Sci USA 101:18036–18041

Hoyoux IJ, Dubois P, Genicot S, Dubail F, Franc JM, Baise Ois E, Feller G, Gerday C (2001) Coldadapted beta-galactosidase from the Antarctic psychrophile *Pseudoalteromonas haloplanktis*. Appl Environ Microbiol 67:1529–1535

Humphry DR, George A, Black GW, Cummings SP (2001) *Flavobacterium frigidarium* sp. nov., an aerobic, psychrophilic, xylanolytic and laminarinolytic bacterium from Antarctica. Int J Syst Evol Microbiol 51:1235–1243

Huston AL (2008) Biotechnological aspects of cold adapted enzymes. In: Margesin R, Schinner F, Marx JC, Gerday C (eds) Psychrophiles: from biodiversity to biotechnology. Springer, Berlin/Heidelberg, pp. 347–364

Irwin JA (2010) Extremophiles and their application to veterinary medicine. Environ Technol 31:857–869

Isaksen GV, Åqvist J, Brandsdal BO (2014) Protein surface softness is the origin of enzyme cold-adaptation of trypsin. PLoS Comput Biol 10(8):e1003813. doi:10.1371/journal.pcbi.1003813

Ishida Y, Tsuruta H, Tsuneta ST, Uno T, Watanabe K, Aizono I (1998) Characteristics of psychrophilic alkaline phosphatase. Biosci Biotechnol Biochem 62:2246–2250

Ito S, Shikata S, Ozaki K, Kawai S, Okamoto K, Inoue S, Takei A, Ohta Y, Satoh T (1989) Alkaline cellulases for laundry detergents production by *Bacillus* sp. KSM 635 and enzymatic properties. Agric Biol Chem 53:1275–1281

Ivanova EP, Gorshkova NM, Bowman JP, Lysenko AM, Zhukova NV, Sergeev AF, Mikhailov VV, Nicolau DV (2004) *Shewanella pacifica* sp. nov., a polyunsaturated fatty acid-producing bacterium isolated from sea water. Int J Syst Evol Microbiol 54:1083–1087

Jensen DB, Vesth TC, Hallin PF, Pedersen AG, Ussery DW (2012) Bayesian prediction of bacterial growth temperature range based on genome sequences. BMC Genomics 13(Suppl 7):S3. doi: 10.1186/1471-2164-13-S7-S3

Jeon CO, Park W, Ghiorse WC, Madsen EL (2004) *Polaromonas naphthalenivorans* sp. nov., a naphthalene-degrading bacterium from naphthalene-contaminated sediment. Int J Syst Evol Microbiol 54:93–97

Jónsdóttir LB, Ellertsson BO, Invernizzi G, Magnúsdóttir M, Thorbjarnardóttir SH, Papaleo E, Kristjánsson MM (2014) The role of salt bridges on the temperature adaptation of aqualysin I, a thermostable subtilisin-like proteinase. Biochim Biophys Acta 1844:2174–2181

Joseph B, Ramteke PW, Thomas G (2008) Cold active microbial lipases: some hot issues and recent developments. Biotechnol Adv 26:457–470

Joseph B, Ramteke PW, Thomas G, Shrivastava N (2007) Standard review. Cold-active microbial lipases: a versatile tool for industrial applications. Biotechnol Mol Biol Rev 2:39–48

Joshi S, Satyanarayana T (2013) Biotechnology of cold-active proteases. Biology (Basel) 2:755–783

Junge K, Eicken H, Deming JW (2003) Motility of *Colwellia psychrerythraea* strain 34H at subzero temperatures. Appl Environ Microbiol 69:4282–4284

Kahlke T, Thorvaldsen S (2012) Molecular characterization of cold adaptation of membrane proteins in the Vibrionaceae core-genome. PLoS ONE 7(12):e51761. doi:10.1371/journal.pone.0051761

Kaksonen AH, Dopson M, Karnachuk O, Tuovinen OH, Puhakka JA (2008) Biological iron oxidation and sulfate reduction in the treatment of acid mine drainage at low temperatures. In: Margesin R, Schinner F, Marx JC, Gerday C (eds) Psychrophiles: from biodiversity to biotechnology. Springer, Berlin/Heidelberg, pp. 429–454

Kalyuzhnyi SV, Gladchenko M, Epov A (2004) Combined anaerobic-aerobic treatment of landfill leachates under mesophilic, submesophilic and psychrophilic conditions. Water Sci Technol 48:311–318

Kamal MZ, Mohammad TAS, Krishnamoorthy G, Rao NM (2012) Role of active site rigidity in activity: MD simulation and fluorescence study on a lipase mutant. PLoS ONE 7(4):e35188. doi:10.1371/journal.pone.0035188

Karasova P, Spiwok V, Mala S, Kralova B, Russel NJ (2002) Beta-galactosidase activity in psychrotrophic microorganisms and their potential use in food industry. Czech J Food Sci 20:43–47

Karr EA, Sattley WM, Jung DO, Madigan MT, Achenbach LA (2003) Remarkable diversity of phototrophic purple bacteria in permanently frozen Antarctic lake. Appl Environ Microbiol 69:4910–4914

Karr EA, Ng JM, Belchik SM, Sattley WM, Madigan MT, Achenbach LA (2006) Biodiversity of methanogenic and other archaea in the permanently frozen lake Fryxwell, Antarctica. Appl Environ Microbiol 72:1662–1666

Katayama T, Tanaka M, Moriizumi J, Nakamura T, Brouchkov A, Douglas TA, Fukuda M, Tomita F, Asano K (2007) Phylogenetic analysis of bacteria preserved in a permafrost ice wedge for 25,000 years. Appl Environ Microbiol 73:2360–2363

Kato Y, Sakala RM, Hayashidani H, Kiuchi A, Kaneuchi C, Ogawa M (2000) *Lactobacillus algidus* sp. nov., a psychrophilic lactic acid bacterium isolated from vacuum-packaged refrigerated beef. Int J Syst Evol Microbiol 50:1143–1149

Kawahara H (2008) Cryoprotectant and ice binding proteins. In: Margesin R, Schinner F, Marx JC, Gerday C (eds) Psychrophiles: from biodiversity to biotechnology. Springer, Berlin/Heidelberg, pp. 229–246

Khachane AN, Timmis KN, Martins dos Santos VAP (2005) Uracil content of 16S rRNA of thermophilic and psychrophilic prokaryotes correlates inversely with their optimal growth temperatures. Nucleic Acids Res 33:4016–4022

Kim H-R, Kim I-H, Hou CT, K-Il K, Shin B-S (2010) Production of a novel cold-active lipase from *Pichia lynferdii* Y-7723. J Agric Food Chem 58:1322–1326

Kitamoto D, Ikegami T, Suzuki GT, Sasaki A, Takeyama Y, Idemoto Y, Koura N, Yanagishita H (2001) Microbial conversion of n-alkanes into glycolipid biosurfactants, annosylerythritol lipids by *Pseudozyma* (*Candida antarctica*). Biotechnol Lett 23:1709–1714

Knoblauch C, Sahm K, Jorgensen BB (1999) Psychrophilic sulfate-reducing bacteria isolated from permanently cold Arctic marine sediments: description of *Desulfofrigus oceanense* gen. nov., sp. nov., *Desulfofrigus fragile* sp. nov., *Desulfofaba gelida* gen. nov., sp. nov., *Desulfotalea psychrophila* gen. nov., sp. nov. and *Desulfotalea arctica* sp. nov. Int J Syst Bact 49:631–643

Kobayashi M, Nagasawa T, Yamada H (1992) Enzymatic synthesis of acrylamide manufacturing process using microorganisms. Trends Biotechnol 10:402–408

Kovacic F, Mandrysch A, Poojari C, Strodel B, Jaeger KE (2016) Structural features determining thermal adaptation of esterases. Protein Eng Des Sel 29:65–76

Kraft B, Engelen B, Goldhammer T, Lin YS, Cypionka H, Könneke M (2013) *Desulfofrigus* sp. prevails in sulfate-reducing dilution cultures from sediments of the Benguela upwelling area. FEMS Microbiol Ecol 84:86–97

Krembs C, Deming JW (2008) The role of exopolymers in microbial adaptation to sea ice. In: Margesin R, Schinner F, Marx JC, Gerday C (eds) Psychrophiles: from biodiversity to biotechnology. Springer, Berlin/Heidelberg, pp. 247–286

Krishnan KP, Sinha RK, Krishna K, Nair S, Singh SM (2009) Microbially mediated redox transformations of manganese (II) along with some other trace elements: a study from Antarctic lakes Polar Biol 32:1765–1778

Kuddus M, Ramteke PW (2012) Recent developments in production and biotechnological applications of cold-active microbial proteases. Crit Rev Microbiol 38:330–338

Kuhn E (2012) Toward understanding life under subzero conditions: the significance of exploring psychrophilic "cold-shock" proteins. Astrobiology 12:1078–1086

Kulakova L, Galkin A, Kurihara T, Yoshimura T, Esaki N (1999) Cold-active serine alkaline protease from the psychrotrophic bacterium *Shewanella* strain Ac10: gene cloning and enzyme purification and characterization. Appl Environ Microbiol 65:611-617

Kurihara T, Esaki N (2008) Proteomic studies of psychrophilic microrganisms. In: Margesin R, Schinner F, Marx JC, Gerday C (eds) Psychrophiles: from biodiversity to biotechnology. Springer, Berlin/Heidelberg, pp. 333–346

Kwon KK, Lee HS, Yang SH, Kim SJ (2005) *Kordiimonas gwangyangensis* gen. nov., sp. nov., a marine bacterium isolated from marine sediments that forms a distinct phyletic lineage (*Kordiimonadales* ord. nov.) in the 'Alphaproteobacteria'. Int J Syst Evol Microbiol 55:2033–2037

Langwaldt JH, Tirola M, Puhakka JA (2008) Microbial adaptation to boreal saturated subsurface: implication in bioremediation of polychlorophenols. In: Margesin R, Schinner F, Marx JC, Gerday C (eds) Psychrophiles: from biodiversity to biotechnology. Springer, Berlin/Heidelberg, pp. 409–428

Larose C, Berger S, Ferrari C, Navarro E, Dommergue A, Schneider D, Vogel TM (2010) Microbial sequences retrieved from environmental samples from seasonal Arctic snow and meltwater from Svalbard, Norway. Extremophiles 14:205–212

Larouche JR, Bowden WB, Giordano R, Flinn MB, Crump BC (2012) Microbial biogeography of arctic streams: exploring influences of lithology and habitat. Front Microbiol 3:309. doi:10.3389/fmicb.2012.00309

Law BA, Goodenough PW (1995) Enzymes in milk and cheese production. In: Tucker GA, Woods LFJ (eds) Enzymes in food processing, 2nd edn. Blackie Academic and Professional, Bishop Briggs/Glasgow, UK, pp. 114–143

Lee HK, Ahn MJ, Kwak SH, Song WH, Jeong BC (2003) Purification and characterization of cold active lipase from psychrotrophic *Aeromonas* sp. LPB 4. J Microbiol 41:22–27

Lee CC, Smith MR, Accinelli R, Williams TG, Wagschal KC, Wong D, Robertson GH (2006) Isolation and characterization of a psychrophilic xylanase enzyme from *Flavobacterium* sp. Curr Microbiol 52:112–116

Lee SG, Koh HY, Han SJ, Park H, Na DC, Kim IC, Lee HK, Yim JH (2010) Expression of recombinant endochitinase from the Antarctic bacterium, *Sanguibacter antarcticus* KOPRI 21702 in *Pichia pastoris* by codon optimization. Protein Expr Purif 71:108–114

Lees RS (1990) Impact of dietary fats on human health. In: RS L, Karel M (eds) Omega-3 fatty acids in health and disease, Series: Food Science and Technology. Marcel Dekker, New York, pp. 1–38

Lettinga G, Rebac S, van Lier J, Zeman G (1999) The potential of sub-mesophilic and/or psychrophilic anaerobic treatment of low strength wastewaters. In: Margesin R, Schinner F (eds) Biotechnological applications of cold adapted organisms. Springer, Berlin/Heidelberg, pp. 221–234

Lettinga G, Rebac S, Zeeman G (2001) Challenge of psychrophilic anaerobic wastewater treatment. Trends Biotechnol 19:363–370

Liebner S, Wagner D (2007) Abundance, distribution and potential activity of methane oxidizing bacteria in permafrost soils from the Lena Delta, Siberia. Environ Microbiol 9:107–117

Liu Q, Liu H, Zhang J, Zhou Y, Xin Y (2013) *Cryobacterium levicorallinum* sp. nov., a psychrophilic bacterium isolated from glacier ice. Int J Syst Evol Microbiol 63:2819–2822

Lo Giudice A, Bruni V, Michaud L (2007) Characterization of Antarctic psychrotrophic bacteria with antibacterial activities against terrestrial microorganisms. J Bas Microbiol 47:496–505

Lorv JSH, Rose DR, Glick BR (2014) Bacterial ice crystal controlling proteins. Scientifica Article ID 976895, 20 pages. http://dx.doi.org/10.1155/2014/976895

Loveland-Curtze J, Miteva V, Brenchley JE (2009) *Herminiimonas glaciei* sp. nov., a novel ultramicrobacterum from 3042 m deep Greenland glacial ice. Int J Syst Evol Microbiol 59:1272–1127

Loveland-Curtze J, Miteva V, Brenchley J (2010) Novel ultramicrobacterial isolates from a deep Greenland icecore represent a proposed new species *Chryseobacterium greenlandense* sp. nov. Extremophiles 14:61–69

Lutz S, Anesio AM, Edwards A, Benning LG (2015) Microbial diversity on Icelandic glaciers and ice caps. Front Microbiol 6:307. doi:10.3389/fmicb.2015.00307

Maccario L, Vogel TM, Larose C (2014) Potential drivers of microbial community structure and function in Arctic spring snow. Front Microbiol 5:413. doi:10.3389/fmicb.2014.00413

Madigan MT, Jung DO (2003) Extremophiles. Worldbook encyclopedia, Science Year 2004 Annual, pp 74–89

Maiangwa J, Ali MS, Salleh AB, Rahman RN, Shariff FM, Leow TC (2015) Adaptational properties and applications of cold-active lipases from psychrophilic bacteria. Extremophiles 19:235–247

Malecki PH, Vorgias CE, Petoukhov MV, Svergun DI, Rypniewski W (2014) Crystal structures of substrate-bound chitinase from the psychrophilic bacterium *Moritella marina* and its structure in solution. Acta Crystallogr D Biol Crystallogr. 70:676–684

Margesin R (2007) Alpine microorganisms: useful tools for low-temperature bioremediation. J Microbiol 45:281–285

Margesin R, Schinner F (1999) Biodegradation of organic pollutants at low temperature. In: Margesin R, Schinner F (eds) Biotechnological applications of cold adapted organisms. Springer, Berlin/Heidelberg/New York, pp. 271–290

Margesin R, Schinner F (2001) Bioremediation (natural attenuation and biostimulation) of diesel-oil contaminated soil in an alpine glacier skiing area. Appl Environ Microbiol 67:3127–3133

Margesin R, Schumann P, Spröer C, Gounot AM (2004) *Arthrobacter psychrophenolicus* sp. nov.,isolated from an alpine ice cave. Int J Syst Evol Microbiol 54:2067–2072

Margesin R, Fonteyne PA, Schinner F, Sampaio JP (2007) *Rhodotorula psychrophila* sp. nov., *Rhodotorula psychrophenolica* sp. nov. and *Rhodotorula glacialis* sp. nov., novel psychrophilic basidiomycetous yeast species isolated from alpine environments. Int J Syst Evol Microbiol 57:2179–2184

Marx JG, Carpenter SD, Deming JW (2009) Production of cryoprotectant extracellular polysaccharide substances (EPS) by the marine psychrophilic bacterium *Colwellia psychrerythraea* strain 34H under extreme conditions. Can J Microbiol 55:63–72

Matsuyama H, Hirabayashi T, Kasahara H, Minami H, Hoshino T, Yumoto I (2006) *Glaciecola chathamensis* sp. nov., a novel marine polysaccharide-producing bacterium. Int J Syst Evol Microbiol 56:2883–2886

Mattes TE, Alexander AK, Richardson PM, Munk AC, Han CS, Stothard P, Coleman NV (2008) The genome of *Polaromonas* sp. strain JS666: insights into the evolution of a hydrocarbon- and xenobiotic-degrading bacterium, and feature of relevance to biotechnology. Appl Environ Microbiol 74:6405–6416

McCammon SA, Innes BH, Bowman JP, Franzmann PD, Dobson SJ, Holloway PE, Skerratt JH, Nichols PD, Rankin LM (1998) *Flavobacterium hibernum* sp. nov., a lactose-utilizing bacterium from a freshwater Antarctic lake. Int J Syst Bacteriol 4:1405–1412

Médigue C, Krin E, Pascal G, Barbe V, Bernsel A, Bertin PN, Cheung F, Cruveiller S, D'Amico S, Duilio A, Fang G, Feller G, Ho C, Mangenot S, Marino G, Nilsson J, Parrilli E, Rocha EP, Rouy Z, Sekowska A, Tutino ML, Vallenet D, von Heijne G, Danchin A (2005) Coping with cold: the genome of the versatile marine Antarctica bacterium *Pseudoalteromonas haloplanktis* TAC125. Genome Res 15:1325–1335

Mereghetti P, Riccardi L, Brandsdal BO, Fantucci P, De Gioia L, Papaleo E (2010) Near native-state conformational landscape of psychrophilic and mesophilic enzymes: probing the folding funnel model. J Phys Chem B 114:7609–7619

Merlino A, Russo Krauss I, Albino A, Pica A, Vergara A, Masullo M, De Vendittis E, Sica F (2011) Improving protein crystal quality by the without-oil microbatch method: Crystallization and preliminary X-ray diffraction analysis of glutathione synthetase from *Pseudoalteromonas haloplanktis*. Int J Mol Sci 12:6312–6319

Methé BA, Nelson KE, Deming JW, Momen B, Melamud E, Zhang X, Moult J, Madupu R, Nelson WC, Dodson RJ, Brinkac LM, Daugherty SC, Durkin AS, DeBoy RT, Kolonay JF, Sullivan SA, Zhou L, Davidsen TM, Wu M, Huston AL, Lewis M, Weaver B, Weidman JF, Khouri H, Utterback TR, Feldblyum TV, Fraser CM (2005) The psychrophilic lifestyle as revealed by the genome sequence of *Colwellia psychrerythraea* 34H through genomic and proteomic analyses. Proc Natl Acad Sci USA 102:10913–10918

Michaux C, Massant J, Kerff F, Frere J-M, Docquier J-D, Vandenberghe I, Samyn B, Pierrard A, Feller G, Charlier P, Van Beeumen J, Wouters J (2008) Crystal structure of a cold-adapted class C b-lactamase. FEBS J 275:1687–1697

Mikhailova G, Likhareva V, Khairullin RF, Lubenets NL, Rumsh LD, Demidyuk IV, Kostrov SV (2006) Psychrophilic trypsin-type protease from *Serratia proteamaculans*. Biochem Sci 71:563–570
Mikhailova AG, Nekrasov AN, Zinchenko AA, Rakitina TV, Korzhenevsky DA, Lipkin AV, Razguljaeva OA, Ovchinnikova MV, Gorlenko VA, Rumsh LD (2015) Truncated variants of *Serratia proteamaculans* oligopeptidase B having different activities. Biochemistry (Mosc). 80:1331–1343
Milne PJ, Hunt AL, Rostoll K, Van Der Walt JJ, Graz CJM (1998) The biological activity of selected cyclic dipeptides. J Pharm Pharmacol 50:1331–1337
Miteva VI (2008) Bacteria in snow and glacier ice. In: Margesin R, Schinner F, Marx JC, Gerday C (eds) Psychrophiles: from biodiversity to biotechnology. Springer, Berlin/Heidelberg, pp. 31–50
Miteva VI, Brenchley JE (2005) Detection and isolation of ultrasmall microorganisms from a 120,000 year old Greenland glacier ice core. Appl EnvMicrobiol 71:7806–7818
Mock T, Thomas DN (2008) Microalgae from Polar Regions: functional genomics and physiology. In: Margesin R, Schinner F, Marx JC, Gerday C (eds) Psychrophiles: from biodiversity to biotechnology. Springer, Berlin/Heidelberg, pp. 347–368
Morgan-Kiss RM, Priscu JC, Pocock T, Gudynaite-Savitch L, Huner NPA (2006) Adaptation and acclimation of photosynthetic microorganisms to permanently cold environments. Microbiol Molec Biol Rev 70:222–252
Morita Y, Nakamura T, Hasan Q, Murakami Y, Yokoyama K, Tamiya E (1997) Cold-active enzymes from cold-adapted bacteria. J Am Oil Chem Soc 74:441–444
Mou Z, Ding Y, Wang X, Cai Y (2014) Comparison of protein-water interactions in psychrophilic, mesophilic, and thermophilic Fe-SOD. Protein Pept Lett 21:578–583
Moyer CL, Morita RY (2007) Psychrophiles and psychrotrophs. In: Encyclopedia of life sciences. John Wiley & Sons, Ltd., New York. www.els.net.
Mukhopadhyay A, Dasgupta AK, Chakrabarti K (2015) Enhanced functionality and stabilization of a cold active laccase using nanotechnology based activation-immobilization. Bioresour Technol. 79:573–584
Murray E, Preston CM, Massana R, Taylor LT, Blakis A, Wu K, Delong EF (1998) Seasonal and spatial variability of bacterial and archaeal assemblages in the coastal waters near Anvers Island, Antarctica. Appl Environ Microbiol 64:2585–2595
Murygina V, Arinbasarov M, Kalyuzhnyi S (2000) Bioremediation of oil polluted aquatories and soils with novel preparation "Rhoder". Biodegradation 11:385–389
Naganuma T, Hua PN, Okamoto T, Ban S, Imura S, Kanda H (2005) Depth distribution of euryhaline halophilic bacteria in Suribati Ike, a meromictic lake in East Antarctica. Polar Biol 28:964–970
Nakagawa T, Ikehata R, Myoda T, Miyaji T, Tomizuka N (2007) Overexpression and functional analysis of cold-active β-galactosidase from *Arthrobacter psychrolactophilus* strain F2. Protein Expr Purif 54:295–299
Nakagawa T, Nagaoka T, Taniguchi S, Miyaji T, Tomizuka N (2004) Isolation and characterization of psychrophilic yeasts producing cold-adapted pectinolytic enzymes. Lett Appl Microbiol 38:383–387
Napolitano MJ, Shain DH (2004) Four kingdoms on glacier ice: convergent energetic processes boost energy levels as temperatures fall. Proc R Soc Lond B 271:273–276
Negoiță TG, Stefanic G, Irimescu Orzan ME, Palanciuc V, Oprea G (2001a) Chemical and biological characterization of soils from the Antarctic East Coast. Polar Biol 24:565–571
Negoiță TG, Stefanic G, Irimescu Orzan ME, Palanciuc V, Oprea G (2001b) Microbial chemical and enzymatic properties in Spitsbergen soils. Polar Forschung 71:41–46
Ni Y, Gu Y, Shi X, Zheng X, Han L, Zhou H, Cheng G (2013) Phylogenetic and physiological diversity of cold-adapted bacteria producing protease from sediments of the bottom layer of the glacier no. 1 in the Tianshan mountains. Wei Sheng Wu Xue Bao 53(2):164–172. [Article in Chinese]
Nichols DS, Bowman J, Sanderson C, Nichols CM, Lewis T, McMeekin T, Nichols PD (1999) Developments with Antarctic microorganisms: culture collections, bioactivity screening, taxonomy, PUFA production and cold-adapted enzymes. Curr Opin Biotechnol 10:240–246
Nichols DS, Sanderson K, Buia A, van de Kamp J, Holloway P, Bowman JP, Smith M, Mancuso C, Nichols PD, McMeekin T (2002) Bioprospecting and biotechnology in Antarctica. In:

Jabour- Green J, HawardM (eds) The Antarctic: past, present and future. Antarctic CRC Research Report #28, Hobart, pp. 85–103

Nielsen TB, Ishii M, Kirk O (1999) Lipases A and B from the yeast *Candida antarctica*. In: Margesin R, Schinner F (eds) Biotechnological applications of cold adapted organisms. Springer, Berlin/Heidelberg/New York, pp. 48–61

Niemann H, Elvert M, Wand U, Samarkin VA, Lehmann MF (2010) First Lipid Biomarker evidence for aerobic methane oxidation in the water column of Lake Untersee (East Antarctica). Geophysical Research Abstracts 12:EGU2010–EGU5921

Nogi Y (2008) Bacteria in deep sea: psychropiezophiles. In: Margesin R, Schinner F, Marx JC, Gerday C (eds) Psychrophiles: from biodiversity to biotechnology. Springer, Berlin/ Heidelberg, pp. 73–82

Nonaka K, Yoon KS, Ogo S (2014) Biochemical characterization of psychrophilic Mn-superoxide dismutase from newly isolated *Exiguobacterium* sp. OS-77. Extremophiles 18:363–373

Nowak M, Olszewski M, Śpibida M, Kur J (2014) Characterisation of single stranded DNA binding proteins from the pychrophilic bacteria *Desulfotalea psychrophila Flavobacterium psychrophilum, Psychrobacter arcticus, Psychrobacter cryohalolentis Psychromonas ingrahamii, Psychroreflexus torquans, Photobacterium profundum*. BMC Microbiology 14:91. doi:10.1186/1471-2180-14-91

Ohgiya S, Hoshino T, Okuyama H, Tanaka S, Ishizaki K (1999) Biotechnology of enzymes from cold adapted microorganisms. In: Margesin R, Schinner F (eds) Biotechnological applications of cold adapted organisms. Springer, Berlin/Heidelberg/New York, pp. 17–35

Oikawa T, Yamamoto N, Shimoke K, Uesato S, Ikeuki T, Fujioka T (2005) Purification, characterization and overexpression of psychrophilic and thermolabile malate dehydrogenase from a novel Antarctic psychrotolerant *Flavobacterium frigidimaris* KUC 1. Biosci Biotechnol Biochem 59:2146–2154

Olivera NL, Sequeiros C, Nievas ML (2007) Diversity and enzyme properties of protease-producing bacteria isolated from sub-Antarctic sediments of Isla de Los Estados, Argentina. Extremophiles 11:517–526

Onofri S, Zucconi L, Selbmann L, de Hoog S, de los Rιos A, Ruisi S, Grube M (2007) Fungal associations at the cold edge of life in algae and cyanobacteria in extreme environments. In: Seckbach J (ed) Cellular origins, life in extreme habitats and astrobiology, vol 11. Springer, Dordrecht, The Netherlands, pp 735–757.

Oswald VF, Chen WT, Harvilla PB, Magyar JS (2014) Overexpression, purification, and enthalpy of unfolding of ferricytochrome c552 from a psychrophilic microorganism. J Inorg Biochem. doi:10.1016/j.jinorgbio.2013.11.002

Ovalle AW (1987) In-place leaching of a block carving mine. In: Cooper WC, Lagos GE, Ugarte G (eds) Copper 87, Hydrometallurgy and electrometallurgy of copper, vol 3. Universidad de Chile, Santiago, Chile, pp. 17–37

Papaleo E, Parravicini F, Grandori R, De Gioia L, Brocca S (2014) Structural investigation of the cold-adapted acylaminoacyl peptidase from *Sporosarcina psychrophila* by atomistic simulations and biophysical methods. Biochim Biophys Acta 1844:2203–2213. doi:10.1016/j.bbapap.2014.09.018

Papaleo E, Pasi M, Riccardi L, Sambi I, Fantucci P, De Gioia L (2008) Protein flexibility in psychrophilic and mesophilic trypsins. Evidence of evolutionary conservation of protein dynamic in trypsin-like serine-proteases. FEBS Lett 582:1008–1018

Park SC, Kim MS, Baik KS, Kim EM, Rhee MS, Seong CN (2008) *Chryseobacterium aquifrigidense* sp. nov., isolated from a water-cooling system. Int J Syst Evol Microbiol 58:607–611

Parrilli E, Papa R, Tutino ML, Sannia G (2010) Engineering of a psychrophilic bacterium for the bioremediation of aromatic compounds. Bioengineered Bugs 3:213–216

Parrilli E, Duillio A, Tutino ML (2008) Heterologous protein expression on psychrophilic hosts. In: Margesin R, Schinner F, Marx JC, Gerday C (eds) Psychrophiles: from biodiversity to biotechnology. Springer, Berlin/Heidelberg, pp. 365–380

Parry BR, Shain DH (2011) Manipulations of AMP metabolic genes increase growth rate and cold tolerance in *Escherichia coli*: Implications for psychrophilic evolution. Mol Biol Evol 28:2139–2145

Parvizpour S, Razmara J, Jomah AF, Shamsir MS, Illias RM (2015) Structural prediction of a novel laminarinase from the psychrophilic *Glaciozyma antarctica* PI12 and its temperature adaptation analysis. J Mol Model. 21:63. doi:10.1007/s00894-015-2617-1

Pathan AA, Bhadra B, Begum Z, Shivaji S (2010) Diversity of yeasts from puddles in the vicinity of midre lovénbreen glacier, arctic and bioprospecting for enzymes and fatty acids. Curr Microbiol. 60:307–314

Pavlova K, Gargova S, Hristozova T, Tankova Z (2008) Phytase from Antarctic yeast strain *Cryptococcus laurentiis* AL 27. Folia Microbiol 23:29–34

Pernthaler J, Glöckner FO, Unterholzner S, Alfreider A, Psenner R, Amann R (1998) Seasonal community and population dynamics of pelagic bacteria and archaea in a high mountain lake. Appl Environ Microbiol 64:4299–4306

Perovich DK, Grenfell TC, Light B, Hobbs PV (2002) Seasonal evolution of the albedo of multi-year Arctic sea-ice. J Geophys Res 107(C10):8044. doi:10.1029/2000JC000438

Phadtare S, Inouye M (2008) Cold shock proteins. In: Margesin R, Schinner F, Marx JC, Gerday G (eds) Psychrophiles: from biodiversity to biotechnology. Springer, Berlin/Heidelberg, pp. 191–210

Poli A, Anzelmo G, Nicolaus B (2010) Bacterial exopolysaccharides from extreme marine habitats: Production, characterization and biological activities. Mar Drugs 8:1779–1802

Prabagaran SR, Manorama R, Delile D, Shivaji S (2007) Predominance of *Roseobacter*, *Sulfitobacter*, *Glaciecola* and *Psychrobacter* in seawater collected from Ushuaia, Argentina, sub Antarctica. FEMS Microbiol Ecol 59:342–355.

Prasad S, Pratibha MS, Manasa P, Buddhi S, Begum Z, Shivaji S (2013) Diversity of chemotactic heterotrophic bacteria associated with arctic cyanobacteria. Curr Microbiol 66:64–71

Price BP (2006) Microbial life in glacial ice and implications for a cold origin of life. FEMS Microbiol Ecol 59:217–231

Prince RC (2005) The microbiology of marine oil spills bioremediation. In: Olivier B, Margot M (eds) Petroleum microbiology. ASM Press, Washington, DC, pp. 317–335

Priscu JC, Christner BC (2004) Earths icy biosphere. In: Bull AT (ed) Microbial diversity and bioprospecting. ASM Press, Washington, DC, pp. 130–145

Priscu JC, Fritsen CH, Adams EE, Giovannoni SJ, Paerl HW, McKay CP et al (1998) Perennial Antarctic lake ice: an oasis for life in a polar desert. Science 280:2095–2098

Purcell AM, Mikucki JA, Achberger AM, Alekhina IA, Barbante C, Christner BC, Ghosh D, Michaud AB, Mitchell AC, Priscu JC, Scherer R, Skidmore ML, Vick-Majors TM, the WISSARD Science (2014) Microbial sulfur transformations in sediments from subglacial Lake Whillans. Front Microbiol. 5:594. doi:10.3389/fmicb.2014.00594

Qoura F, Elleuche S, Brueck T, Antranikian G (2014) Purification and characterization of a cold-adapted pullulanase from a psychrophilic bacterial isolate. Extremophiles 18:1095–1102

Rabus R, Ruepp A, Frickey T, Rattei T, Fartmann B, Stark M, Bauer M, Zibat A, Lombardot T, Amann J, Gellner K, Teeling H, Leuschner WD, Glockner F-O, Lupas AN, Amann R, Klenk H-P (2004) The genome of *Desulfotea psychrophila* a sulphate reducing bacterium from permanently cold Arctic sediment. Environ Microbiol 6:887–902

Raja A, Prabakaran P, Gjalakshmi P (2010) Isolation and screening of antibiotic producing actinomycetes and its nature from Rothang Hill soil against Viridans *Streptococcus* sp. Res J Microbiol 5:44–49

Ramaiah N (1994) Production of certain hydrolytic enzymes by psychrophilic bacteria from the Antarctic krill, zooplankton and seawater. In: Ninth Indian Expedition to Antarctica, Scientific Report, National Institute of Oceanography, Department of Ocean Development, Goa, India Technical Publication, vol 6. Scientific report of Ninth Indian Expedition to Antarctica, New Delhi, pp. 107–114

Ramli ANM, Mahadi NM, Rabu A, Murad AM, Bakar FD, Illias RM (2011) Molecular cloning, expression and biochemical characterisation of a cold-adapted novel recombinant chitinase from *Glaciozyma antarctica* PI12. Microb Cell Fact 10:94. doi:10.1186/1475-2859-10-94

Ramteke PW, Joseph B, Kuddus M (2005) Extracellular lipases from anaerobic microorganism of Antarctic. Ind J Biotechnol 4:293–294

Ramya LN, Pulicherla KK (2015) Molecular insights into cold active polygalacturonase enzyme for its potential application in food processing. J Food Sci Technol. 52:5484–5496

Rashidah A, Sabrina S, Ainihayati A, Shanmugapriya P, Nazalan N, Razip S (2007) Psychrophilic enzymes from the Antarctic isolates. In: Proceeedings of the international symposium Asian collaboration in IPY (2007–2008) Science Council of Japan, Tokyo 1st March 2007. National Institute of Polar Research, Tokyo, Japan, pp. 116–119

Rasol R, Rashidah AR, Nazuha RS, Smykla J, Maznah WO, Alias SA (2014) Psychrotrophic lipase producers from arctic soil and sediment samples. Pol J Microbiol 63:75–82

Raymond JA, Fritsen C, Shen K (2007) An ice-binding protein from an Antarctic sea ice bacterium. FEMS Microbiol Ecol 61:214–221

Rendleman JA Jr (1996) Enzymatic conversion of malto-oligosaccharides and maltodextrin into cyclodextrin at low temperature. Biotechnol Appl Biochem 24:129–137

Riley M, Staley JT, Danchin A, Wang TZ, Brettin TS, Hauser LJ, Land M, Thompson LS (2008) Genomics of an extreme psychrophile, *Psychromonas ingrahamii*. BMC Genomics 9:210

Rippa V, Papa R, Giuliani M, Pezzella C, Parrilli E, Tutino ML, Marino G, Duilio A (2012) Regulated recombinant protein production in the Antarctic bacterium *Pseudoalteromonas haloplanktis* TAC125. Methods Mol Biol. 824:203–218

Rivkina E, Gilichinsky D, Wagener S, Tiedje J, McGrath J (1998) Biogeochemical activity of anaerobic microorganisms from buried permafrost sediments. Geomicrobiol J 15:187–193

Rivkina EM, Friedmann EI, McKay CP, Gilichinsky DA (2000) Metabolic activity of permafrost bacteria below the freezing point. Appl Environ Microbiol 66:3230–3233

Rodrigues-Diaz F, Ivanova N, He Z, Huebner M, Zhou J, Tiedje JM (2008) Architecture of thermal adaptation in an *Exiguobacterium sibiricum* strain isolated from 3 million year old permafrost: a genome and transcriptome approach. BMC Genomics 9:547

Rogers SO, Shtarkman YM, Koçer ZA, Edgar R, Veerapaneni R, D'Elia T (2013) Ecology of subglacial Lake Vostok (Antarctica), based on metagenomic/metatranscriptomic analyses of accretion ice. Biology 2:629–650

Rossi G (1999) Biohydrometallurgical processes and temperature. In: Margesin R, Schinner F (eds) Biotechnological applications of cold adapted organisms. Springer, Berlin/Heidelberg/New York, pp. 291–308

Rossi M, Buzzini P, Cordisco L, Amaretti A, Sala M, Raimondi S, Ponzoni C, Pagnoni UM, Matteuzzi D (2009) Growth, lipid accumulation and fatty acid composition in obligate psychrophilic, facultative psychrophilic, and mesophilic yeasts. FEMS Microbiol Ecol 69:363–337

Russell NJ (2008) Membrane components and cold sensing. In: Margesin R, Schinner F, Marx JC, Gerday C (eds) Psychrophiles: from biodiversity to biotechnology. Springer, Berlin/Heidelberg, pp. 177–190

Russell NJ, Cowan DA (2005) Handling of psychrophilic microorganisms. In: Rainey FA, Oren A (eds) Extremophiles. Methods in Microbiology 35:371–393

Russell NJ, Nichols DS (1999) Polyunsaturated fatty acids in marine bacteria - a dogma rewritten. Microbiology 145:767–779

Rybalka N, Andersen RA, Kostikov I, Mohr KI, Massalski A, Olech M, Friedl T (2008) Testing for endemism, genotypic diversity and species concepts in Antarctic terrestrial microalgae of the *Tribonemataceae* (*Stramenopiles*, *Xanthophyceae*). Environ Microbiol Rep 11:554–565

Rysgaard S, Glud RN, Sejr MK, Blicher ME, Stahl HJ (2008) Denitrification activity and oxygen dynamics in Arctic sea ice. Polar Biol 31:527–537

Sahay S, Hamid B, Singh P, Ranjan K, Chauhan D, Rana RS, Chaurse VK (2013) Evaluation of pectinolytic activities for oenological uses from psychrotrophic yeasts. Lett Appl Microbiol. 57:115–121

Sanchez LA, Gomez FF, Delgado OD (2009) Cold-adapted microorganisms as a source of new antimicrobials. Extremophiles 13:111–120

Santos AF, Pires F, Jesus HE, Santos ALS, Peixoto R, Rosado AS, D'Avila-Levy CM, Branquinha MH (2015) Detection of proteases from *Sporosarcina aquimarina* and *Algoriphagus antarcticus* isolated from Antarctic soil. Anais da Academia Brasileira de Ciências 87:1678–2690

Sarmiento F, Peralta R, Blamey JM (2015) Cold and hot extremozymes: industrial relevance and current trends. Front Bioeng Biotechnol 3:148. doi:10.3389/fbioe.2015.00148

Sattley WM, Madigan MT (2006) Isolation, characterization and ecology of cold-active, chemolithotrophic, sulfur-oxidizing bacteria from perennially ice-covered Lake Fryxell, Antarctica. Appl Environ Microbiol 72:5562–5568

Sattley WM, Madigan MT (2007) Cold-active acetogenic bacteria from surcial sediments of perennially ice-covered Lake Fryxell, Antarctica. FEMS Microbiol Lett 272:48–54

Saunders NFW, Thomas T, Curmi PMG, Mattick JS, Kuczek E, Slade R, Davis J, Franzmann PD, Boone D, Rusterholtz K, Feldman R, Gates C, Bench S, Sowers K, Kadner K, Aerts A, Dehal P, Detter C, Glavina T, Lucas S, Richardson P, Larimer F, Hauser L, Land M, Cavicchioli R (2003) Mechanisms of thermal adaptation revealed from the genomes of the Antarctic Archaea *Methanogenium frigidum* and *Methanococcoides burtonii*. Genome Res 13:1580–1588

Sælensminde G, Halskau Ř Jr, Jonassen I (2009) Amino acid contacts in proteins adapted to different temperatures: hydrophobic interactions and surface charges play a key role. Extremophiles 13:11–20

Säwström C, Laybourn-Parry J, Granéli W, Anesio AM (2007) Heterotrophic bacterial and viral-dynamics in Arctic freshwaters: results from a field study and nutrient-temperature manipulation experiments. Polar Biol 30:1407–1415

Seo HJ, Bae SS, Lee J-H, Kim S-J (2005) *Photobacterium frigidiphilum* sp. nov., a psychrophilic, lipolytic bacterium isolated from deep-sea sediments of Edison Seamount. Int J Syst Evol Microbiol 55:1661–1666

Sheridan PP, Brenchley JE (2000) Characterization of a salt-tolerant family 42 beta-galactosidase from a psychrophilic Antarctic *Planococcus* isolate. Appl Environ Microbiol 66:2438–2444

Siddiqui KS, Cavicchioli R (2006) Cold-adapted enzymes. Annu Rev Biochem 75:403–433

Siddiqui KS (2015) Some like it hot, some like it cold: Temperature dependent biotechnological applications and improvements in extremophilic enzymes. Biotechnol Adv. 33:1912–1922

Siddiqui KS, Poljak A, Guilhaus M, De Francisci D, Curmi PM, Feller G, D'Amico S, Gerday C, Uversky VN, Cavicchioli R (2006) Role of lysine versus arginine in enzyme cold-adaptation:modifying lysine to homo-arginine stabilizes the cold-adapted alpha-amylase from *Pseudoalteromonas haloplanktis*. Proteins 1:486–501

Siglioccolo A, Gerace R, Pascarella S (2010) "Cold spots" in protein cold adaptation: Insights from normalized atomic displacement parameters (B′-factors). Biophys Chem 153:104–114

Signori CN, Thomas F, Enrich-Prast A, Pollery RCG, Sievert SM (2014) Microbial diversity and community structure across environmental gradients in Bransfield Strait Western Antarctic Peninsula. Front Microbiol 5:647. doi:10.3389/fmicb.2014.00647

Singh L, Ramana Venkata K (1998) Isolation and characterization of psychrotrophic Antarctic bacteria from blue-green algal mats and their hydrolytic enzymes. Scientific report; Fourteenth Indian expedition to Antarctica. Vol. 12:199–206

Singh SM, Puja G, Bhat DJ (2006) Psychrophilic fungi from Schirmacher Oasis, East Antarctica. Curr Sci 90:1388–1392

Singh P, Singh SM, Roy U (2015) Taxonomic characterization and the bio-potential of bacteria isolated from glacier ice cores in the high Arctic. J Basic Microbiol. 56:275–285. doi:10.1002/jobm.201500298

Singh P, Singh SM, Tsuji M, Prasad GS, Hoshino T (2014) *Rhodotorula svalbardensis* sp. nov., a novel yeast species isolated from cryoconite holes of Ny-Alesund, Arctic. Cryobiology 68:122–128

Sizova M, Panikov N (2007) *Polaromonas hydrogenivorans* sp. nov., a psychrotolerant hydrogen-oxidizing bacterium from Alaskan soil. Int J Syst Evol Microbiol 57:616–619

Skidmore ML, Foght JM, Sharp MJ (2000) Microbial life beneath a high arctic glacier. Appl Environ Microbiol 66:3214–3220

Smith DW, Emde KME (1999) Effectiveness of wastewater lagoons in cold regions. In: Margesin R, Schinner F (eds) Biotechnological applications of cold adapted organisms. Springer, Berlin/Heidelberg/New York, pp. 235–256

Somkutl GA, Holsinger VH (1997) Microbial technologies in the production of low-lactose dairy foods. Food Sci Technol Int J 3:163–169

Sonan GK, Receveur-Brechot V, Duez C, Aghari N, Czjzek M, Haser R, Gerday C (2007) The linker region plays a key role in the adaptation to cold of the cellulase from an Antarctic bacterium. Biochem J 407:293–302

Spring S, Merkhoffer B, Weiss N, Kroppenstedt RM, Hippe H, Stackebrandt E (2003) Characterization of novel psychrophilic clostridia from an Antarctic microbial mat: description of *Clostridium frigoris* sp. nov., *Clostridium lacusfryxellense* sp. nov., *Clostridium bowmanii* sp. nov. and*Clostridium psychrophilum* sp. nov. and reclassification of *Clostridium laramiense* as *Clostridium estertheticum* subsp. *laramiense* subsp. nov. Int J Syst Evol Microbiol 53:1019–1029

Stefanic G (1994) Biological definition, quantifying method and agricultural interpretation of soil fertility. Rom Agric Res 2:107–116

Steven B, Briggs G, McKay CP, Pollard WH, Greer CW, Whyte LG (2007) Characterization of the microbial diversity in a permafrost sample from the Canadian high Arctic using culture-dependent and culture-independent methods. FEMS Microbiol Ecol 59:513–523

Stibal M, Gözdereliler E, Cameron KA, Box JE, Stevens IT, Gokul JK, Schostag M, Zarsky JD, Edwards A, Irvine-Fynn TDL, Jacobsen CS (2015) Microbial abundance in surface ice on the Greenland Ice Sheet. Front Microbiol 6:225. doi:10.3389/fmicb.2015.00225

Struvay C, Feller G (2012) Optimization to low temperature activity inpsychrophilic enzymes. Int J Mol Sci 13:11643–11665

Sundareswaran VR, Singh AK, Dube S, Shivaji S (2010) Aspartate aminotransferase is involved in cold adaptation in psychrophilic *Pseudomonas syringae*. Arch Microbiol. 192:663–672

Suzuki T, Nakayama T, Kurihara T, Nishino T, Esaki N (2001) Cold active lipolytic activity of psychrotrophic *Acinetobacter* sp. strain no. 6. J Biosci Bioeng 92:144–148

Sweet CR, Watson RE, Landis CA, Smith JP (2015) Temperature-dependence of lipid A acyl structure in *Psychrobacter cryohalolentis* and arctic isolates of*Colwellia hornerae* and *Colwellia piezophila*. Mar Drugs 13:4701–4720

Tajima T, Hamada M, Nakashimada Y, Kato J (2015) Efficient aspartic acid production by a psychrophile-based simple biocatalyst. J Ind Microbiol Biotechnol 42:1319–1324

Tajima T, Fuki K, Kataoka N, Kudou D, Nakashimada Y, Kato J (2013) Construction of a simple biocatalyst using psychrophilic bacterial cells and its application for efficient 3-hydroxypropionaldehyde production from glycerol. AMB Express 3:69 http://www.amb-express.com/content/3/1/69

Takeuchi N, Kohshima S (2004) A snow algal community on Tyndall glacier in the Southern Patagonia Icefield, Chile. Arct Antarct Alp Res 36:92–99

Tamaki H, Hanada S, Kamagata Y, Nakamura K, Nomura N, Nakano K, Matsumura M (2003) *Flavobacterium limicola* sp. nov., a psychrophilic, organic-polymer-degrading bacterium isolated from freshwater sediments. Int J Syst Evol Microbiol 53:519–526

Tang MA, Motoshima H, Watanabe K (2014) Cold adaptation: structural and functional characterizations of psychrophilic and mesophilic acetate kinase. Protein J 33:313–322

Tashyrev OB (2009) The complex researches of structure and function of Antarctic terrestrial microbial communities. Ukr Antarctic J 8:343–357

Tebo BM, Davis RE, Anitori RP, Connell LB, Schiffman P, Staudigel H (2015) Microbial communities in dark oligotrophic volcanic ice cave ecosystems of Mt. Erebus, Antarctica. Front Microbiol 6:179. doi:10.3389/fmicb.2015.00179

Thomas-Hall S, Watson K (2002) *Cryptococcus nyarrowii* sp. nov., a basidiomycetous yeast from Antarctica. Int J Syst Evol Microbiol 52:1033–1038

Thomas-Hall S, Hall R, Turchetti B, Buzzini P, Branda E, Boekhout T, Theelen B, Watson K (2010) Cold-adapted yeasts from Antarctica and the Italian Alps – description of three novel species: *Mrakia robertii* sp. nov., *Mrakia blollopis* sp. nov. and *Mrakiella niccombsii* sp. nov. Extremophiles 14:47–59

Tian F, Yu Y, Chen B, Li H, Yao YF, Guo XK (2009) Bacterial, archaeal and eukaryotic diversity in Arctic sediment as revealed by 16S rRNA and 18S rRNA gene clone libraries analysis. Polar Biol 32:93–103

Tkaczuk KL, Buinicki JM, Bialkowska A, Bielecki S, Turkievicz M, Cieslinski H, Kur J (2005) Molecular modelling of a psychrophilic beta-galactosidase. Biocat Biotransf 23:201–209

Tuorto SJ, Darias P, McGuinness LR, Panikov N, Zhang T, Haggblom MM, Kerkhof LJ (2014) Bacterial genome replication at subzero temperatures in permafrost. The ISME J 8:139–149

Turkiewicz M, Pazgier M, Donachie SP, Kalinowska H (2005) Invertase and glucosidase production by the endemic Antarctic marine yeast *Leucosporidium antarcticum*. Pol Polar Res 26:125–136.

Tutino ML, Duilio A, Parrilli R, Remaut E, Sannia G, Marino G (2001) A novel replication element from an Antarctic plasmid as a tool for the expression of proteins at low temperature. Extremophiles 5:257–264

Vishniac HS, Kurtzman CP (1992) *Cryptococcus antarcticus* sp. nov., and *Cryptococcus albidosimilis* sp. nov., basidioblastomycetes from Antarctic soils. Int J Syst Bact 42:547–553

Vorobyova E, Soina V, Gorlenko M, Minkovskaya M, Zalinova N, Mamukelashvili A, Gilichinsky D, Rivkina E, Vishnivetskaya T (1997) The deep cold biosphere: facts and hypothesis. FEMS Microbiol Rev 20:277–290

Voytek MA, Priscu JC, Ward BB (1999) The diversity and abundance of ammonia oxidizing bacteria in lake of the McMurdo Dry Valley, Antarctica. Hydrobiologia 401:113–130

Wagner-Döbler I, Rheims H, Felske A, El-Ghezal A, Flade-Schroder D, Laatsch H, Lang S, Pukall R, Tindall BJ (2004) *Oceanibulbus indolifex* gen. nov., sp. nov., a North Sea alphaproteobacterium that produces bioactive metabolites. Int J Syst Evol Microbiol 54:1177–1184

Wang Q, Zhang C, Hou Y, Lin X, Shen J, Guan X (2013) Optimization of cold-active lipase production from psychrophilic bacterium *Moritella* sp. 2-5-10-1 by statistical experimental methods. Biosci Biotechnol Biochem 77:17–21

Wang X, Zhao Y, Tan H, Chi N, Zhang Q, Du Y, Yin H (2014) Characterisation of a chitinase from *Pseudoalteromonas* sp. DL-6, a marine psychrophilic bacterium. Int J Biol Macromol. 70:455–462

Wartiainen I, Hestnes AG, McDonald IR, Svenning Mette M (2006) *Methylocystis rosea* sp. nov., a novel methanotrophic bacterium from Arctic wetland soil, Svalbard, Norway (786°N). Int J Syst Microbiol 56:541–547

Wells LE (2008) Cold active viruses. In: Margesin R, Schinner F, Marx JC, Gerday C (eds) Psychrophiles: from biodiversity to biotechnology. Springer, Berlin/Heidelberg, pp. 157–176

Wharton RA Jr, McKay CP, Simmons GM Jr, Parker CB (1985) Cryoconite holes on glaciers. Bioscience 35:499–503

Whyte LG, Hawari J, Zhou E, Bourbonnere L, Inniss WE, Greer CW (1998) Biodegradation of variable-chain-length alkanes at low temperatures by a psychrotrophic *Rhodococcus*. Appl Environ Microbiol 64:2578–2584

Wi AR, Jeon SJ, Kim S, Park HJ, Kim D, Han SJ, Yim JH, Kim HW.(2014) Characterization and a point mutational approach of a psychrophilic lipase from an arctic bacterium, *Bacillus pumilus*. Biotechnol Lett 36:1295–1302

Wierzchos J, de los Rios A, LG S, Ascaso C (2004) Viability of endolithic microorganisms in rocks from the McMurdo Dry Valleys of Antarctica established by confocal and fluorescence microscopy. J Microsc Oxford 216:57–61

Wilhelm RC, Niederberger TD, Greer C, Whyte LG (2011) Microbial diversity of active layer and permafrost in an acidic wetland from the Canadian High Arctic. Can J Microbiol 57:303–315

Wilmes B, Kock H, Glagla S, Albrecht D, Voigt B, Markert S, Gardebrecht A, Bode R, Danchin A, Feller G, Hecker M, Schweder T (2011) Cytoplasmic and periplasmic proteomic signatures of exponentially growing cells of the psychrophilic bacterium *Pseudoalteromonas haloplanktis* TAC125. Appl Environ Microbiol 77:1276–1283

Woo JH, Hwang YO, Kang SG, Lee HS, Cho JC, Kim SJ (2007) Cloning and characterization of three epoxide hydrolases from a marine bacterium, *Erythrobacter litoralis* HTCC2594. Appl Microbiol Biotechnol 76:365–375

Xu Y, Wang D, Shi X, Zheng X, Zhou H, Liu Y, Ni Y (2011) Selective isolation and diversity of cold-adapted lipase-producing strains from permafrost soil at the terminus of a glacier in the Tianshan Mountains. Wei Sheng Wu Xue Bao 51:233–240

Yakimov MM, Giuliano L, Gentile G, Crisafi E, Chernikova TN, Abraham W-R, Lunsdorf H, Timmis KN, Golyshin PN (2003) *Oleispira antarctica* gen. nov., sp. nov., a novel hydrocarbonoclastic marine bacterium isolated from Antarctic coastal sea water. Int J Syst Evol Microbiol 53:779–785

Yan BQ, Chen XL, Hou XY, He H, Zhou BC, Zhang YZ (2009) Molecular analysis of the gene encoding a cold-adapted halophilic subtilase from deep-sea psychrotolerant bacterium *Pseudoalteromonas* sp. SM9913: cloning, expression, characterization and function analysis of the C-terminal PPC domains. Extremophiles 13:725–733

Yang C, Wang F, Hao J, Zhang K, Yuan N, Sun M (2010) Identification of a proteolytic bacterium, HW08, and characterization of its extracellular cold-active metalloprotease Ps5. Biosci Biotechnol Biochem 74:1220–1225

Yang SH, Lee JH, Ryu JS, Kato C, Kim SJ (2007) *Shewanella donghaensis* sp. nov., a psychrotrophic, piezosensitive bacterium producing high levels of polyunsaturated fatty acid, isolated from deep sea sediments. Int J Syst Evol Microbiol 57:208–212

Yu Y, Li HR, Zeng YX, Chen B (2011) Bacterial diversity and bioprospecting for cold-active hydrolytic enzymes from culturable bacteria associated with sediment from Nella Fjord, EasternAntarctica. Mar Drugs 9:184–195

Yu Z-C, Tang B-L, Zhao D-L, Pang X, Qin Q-L, Zhou B-Cheng et al (2015) Development of a cold-adapted *Pseudoalteromonas* expression system for the *Pseudoalteromonas* proteins intractable for the *Escherichia coli* system. PLoS ONE 10(9):e0137384. doi:10.1371/journal.pone.0137384.

Yu D, Margesin R (2014) Partial characterization of a crude cold-active lipase from *Rhodococcus cercidiphylli* BZ22. Folia Microbiol (Praha) 59:439–445

Yu Y, Li H, Zeng Y, Chen B (2009) Extracellular enzymes of cold-adapted bacteria from Arctic sea ice, Canada Basin. Polar Biol 32:1539–1547

Yumoto I, Nakamura A, Iwata H, Kojima K, Kusumoto K, Nodasaka Y, Matsuyama H (2002) *Dietzia psychralkaliphila* sp. nov., a novel, facultatively psychrophilic alkaliphile that grows on hydrocarbons. Int J Syst Evol Microbiol 52:85–90

Yumoto I, Hirota K, Sogabe Y, Nodasaka Y, Yokota Y, Hoshino T (2003) *Psychrobacter okhotskensis* sp. nov., a lipase-producing facultative psychrophile isolated from the coast of the Okhotsk Sea. Int J Syst Evol Microbiol 53:1985–1989

Zachariassen KE, Lundheim R (1999) Application of antifreeze proteins. In: Margesin R, Schinner F (eds) Biotechnological applications of cold adapted organisms. Springer, Berlin/Heidelberg, pp. 319–332

Zakaria MM, Ashiuchi M, Yamamoto S, Yagi T (1998) Optimization for beta-mannanase production of a psychrophilic bacterium, Flavobacterium sp. Biosci Biotechnol Biochem 62:655–660

Zakhia F, Jungblut AD, Taton A, Vincent WF, Wilmotte A (2008) Cyanobacteria in cold ecosystems. In: Margesin R, Schinner F, Marx JC, Gerday C (eds) Psychrophiles: from biodiversity to biotechnology. Springer, Berlin/Heidelberg, pp. 121–135

Zhang S-C, Sun M, Li T, Wang Q-H, Hao J-H et al (2011) Structure analysis of a new psychrophilic marine protease. PLoS ONE 6(11):e26939. doi:10.1371/ journal.pone.0026939

Halophilic Microorganisms from Man-Made and Natural Hypersaline Environments: Physiology, Ecology, and Biotechnological Potential

8

Madalin Enache, Gabriela Teodosiu, Takashi Itoh, Masahiro Kamekura, and Helga Stan-Lotter

Contents

M. Enache (✉) • G. Teodosiu
Institute of Biology of the Romanian Academy,
Splaiul Independentei 296, P.O. Box 56-53, Bucharest, Romania
e-mail: madalin.enache@ibiol.ro

T. Itoh
Japan Collection of Microorganisms, RIKEN BioResource Center, Saitama 351-0198, Japan

M. Kamekura
Halophiles Research Institute, 677-1 Shimizu, Noda 278-0043, Japan

H. Stan-Lotter
Department of Molecular Biology, University of Salzburg, 5020 Salzburg, Austria

H. Stan-Lotter, S. Fendrihan (eds.), *Adaption of Microbial Life to Environmental Extremes*, DOI 10.1007/978-3-319-48327-6_8

8.1 Introduction

What are hypersaline environments? Geologists or geochemists define saline lakes *sensu lato* as bodies of water with salinity more than 3 g/l (0.3 %), while those *sensu stricto* (hypersaline) are bodies of water that exceed the modest 35 g/l (3.5 %) salt of oceans (Williams 1998). Many microbiologists use the term hypersaline to denote the well-known salt lakes, such as the Dead Sea and the Great Salt Lake, or crystallizer ponds of solar salterns, environments almost saturated with salt.

A lot of hypersaline environments are found in nature throughout the world, natural or man-made (Javor 1989a). Rock salt from salt deposits has been a source of sodium chloride for human beings from prehistoric period (Multhauf 1978). The subterranean salt deposit of the Mediterranean Sea is the result of Messinian Salinity Crisis 5.96–5.33 My ago (Duggen et al. 2003). Strong brines occur in both marine-derived (thalassic) and nonmarine (athalassic) systems. In arid coastal zones, large scale sabkhas, salt flats, or strong brines of tiny scales are observed (Javor 1989b). These hypersaline environments are too harsh for normal life to exist, but a variety of microbes, both Bacteria and Archaea, survive.

8.2 Halophilic Microorganisms

When microbiologists are confronted with the question of how to define halophilic microorganisms, an answer is the classical definition by Kushner (1985), who categorized them into slight, moderate, and extreme halophiles, depending on the NaCl concentration that supported optimal growth. The term "halotolerant" is generally used for organisms that are able to grow in media without added NaCl but are able to grow at high salt concentration, while "halophilic" is for those that require addition of NaCl or other salts to media for their growth. The definition seems clear, but the fact is that many "halotolerant" microorganisms are able to grow at higher salt concentrations than some "halophilic" microorganisms.

Another answer is an operational definition by Oren (2002): "microorganisms that are able to grow well above 100 g/l salt." Since the time Oren compiled the data, numerous papers on the "halophilic microorganisms" have been published proposing new genera and species, both Bacteria and Archaea. In this chapter, we arbitrarily define halophilic microorganisms as those that require NaCl higher than 30 g/l for growth and are able to grow well above 200 g/l, irrespective of the salt concentration of the origin of isolation. Although most origins are hypersaline, above 100 g/l salt as defined by Oren, sometimes data of salt concentration of bodies of water are not described in the original papers. When the sources are soil samples (Ventosa et al. 2008) or materials collected on seashores, leaves of plants, or birds (Brito-Echeverría et al. 2009; Yim et al. 2015), it is difficult to measure the exact salt concentration of the very small niches in those samples where halophilic microorganisms are thriving.

8.2.1 Haloarchaea

Haloarchaea, members of the family *Halobacteriaceae*, are a group of extremely halophilic microorganisms, forming a part of the domain Archaea. The early investigations on the general bacteriology of red-pigmented halophilic Bacteria as the cause of the reddening of salted hides (red heat) and fish (red eye) began in the 1920s. The scientists came to the conclusion that marine or solar salts and rock salts are contaminated by the "halophilic Bacteria" (Anderson 1954; Juez 1988). The eighth edition of Bergey's Manual of Determinative Bacteriology placed rod-shaped extreme halophiles in the genus *Halobacterium* with two species, *Halobacterium salinarum* and *Halobacterium halobium,* and coccoid extreme halophiles in the genus *Halococcus*, with only the species *Halococcus morrhuae* (Gibbons 1974). Soon, *Hbt. halobium* was suggested to be a member of Archaebacteria by Magrum et al. (1978). Tindall et al. (1980) were the first to isolate alkaliphilic haloarchaea that grow only in media of pH higher than 7.5, and they introduced the novel genera *Natronobacterium* and *Natronococcus* (Tindall et al. 1984). Two more genera, *Haloarcula* and *Haloferax*, were proposed to accommodate some new isolates and several species of the genus *Halobacterium* (Torreblanca et al. 1986). Since then, numerous strains have been isolated from hypersaline environments distributed all over the world. Neutrophilic strains were differentiated at the generic level based on physiological characteristics and on the presence or absence of specific membrane lipids, phosphatidylglycerosulfate (PGS) and glycolipids (Kates 1995; Grant et al. 2001). Alkaliphilic strains, on the other hand, were devoid of detectable amount of glycolipids. The introduction of the PCR technique (Saiki et al. 1988) made analysis of 16S rRNA gene sequences easy in microbial taxonomy (Weisburg et al. 1991). The first phylogenetic tree of haloarchaea was reconstructed in 1993 using 19 sequences available at that time, demonstrating that species of the genera *Halobacterium, Halococcus, Haloarcula, Haloferax*, and *Natronobacterium* formed coherent clusters (Kamekura and Seno 1993).

At present, haloarchaeal strains are classified in 215 species of 53 genera (as of March 2016) (http://www.bacterio.net; Enache et al. 2007b; Burns et al. 2010; Minegishi et al. 2010; Shimane et al. 2010). Table 8.1a shows the geographical distribution of most of the representative strains. All strains of halophilic Archaea require NaCl higher than at least 4.7 % (0.8 M) and are able to grow up to at least 23 % (4.0 M) NaCl. Although the majority are neutrophiles, alkaliphilic species have been isolated from soda lakes or Wadi Natrun in Kenya, China, India, Egypt, etc. (Horikoshi 1999; Rees et al. 2004). They are accommodated in the genera *Natronobacterium, Natronococcus, Natronomonas* (Kamekura et al. 1997), *Halalkalicoccus* (Xue et al. 2005), and *Natronolimnobius* (Itoh et al. 2005). The genera *Halobiforma, Halorubrum, Natrialba*, and *Natronorubrum* consist of both neutrophilic and alkaliphilic species.

Halophilic Archaea are believed to survive for long times in fluid inclusions of halite (Grant et al. 1998; Kunte et al. 2002; Stan-Lotter et al. 2003; Park et al. 2009; Gramain et al. 2011; Lowenstein et al. 2011; Fendrihan et al. 2012; Jaakkola et al.

2016). Recently, Fendrihan et al. (2009b) suggested the use of Raman spectroscopy as a potential method for the detection of extremely halophilic Archaea embedded in halite.

8.2.2 Halophilic Bacteria

Table 8.1b is a summary of saline and hypersaline environments and the halophilic Bacteria isolated from these sites. The range of salt concentrations, mostly NaCl or

Table 8.1a Halophilic Archaea and their distribution around the world

Country	Saline environment	Archaea
Algeria	Ezzemoul sabkha	*Halorubrum ezzemoulense*
Antarctica	Deep Lake	*Halorubrum lacusprofundi*
Argentine	Salt flats (not specified)	*Haloarcula argentinensis* *Halomicrobium mukohataei*
Australia	Hamelin Pool, Shark Bay, Western Australia	*Halococcus hamelinensis* *Haloferax elongans* *Haloferax mucosum*
	Cheetham Salt Works, Geelong, Victoria	*Halonotius pteroides* *Haloquadratum walsbyi* *Halorubrum coriense* *Natronomonas moolapensis*
Austria	Salt mine, Altaussee	*Halobacterium noricense*
	Rock salt, Bad Ischl salt mine	*Halococcus dombrowskii* *Halococcus salifodinae*
Canada	Salted cowhide	*Halobacterium salinarum* [a]
Chile	Lake Tebenquiche, Atacama Saltern	*Halomicrobium katesii* *Halorubrum tebenquichense*
China	Saline soil, Daqing, Heilongjiang Province	*Haloterrigena daqingensis*
	Fuqing solar saltern, Fujian Province	*Halorubrum litoreum*
	Rudong marine solar saltern, Jiangsu Province	*Haladaptatus litoreus* *Halogeometricum rufum* *Halogranum rubrum* *Halopelagius inordinatus* *Haloplanus vescus* *Halosarcina limi*
	Solar saltern, Zhoushan Archipelago, Zhejiang Province	*Haloferax larsenii*
	Sea salt, Qingdao, Shandong Province	*Halococcus qingdaonensis*
	Gangxi marine solar saltern	*Salinigranum rubrum*
	Marine solar saltern near city of Shanya, Hainan Province	*Halomicroarcula salina*
	Yuncheng salt lake in Shanxi Province	*Halorussus amylolyticus*
	Brine from marine solar saltern Tainan in Eastern China	*Halobellus clavatus*

Table 8.1a (continued)

Country	Saline environment	Archaea
China (Inner Mongolia Autonomous Region)	Baerhu Soda Lake	*Natronolimnobius baerhuensis* *Natronolimnobius innermongolicus*
	Lake Bagaejinnor	*Halorubrum kocurii*
	Lake Chagannor, 17 °C, pH 10.5	*Halorubrum luteum* *Natronorubrum sediminis*
	Chahannao soda lake	*Halobiforma nitratireducens* *Natrialba chahannaoensis*
	Lake Ejinor	*Halorubrum ejinorense* *Halorubrum orientale* *Halovivax asiaticus* *Natrinema ejinorense*
	Jilantai salt lake	*Halobacterium jilantaiense*
	Lake Shangmatala	*Halopiger xanaduensis*
	Lake Xilinhot	*Halostagnicola larsenii* *Haloterrigena salina* *Halovivax ruber*
	Unnamed soda lake, Hulunbuir prefecture	*Natrialba hulunbeirensis*
China (Tibet Autonomous Region)	Bange salt-alkaline lake, pH 10	*Natronorubrum bangense* *Natronorubrum tibetense*
	Lake Zabuye, pH 9.4	*Halalkalicoccus tibetensis* *Halorubrum tibetense*
China (Xinjiang Uygur Autonomous Region)	Aibi (or Ebinur) salt lake	*Haloarcula amylolytica* *Halorubrum lipolyticum* *Haloterrigena limicola* *Haloterrigena longa* *Haloterrigena saccharevitans* *Natrinema versiforme* *Natronorubrum aibiense*
	Aiding salt lake	*Halorubrum aidingense* *Natronorubrum sulfidifaciens*
	Xiao-Er-Kule Lake	*Halorubrum xinjiangense*
	Saline lake (sampling site not specified)	*Halorubrum alkaliphilum*
	Ayakekum salt lake, Altun Mountain, pH 7.8	*Halobiforma lacisalsi* *Halorubrum arcis* *Natrinema altunense*
Egypt	Brine pool, Sinai	*Haloarcula quadrata*
	Saline soil, Aswan	*Halobiforma haloterrestris* *Halopiger aswanensis* *Natrialba aegyptiaca*
	Solar saltern, Alexandria	*Haloferax alexandrinus*
	Wadi Natrun	*Natronomonas pharaonis*
France	Solar salt "Sel marin de Guerande"	*Halomicroarcula pellucida*
Iran	Mud of hypersaline lake Aran-Bidgol (35° 70′ N, 51° 39′ E)	*Halovivax limisalsi*

(continued)

Table 8.1a (continued)

Country	Saline environment	Archaea
	Brine of hypersaline lake Aran-Bidgol	*Halorubrum halodurans*
	Brine of hypersaline lake Aran-Bidgol	*Halopenitus persicus*
Israel/Jordan	The Dead Sea	*Haloarcula marismortui* *Halobaculum gomorrense* *Halococcus morrhuae* [b] *Haloferax volcanii* *Haloplanus natans* *Halorubrum sodomense*
Italy	"Red heat" in salted hides	*Natrinema pellirubrum* [c]
	Solar salt, Trapani, Sicily	*Halorubrum trapanicum* [c]
Japan	Solar salt, Niigata	*Natronoarchaeum mannanilyticum*
	Salt field, Ishikawa	*Haloarcula japonica*
	Sea sand (sampling site not specified)	*Natrialba asiatica*
	Commercial salt, also from Australia and Bolivia	*Halocalculus aciditolerans*
	Commercial salt from seawater of Yonaguni island, Okinawa	*Salarchaeum japonicum*
Kenya	Lake Magadi	*Halorubrum vacuolatum* *Natrialba magadii* *Natronobacterium gregoryi* *Natronococcus amylolyticus* *Natronococcus occultus*
Mexico	Solar saltern, Baja California	*Halorubrum chaoviator*
	Former lake Texcoco	*Natronobacterium texcoconense*
Pakistan	Commercial rock salt	*Halorubrum gandharaense*
Philippines	Solar salt (sampling site not specified)	*Halarchaeum acidiphilum*
	Commercial salt from seawater in the Philippines, also Indonesia (Bali) and Japan (Okinawa)	*Halarchaeum rubridurum*
Puerto Rico	Solar saltern, Cabo Rojo	*Halogeometricum borinquense* *Haloterrigena thermotolerans*
Red Sea	Shaban Deep, brine-sediment interface (depth of 1447 m, pH 6.0)	*Halorhabdus tiamatea*
Romania	Telega Lake, Prahova	*Haloferax prahovense*
South Korea	Jeotgal (salty condiment)	*Haladaptatus cibarius* *Halalkalicoccus jeotgali* *Halorubrum cibi* *Haloterrigena jeotgali* *Natronococcus jeotgali*

Table 8.1a (continued)

Country	Saline environment	Archaea
Spain	Fuente de Piedra salt lake, Malaga	*Haloterrigena hispanica*
	San Fernando solar saltern, Cadiz	*Halococcus saccharolyticus*
	Santa Pola solar saltern, Alicante	*Haloarcula hispanica* *Haloferax gibbonsi* *Haloferax lucentense* *Haloferax mediterranei*
	Isla Bacuta saltern in Huelva	*Halovenus salina*
Taiwan	Solar salt (sampling site not specified)	*Natrialba taiwanensis*
Thailand	Fermented salty foods (kapi, nam pla, Pla-ra)	*Halobacterium piscisalsi* *Halococcus thailandensis* *Natrinema gari*
Turkmenistan	Saline soil	*Halorubrum distributum* *Halorubrum terrestre* *Haloterrigena turkmenica*
USA	The Great Salt Lake, Utah	*Halorhabdus utahensis*
	Death Valley, California	*Haloarcula vallismortis*
	Saltern, San Francisco Bay, California	*Haloferax denitrificans* *Halorubrum saccharovorum*
	Cargill Solar Salt Plant, Newark, California	*Halorubrum californiense*
	Rock salt crystals, Carlsbad, New Mexico	*Halosimplex carlsbadense*
	Zodletone Spring, Oklahoma	*Haladaptatus paucihalophilus* *Haloferax sulfurifontis* *Halosarcina pallida*

Natrinema pallidum NCIMB 777 was isolated from salted cod fish, but the site of isolation is not clear

[a]Lochhead (1934)

[b]Kocur and Hodgkiss (1973)

[c]Online catalogue of NCIMB

a mixture of salt in some cases, that permitted growth and optimum concentration are also indicated. The species listed in Table 8.1b belong to the following classes: *Cyanobacteria* of the phylum *Cyanobacteria*, *Alphaproteobacteria* and *Gammaproteobacteria* of the phylum *Proteobacteria*, *Clostridia* and *Bacilli* of the phylum *Firmicutes*, *Actinobacteria* of the phylum *Actinobacteria*, and *Flavobacteria* of the phylum *Bacteroidetes*. References for each species are not given in the list because of the huge number which would have to be cited. For example, the genus *Halomonas* and the genus *Marinobacter* (both of the phylum *Gammaproteobacteria*) contain 90 and 39, respectively, validly published species (as of March 2016). Readers are recommended to consult the very useful website "List of Prokaryotic names with Standing in Nomenclature," maintained by Aidan C. Parte, Curator, at http://www.bacterio.net for relevant papers.

Table 8.1b Halophilic Bacteria and their distribution around the world

Country	Saline environment	Bacteria	Range of NaCl (%)	Optimum NaCl (%)
Algeria	Ezzemoul sabkha	*Halomonas sabkhae*	5–25	7.5
		Salicola salis	10–25	15–20
Canada	Contaminant on agar plate	*Actinopolyspora halophile*	10–33	15–20
Chile	Solar saltern, Cahuil, Pichilemu	*Halomonas nitroreducens*	3–20	5–7.5
China	Xiaochaidamu salt lake, Qinghai province	*Gracilibacillus halophilus*	7–30	15
China (Inner Mongolia Autonomous Region)	Lake Chagannor, 17 °C, pH 10.5	*Bacillus chagannorensis*	3–20 (salts)	7 (salts)
		Salsuginibacillus kocurii	3–20 (salts)	10 (salts)
	Lake Shangmatala	*Aquisalibacillus elongatus*	3–20	10
	Lake Xilinhot	*Virgibacillus salinus*	3–20	10
	Xiarinaoer soda lake	*Salsuginibacillus halophilus*	9–30	19
China (Xinjiang Uygur Autonomous Region)	Aiding salt lake	*Alkalibacillus salilacus*	5–20	10–12
		Bacillus aidingensis	8–33	12
		Lentibacillus halodurans	5–30	8–12
		Prauserella sedimina	5–20	10
		Salinibacillus aidingensis	5–20	10
	Qijiaojing Lake	*Haloechinothrix alba*	9–23	15
	Saline lake (sampling site not specified)	*Bacillus salarius*	3–20	10–12
		Lentibacillus lacisalsi	5–25	12–15
		Saccharopolyspora halophilia	3–20	10–15
	Saline soil (sampling site not specified)	*Alkalibacillus halophilus*	5–30	10–20
		Nocardiopsis salina	3–20	10
		Prauserella halophila	5–25	10–15
		Saccharomonospora	5–20	10
		paurometabolica	5–25	10–15
		Streptomonospora alba	5–20	10
		Streptomonospora amylolytica	5–25	10
		Streptomonospora flavalba	5–20	10
		Streptomonospora halophila		
Congo	Oil-well head sample	*Halanaerobium congolense*	4–24	10
Egypt	Wadi Natrun	*Natranaerobius thermophilus*	18–29	19–23
		Natranaerobius trueperi	19–31	22
		Natronovirga wadinatrunensis	19–31	23
		Thiohalospira alkaliphila	3–23	12
France	Salin-de-Giraud saltern, Camargue	*Halanaerobacter salinarius*	5–30	14–15
		Halorhodospira neutriphila	6–30	9–12
		Thiohalocapsa halophila	3–20	7
Greece	Saltworks, Mesolongi	*Bacillus halochares*	6–23	15

Table 8.1b (continued)

Country	Saline environment	Bacteria	Range of NaCl (%)	Optimum NaCl (%)
Iran	Lake Aran-Bidgol	*Lentibacillus persicus*	3–25	7.5–10
	Howz Soltan Lake	*Bacillus persepolensis*	5–20	10
Iraq	Saline soil	*Actinopolyspora iraqiensis*	5–20	10–15
Israel/Jordan	The Dead Sea	*Rhodovibrio sodomensis*	6–20	12
		Salisaeta longa	5–20	10
		Selenihalanaerobacter shriftii	10–24	21
		Virgibacillus marismortui	5–25	10
Japan[a]	Solar salt	*Nesterenkonia halobia*	3–25	–
	Salted foods	*Chromohalobacter japonicus*	5–25	7.5–12.5
		Halanaerobium fermentans	7–25	10
Kenya	Lake Magadi	*Natroniella acetigena*	10–26	12–15
Kuwait	Salt marsh soil	*Saccharomonospora halophila*	10–30	–
Mexico	Solar saltern, Baja California	*Halospirulina tapeticola*	3–20	10
	Brine water, Gulf of Mexico	*Halanaerobium acetethylicum*	5–22	10
Mongolia	Barun-Davst-Nur	*Halovibrio denitrificans*	12–30	12–15
Nauru (South Pacific)	Seashore wood	*Salimicrobium halophilum*	3–30	–
Peru	Maras salterns, Andes	*Salicola marasensis*	10–30	15
Portugal	Terminal pond of a saltern	*Rhodovibrio salinarum*	3–24	9–15
Puerto Rico	Black mangrove, solar saltern of Cabo Rojo	*Halobacillus mangrove*	5–20	10
Russia	Kulunda Steppe, Altai	*Halospina denitrificans*	12–30	15–18
		Methylohalomonas lacus	3–23	12
		Thiohalorhabdus denitrificans	9–23	18
		Thiohalospira halophila	12–30	15–18
		Thiomicrospira halophila	3–20	9
Senegal	Retba Lake	*Halanaerobium lacusrosei*	7.5-saturated	18–20
South Korea	Byunsan solar saltern, Yellow Sea	*Lentibacillus salinarum*	3–24	10–12
	Solar saltern, Yellow Sea	*Alkalibacillus flavidus*	4–26	10
	Seawater, Yellow Sea	*Salinisphaera dokdonensis*	4–21	10
	Kunsan solar saltern	*Nocardiopsis kunsanensis*	3–20	10
	Jeotgal (salty condiment)	*Lentibacillus jeotgali*	3–20	10–15

(continued)

Table 8.1b (continued)

Country	Saline environment	Bacteria	Range of NaCl (%)	Optimum NaCl (%)
Spain	Soil from Fuente de Piedra, saline wetland, Malaga	*Halomonas fontilapidosi*	3–20	5–7.5
	Cabo de Gata solar saltern, Almeria	*Halomonas almeriensis*	5–25	7.5
		Kushneria indalinina	3–25	7.5–10
		Salinicola halophilus	3–25	7.5–10
	Mallorca solar saltern, Balearic Islands	*Salinibacter ruber*	15–33 (salts)	20–30 (salts)
	Santa Pola solar saltern, Alicante	*Halomonas cerina*	7.5–20	7.5–10
		Virgibacillus salexigens	7–20 (salts)	10 (salts)
Thailand	Fermented salty foods (kapi, nam pla, Pla-ra)	*Lentibacillus halophilus*	12–30	20–26
		Lentibacillus juripiscarius	3–30	10
		Lentibacillus kapialis	5–30	15
Tunisia	Chott el Guettar	*Halothermothrix orenii*	4–20	10
	Chott el Djerid	*Halanaerobaculum tunisiense*	14–30	20–22
Ukraine	Lake Sivash, Crimea	*Halanaerobium saccharolyticum*	3–30	10
		Halocella cellulosilytica	5–20	15
		Orenia sivashensis	5–25	7–10
USA	The Great Salt Lake, Utah	*Halomonas variabilis*	7–29	9
	Death Valley, California	*Actinopolyspora mortivallis*	5–30	10–15
	Saltern, San Francisco Bay, California	*Halanaerobacter chitinivorans*	3–30	12–18
	Evaporated seawater, Oregon	*Rhodothalassium salexigens*	5–20	12
	Saline oil field brine, Oklahoma	*Arhodomonas aquaeolei*	6–20	15
		Halanaerobium salsuginis	6–24	9
	Searles Lake, California	*Halarsenatibacter silvermanii*	20 - saturated	Saturated

[a]The following species have been isolated from ordinary, nonsaline soil samples taken in Japan: *Alkalibacillus silvisoli* 5–25 % (10–15 %), *Geomicrobium halophilum* 5–25 % (10–15 %), *Halalkalibacillus halophilus* 5.0–25 % (10–15 %)

8.2.3 Romanian Hypersaline Environments

Hypersaline environments are widely distributed also in Romania, either in solid or liquid forms: salt lakes and salt mines located in Prahova County, the Techirghiol Lake nearby to Black Sea coast, the Balta Albă Lake in Buzău County, etc. (Fig. 8.1). Some of these environments have been well described some time ago (Brosteanu 1901) and today also constitute an attractive research area either for geologists and biologists (Sencu 1968; Faghi et al. 1999; Teodosiu et al. 1999; Har et al. 2006).

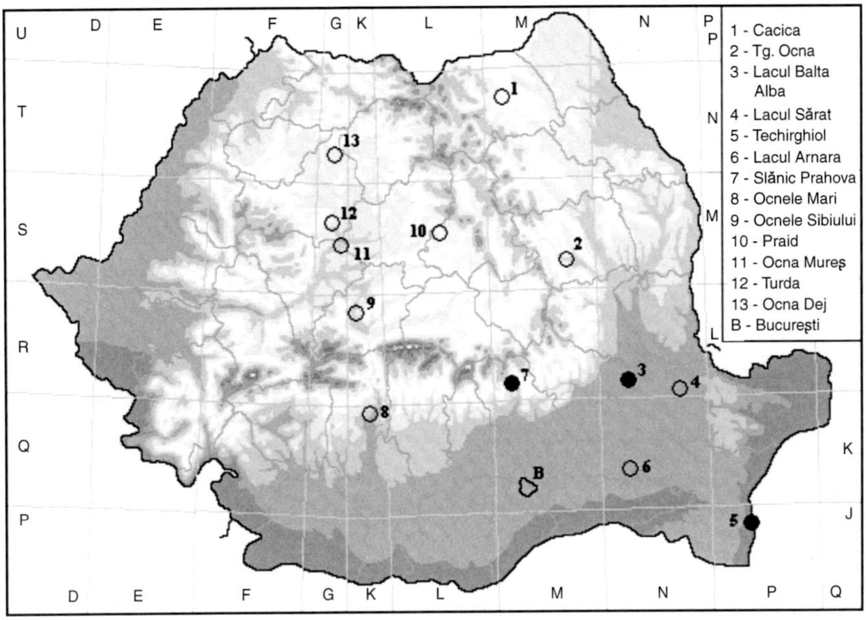

Fig. 8.1 Geographical positions of major hypersaline environments in Romania. The filled points represent the examined areas described in the text. (*1*) salt mine at Cacica; (*2*) salt mine at Targu; (*3*) salt lake (Lacul Balta Albă); (*4*) hypersaline salt lake (Lacul Sarat); (*5*) salty therapeutic mud lake at Techirghiol (near the coast of the Black Sea); (*6*) haloalkaline lake (Lacul Amara); (*7*) lakes and salt mine of Slănic (Slănic Prahova); (*8*) salt mine at Mari (Ocnele Mari); (*9*) lakes and salt mine at Sibiu (Ocnele Sibiu); (*10*) salt mine at Praid (Praid); (*11*) salt mine at Mures (Ocna Mures); (*12*) salt mine at Turda (Turda); (*13*) salt mine at Dej (Ocna Dej); B, Bucharest, capital of Romania

Salt exploitation in the Romanian Carpathian area has been conducted by various methods from antiquity until today (Drăgănescu and Drăgănescu 2001), due to the presence of lots of salt massifs with characteristics that supported their continued use such as surface proximity, superior purity of NaCl, or large reserves. In some places, the massifs are located at the surface as small salt mountains, for example, at Slănic, Praid, Sărata Monteoru (Fig. 8.1). In the Slănic Prahova area, for example, salt exploitation started in 1685 by using bell-type exploitation technology. After the eighteenth century, some areas of exploitation were abandoned, resulting in various man-made salt lakes with different depth (varying from 3 to 40 m, see Table 8.2) and width, along to the left or right side of the Slănic Valley (Drăgănescu 1990). On the other hand, in Telega, where similar technologies were used, salt extraction started before 1685, and relinquished exploitation areas resulted in various man-made salt lakes, as shown in Table 8.3 (Gâştescu 1971; Enache et al. 2008a).

The salt deposit in Slănic Prahova formed in the Neogene period (24 My ago) and is located at around 100 km north of Bucharest in a sub-Carpathian hillock area (Figs. 8.1 and 8.2). This salt deposit has various lengths and thicknesses as described previously (Drăgănescu 1990; Enache et al. 2008a). Investigations for the presence of halophilic microorganisms were conducted in several salt lakes which formed in

Table 8.2 Chemical features of examined lakes and colony-forming units (c.f.u.), as determined on agar plates containing 12.5 % NaCl and 16 % $MgCl_2 \times 6H_2O$, from surface water samples

Lake	Maximum depth (m)	pH	Density	Chloride content (g/l)	c.f.u./ml
Red Bath	3	7.9	1.06	74.9	2500
Green Bath	40	9.0	1.10	138.5	1050
Shepherd Bath	7.25	8.7	1.07	97.4	2400
Bride Cave	32	8.3	1.20	254.6	750
Techirghiol	9	7.0	-	60	-
Telega (Palada)	36	8.3	1.15	161	1100
White Bath	-	8.6	1.10	44	890

Table 8.3 Characteristics of salt lakes from Telega (Gâştescu 1965)

Lake	Height above the sea (m)	Surface (m^2)	Maximum width (m)	Maximum depth (m)
Doftana	413	9200	84	26
Central Bath	414	1344	38	45
Sweet Lake	424	1480	35	21
Stavrică	415	1740	52	107.5
Mocanu	415	630	22	14
Palada (Telega)	416	1416	30	36

the relinquished salt mine exploitations in this salt deposit (see Sect. 2.4). These lakes are known today as Bride Cave, Red Bath, Green Bath, and Shepherd Bath.

The salt deposit of Telega (Doftana-Telega), which formed also in the Neogene period, is located at a similar distance and position from Bucharest like the Slănic deposit and is a mixture of crystals with colors varying from white to gray and swarthy, having a surface of 2.1 square kilometers. Various salt lakes which resulted in the mouth of an abandoned salt mine in this deposit are detailed in Table 8.3.

Another hypersaline environment with large reserves located at Ocnele Mari (Vâlcea County) has been exploited by dissolving salt, and the brine was channeled directly into a petrochemical plant by pipelines for the production of various chemicals (NaOH, Na_2CO_3, HCl, Cl_2, NaOCl, etc.). Consequently, a large subterranean void resulted which nowadays is under control to preserve collapsing. A huge man-made salt lake, estimated as larger than 50 ha, is expected to be present. This man-made hypersaline environment will constitute a further subject for halophilic research.

Another examined salt lake is Techirghiol Lake, located near the Black Sea coast. This lake is characterized by variable concentrations of salt, ranging from negligible to 60 g/L. The lake is relatively well known for its biological communities and saline regimes, and has been well investigated over time, either because of the important therapeutic properties of the mud present in the lake or the dynamics of moderately halophilic microbial communities.

Another hypersaline environment, recently investigated by our team, is represented by the Balta Albă salt lake, located near the town of Râmnicu Sărat. The area is characterized by the presence of various salty (not extremely) soils. Various

Fig. 8.2 Salt massifs in the Slănic Prahova (**a**, **b**) and Praid (**c**) areas. The red color may possibly be due to the presence of microorganisms or their remnants inside of the salt massif in Slănic (**b**)

hypotheses have been elaborated on the origin of a salt lake in this area, arguing for either an ancient origin or the consequence of intense evaporation processes which took place in this area. Some geological data supported the latter hypothesis (Gâştescu 1965). The salinity of this lake was 40 g/L in our surface sample water.

8.2.4 Halophiles from Salty Environments in Romania

The chloride concentration of water samples taken from these salt lakes ranged from 44 to 254.6 g/l (Table 8.2) and pH values between 7.0 and 9.0. A number of these environments have been found to contain halophilic microorganisms able to grow in the presence of high concentrations of NaCl and $MgCl_2 \times 6H_2O$. The numbers of colonies ranged from 750 c.f.u./ml in water samples from Bride Cave until to around 2500 c.f.u./ml in Red Bath (Table 8.1a and 8.1b). In other lakes, the colony numbers were relatively low, when compared to those reported for various salt lakes with similar salt concentrations (Enache et al. 2008b). Investigations have been conducted (Enache et al. 1999b, 2000a, b), and a new species *Haloferax prahovense* was proposed (Enache et al. 2007a).

The diversity of halophiles, mainly in crystallizer ponds of solar salterns, remains one of the important research targets, although the constant NaCl concentrations of these artificial hypersaline environments over the years could be considered as a selective pressure for halophilic microorganisms. Due to various climatic and other natural conditions, different salt concentrations could be found in the salt lakes. Although various molecular techniques were applied to understand the ecological principles of salt lakes, the microorganisms which could be isolated and cultivated in the laboratory appear to play the most important role in the ecological economy of these lakes.

The number of colony-forming units (c.f.u.) decreased with increasing chloride and sodium concentrations in the examined salt lakes, but the colonies assigned to be Archaea had no apparent correlation with the concentrations of these two ions (Enache et al. 2008b). The predominant presence of *Haloferax* species in these lakes (Enache et al. 2008a, b) suggests that members of this genus play an important role in the ecology of salt lakes, even though the largest numbers of species, which have been identified in hypersaline environments, are of the genus *Halorubrum.* On the other hand, the microbiota of subterranean rock salt from the Slănic area are characterized by the presence of some *Halorubrum* and *Haloarcula* species, and some isolates appear to be closely related to *Hbt. noricense*, a strain isolated from an ancient evaporite formed in the Permo-Triassic period (Gruber et al. 2004). The *Haloferax* members, observable in hypersaline lakes located at the surface of the Slănic salt deposit, were identified also in subterranean rock salt, but with decreasing numbers (Enache, 2015).

Recent publications described the isolation of extremely halophilic Archaea from four hypersaline meromictic lakes in the Transylvanian Basin close to the salt mines of Turda (# 12 in Fig. 8.1), Ocnele Sibiu (# 9 in Fig. 8.1), and Tarva (east of the town of Târgu Mures) (Baricz et al. 2014, 2015). *Haloferax* was the most frequently isolated genus, followed by *Halobacterium* and *Halorubrum*, and less abundant species of the genera *Halomicrobium, Haloarchaeobius, Halopenitus, Halarchaeum, Halorubellus, Halovenus, Haladaptatus, Haloarcula, Halococcus, Haloterrigena, Halopelagius, Natronococcus, Natronorubrum,* and *Natrinema* were also detected.

8.3 Biotechnological Potential

The physiological and biochemical features specific for halophilic Archaea and Bacteria as well as their capacity to produce biopolymers, enzymes, and osmoprotectors of industrial interest, with properties superior to those synthesized by non-halophilic species, make them a promising group of unlimited biotechnological potential (Rodriguez-Valera 1992). In the following sections, we would like to describe the potential for biotechnological application of halophilic microorganisms, including those isolated from Romanian hypersaline environments. Readers are also advised to consult previous general reviews (Rodriguez-Valera 1992; Ventosa et al. 1998; Horikoshi 1999; Mellado and Ventosa 2003; Ventosa 2004; Borgne et al. 2008; Oren 2010; Fathepure 2014; Charlesworth and Burns 2015; Enache et al. 2015).

8.3.1 Bioremediation

Biodegradation of organic pollutants by halophilic Bacteria and Archaea has been reviewed by Borgne et al. (2008). A few moderately halophilic Bacteria were found to show the capacity, to various degrees, of decontaminating the pesticide dichlorvos ($C_4H_7O_4Cl_2P$) from saline environments (Oncescu et al. 2007). These strains, which were isolated from Shepherd Bath, Telega (Palada), and Techirghiol lakes, were tentatively assigned as *Halomonas* species according to their preliminary biochemical and physiological characterization. The kinetics of dichlorvos degradation appeared to follow the mechanism of "compatible solutes," used by large numbers of organisms to cope with osmotic stress generated by the presence of high concentrations of salt. Considerable efforts are still necessary in order to estimate the true potential of these halophilic microorganisms to be applied in environmental processes and in the remediation of contaminated hypersaline ecosystems. This effort should also be focused on basic research to understand the overall degradation mechanism, to identify the enzymes involved in the degradation process and the metabolic regulation.

Contamination of hypersaline environments with petroleum compounds is another important area for bioremediation with halophilic microorganisms (reviewed by Fathepure 2014). Due to their magnitude – e.g., hundreds of kilometers of coastline in oil-producing countries all over the world – considerable economic aspects are at stake. Oil fields pose special problems due to their sheer numbers and due to the high salinity which is present in the brackish "produced water," generated during oil and natural gas extraction. Biodegradation of hydrocarbons in moderate to high salinity environments has been demonstrated with halophilic Bacteria and Archaea which possess the metabolic capacity to degrade a variety of aliphatic and aromatic hydrocarbons in varying salinities. Though progress has been made in understanding the diversity of microorganisms responsible for hydrocarbon degradation under aerobic conditions, similar information under anaerobic conditions is lacking. A better knowledge of the diversity of catabolic pathways should lead to the development of robust bioremediation processes for hypersaline environments (Fathepure 2014).

8.3.2 Nanobiotechnology

Nanobiotechnology applies biological materials for the synthesis of nanomaterials and design of nano-devices. Although nanoparticles have been synthesized by physical and chemical methods, these processes have several disadvantages. Besides using up much energy and requiring often expensive instrumentation, contamination from the toxic precursors and the generation of hazardous by-products are major drawbacks (Srivastava and Kowshik 2015; Srivastava et al. 2014). Therefore biological resources have been explored. Halophilic microorganisms are known to encounter metals in their natural habitat and could be exploited for the synthesis of metal-based nanoparticles. Srivastava et al. (2013) obtained silver nanoparticles with the haloarchaea *Halococcus salifodinae* BK_3, which showed good antibacterial activity against both Gram-positive and Gram-negative bacteria.

Enache et al. (2015) described the responses of several moderately halophilic Bacteria such as *Virgibacillus halodenitrificans* and *Bacillus subtilis* to the presence of various nanomaterials like titanium nanotubes or silica microtubes in their culture medium. The experiments revealed an antimicrobial action of the titanate nanostructures. Although a bacteriocin (halocin)-like attack mechanism is likely, further investigation will be necessary concerning the interaction between these nanomaterials and halophiles (Merciu et al. 2009). If silica nanostructures were used, which have spherical shapes or were doped with platinum, a transient positive effect on the growth of tested halophilic Bacteria was observed. This might have been due to the excessive synthesis of exopolysaccharides.

8.3.3 S-layers

Crystalline cell surface layers are commonly observed cell envelope structures of several Bacteria and Archaea, and they have numerous applications in biotechnology and nanotechnology (see review by Sleytr et al. 2014). The extremely halophilic Archaea lack a peptidoglycan component in their cell wall and contain simple S-layers external to the cell membrane (Trachtenberg et al. 2000; Eichler 2003). S-layers consist of identical protein or glycoprotein subunits and completely cover the cell surface during all stages of growth and division. Most S-layers are 5–15 nanometers in thickness and possess pores of identical size and morphology in the 2–8 nm range (Sleytr and Sara 1997; Schuster et al. 2005).

Isolated S-layer subunits of various microorganisms have the intrinsic ability to self-assemble into highly defined monomolecular arrays either in suspension, at air/water interfaces or liquid/surface interfaces, including lipid films, liposomes, and solid supports such as silicon wafers (Schuster and Sleytr 2000; Schuster et al. 2005). The relative simplicity, regularity, and symmetry within the monolayer plane of the S-layer make it an attractive subject for nano-biotechnological studies with targets for medical applications (Sleytr et al. 1999; Trachtenberg et al. 2000).

A striking feature of many S-layers of Bacteria and Archaea is their excellent antifouling property. Even when cells were harvested from complex environments or growth media containing a great variety of macromolecular components, the S-layer lattices were never masked by adsorbed molecules. Detailed studies on molecular interactions and permeability using isolated S-layers or S-layer ultrafiltration membranes confirmed that the surface of the lattice is charge neutral, preventing nonspecific binding of molecules and pore blocking. Recently, the unique antifouling properties of S-layers were successfully exploited for coating microfluidic channels in lab-on-a-chip devices (Sleytr et al. 2014, and references therein).

Our investigations on S-layers were carried out using a halophilic archaeon, *Haloferax* sp. strain GR 2 (deposited as JCM 13922), isolated from the Bride Cave Lake in Prahova County. Preliminary investigations related to the binding of S-layer to some porous silicon substrates were performed. The biochemical characterizations by protein content and chemical treatment had demonstrated the presence of S-layer in the isolated strain. Transmission electron microscopic examination of the isolated S-layer showed the existence of the monomolecular crystalline lattice with a highly ordered arrangement in the dense form, while in relaxed form after treatment with 4 M urea (Dumitru et al. 2007). The S-layer proteins attached to both hydrophilic and hydrophobic surfaces of all plates of porous silicon, which were investigated, but it seemed that the hydrophobic surfaces were more favorable. Thus, the treatment of silicon plates with hexamethyldisilazane, which promotes the hydrophobicity and organic character of the porous silicon surface, increased the amount of attached S-layer protein (Sleytr et al. 1999; Dumitru et al. 2007; Kleps et al. 2009).

8.3.4 Extracellular Enzymes

A great deal of information on eukaryotic and bacterial halophilic enzymes is currently available, for example, on amylases, lipases, nucleases, nucleotidases, proteases, etc. (Ventosa et al. 2005; Oren 2010; Enache et al. 2015).

For the production of Thai fish sauce (*nam pla*), a condiment similar to *Garum* or *Liquamen* of the ancient Roman society, prepared from fish in concentrated brine, proteases of haloarchaea which are present in solar salt, plays an important role in the degradation of fish protein into amino acids (Thongthai et al. 1992). Further studies on haloarchaeal enzymes are expected to contribute to the elucidation of the properties of these extracellular enzymes.

Amylases were produced by some strains isolated from Bride Cave and Techirghiol lakes (Enache and Faghi 1999; Enache et al. 2009). The enzyme of *Hfx.* sp. GR1 purified by ethanol precipitation and differential chromatography showed maximum activity at pH 6.5 and 50 °C in the presence of 3.5 M NaCl and lost its activity below 1.5 M NaCl (Enache et al. 2001). Amylases produced by strains of *Haloferax* and *Halorubrum* isolated from Techirghiol Lake, a low-salt environment, showed higher activity with increasing concentration of $MgCl_2$ in the presence of the relatively low 2.1 M NaCl, but activity decreased with increasing Mg concentrations at the higher concentration of 3.4 M NaCl (Enache et al. 2009).

Extracellular lipase activity was detected in some strains isolated from lakes Shepherd Bath, Green Bath, Red Bath and Bride Cave. Among them, the enzyme produced by *Hfx.* sp. GR1 was influenced by NaCl concentrations in the growth media and had a maximum activity at 3 M NaCl. The activity was lost at NaCl concentrations below 2.5 M (Enache et al. 2004a).

The molecular mechanisms of the adaptation of enzyme proteins to high salt have been described in detail (Vellieux et al. 2007; Yamamura et al. 2009).

8.3.5 Halocins

The halocins, proteinaceous antibiotics, which are haloarchaeal equivalents of bacteriocins, were first discovered by Rodriguez-Valera et al. (1982). The wide variety of activity spectra detected for halocins (H1, H4, H6, S8, C8 etc.) may imply that a great number of different halocins are produced and probably show various mechanisms of action (O'Connor and Shand 2002, Torreblanca et al. 1994). Halocins were also detected in some strains isolated from Romanian salt lakes, and they showed a variety of action spectra (Enache et al. 1999a). When compared as halocin producers and targets, some strains showed identical patterns, supporting the tight clustering of strains in the phylogenetic tree reconstructed from 16S rRNA gene sequences (Enache et al. 2004b, 2008b).

An interesting and relevant use of halocin-producing strains is in the textile industry, where considerable amounts of salt are used in the tanning process. These conditions allow halophiles including some haloarchaea to grow which in turn can damage the product, and halocins have been used to prevent this unwanted growth (Charlesworth and Burns 2015).

Halocin 6 (H6) is a protein of 32 kDa produced by *Hfx. gibbonsii* SH7 and blocks the Na^+/H^+ antiporter in sensitive strains. Quite interestingly, H6 has been shown to inhibit in vitro the Na^+/H^+ exchanger of mammalian cells and to exert in vivo cardioprotective effects against ischemia and reperfusion injury (Lequerica et al. 2006). Halocin production is nearly universal, and relatively few have been fully characterized, in particular their molecular diversity and differences in activity, and it is thought possible that other therapeutic treatments could result from further study (Charlesworth and Burns 2015).

8.3.6 Exopolysaccharides

Halophilic microorganisms are able to synthesize EPS, which are biopolymers combining, in an excellent manner, the rheological properties (high viscosity and pseudoplasticity) with a remarkable resistance to extreme salinity, temperature, and pH values, conditions which are encountered in several industrial processes and which make usage of biopolymers produced by non-halophilic microorganisms impossible (Rodriguez-Valera 1992; Ventosa et al. 1998). The physical and chemical properties of the EPS produced by halophilic microorganisms enable their utilization in the food, textile, and dye industries, also for the production of pharmaceuticals and cosmetic products, in the oil-extracting industry as well as for processes for the removal of toxic compounds.

The biosynthetic activity for EPS was detected in some *Haloferax* strains isolated from Telega (Palada) Lake such as *Hfx. prahovense* and *Hfx.* sp. TL5. The optimal conditions for EPS production by *Hfx. prahovense* were the same as those resulting in highest cell growth. The maximum EPS yield (0.475 g%) was obtained in medium with 3 % glucose as single carbon source at 2 M NaCl, under stirring at 200 rpm, at 37 °C, after 7 days of incubation. The strain produced EPS also in media with galactose, lactose, maltose, sucrose, or fructose as carbon source. Higher salt concentrations (5 M) and higher temperature (45 °C) had an inhibitory effect on both growth and EPS synthesis. Synthesis of EPS started during the early exponential growth phase, increased concomitantly with a rise in the number of viable cells, and then decreased after 7 days of cultivation. The monomer composition of the EPS from *Hfx. prahovense* was similar to the composition of EPS synthesized by halophilic Archaea of the genus *Haloferax* (Anton et al. 1988). The polymer of *Hfx. prahovense* was a heteropolysaccharide containing mainly glucose, fructose, galactose, and mannose as was observed by TLC. Differential scanning calorimetry revealed that the polymer was stable up to 207 °C; the chemical composition observed by TLC was confirmed by FTIR investigations. FTIR also showed the presence of uronic acids and sulfate in the polymer (Popescu et al. unpublished results). A similar highly thermostable EPS was isolated and characterized from cultures of some moderately halophilic Bacteria isolated from Shepherd Bath (Cojoc et al. 2009).

8.3.7 Resistance to Heavy Metals

Several halophilic Archaea (Dumitru et al. 2002) and Bacteria (Enache et al. 2000c) isolated in Romania were shown to be resistant to heavy metals. The data suggested that moderately halophilic Bacteria exhibited a higher tolerance to metallic ions as compared to halophilic Archaea. The investigated haloarchaeal strains were susceptible to Zn and Hg but moderately resistant to Cr and Ni, being classified as tolerant according to the criteria proposed by Nieto (1991).

The metal tolerance level of our isolate *Haloferax* sp. TL5 (assigned as a strain of *Hfx. prahovense*, see Enache et al. 2008a) was compared with that of *Hfx. mediterranei*. Strain *Hfx.* sp. TL5 showed a similar behavior to the collection strain *Hfx. mediterranei*; both strains tolerated 5.0 mM Cr and 2.5 mM Ni and Pb. Strain *Hfx.* sp. TL5 had a higher susceptibility for Zn ions, compared with *Hfx. mediterranei*.

We also measured the capacity of these strains to reduce the concentration of several heavy metal ions from media. Both strains showed the capacity to reduce the concentration of Pb, Cr, Zn, and Ni ions from media with high salinity (Popescu and Dumitru 2009). The two strains produced higher cell densities when grown in media with metal ions and 2–2.5 % glucose than in media without glucose. This suggests that EPS synthesized in the presence of glucose may protect the cells against the toxicity of heavy metals. The two *Haloferax* strains showed the same capacity to reduce the concentration of Pb ion; for example, the initial concentration of 331 mg Pb/l was reduced to 5 mg/l after 10 days of cultivation. *Hfx.* sp. TL5 has a higher biosorption capacity of Cr and Ni ions from medium with or without glucose than *Hfx. mediterranei*. *Hfx. mediterranei* presented a higher removal activity of Zn ion from media with or without glucose than *Hfx.* sp. TL5 (Popescu and Dumitru 2009). The results revealed that the synthesis of EPS enhanced the reduction activity of Cr, Zn, and Ni by the haloarchaeal strains which were investigated. The anionic nature of EPS synthesized by *Haloferax* strains, based on their high sulfate and uronic-acid contents (Rodriguez-Valera 1992; Mellado and Ventosa 2003), is similar to that of EPS synthesized by other halophilic microorganisms and may be responsible for the capacity of these strains to bind and remove heavy metals from solutions with high NaCl concentrations. This property would make these biopolymers a viable alternative to the more aggressive physical and chemical methods, and they could be used as bio-adsorbents in polluted hypersaline environments.

8.3.8 Therapeutical Value

The salt lakes in Romania have been used also for various economical, recreational, and therapeutic purposes. The therapeutic use of the mud from salt lakes started in 1840 (Bulgăreanu 1993). Although attributed to the accumulation of sapropelic material, the mechanisms of mud formation and their microbiota are poorly understood. A few attempts in our laboratory to elucidate the mechanisms, using the mud from Techirghiol Lake, were fruitless. An industry has been developed for the exploitation of mud mainly from Balta Albă and Techirghiol lakes.

In Middle and Eastern Europe, the advantageous effect of some natural salt caves on lung diseases has been known since the nineteenth century; possibly, salt miners knew it far earlier based on the observation that injured animals went to caves for recovering. The beneficial effect of salt (speleotherapy) was reported by the Polish doctor F. Bochkowsky in 1843. Speleotherapy is based on the in-patient treatment in salt caves possessing a specific microclimate. The effects of salt mine treatment on health in the village of Solotvino in the Carpathian Mountains have been investigated by Russian scientists (Simyonka 1989). Several salt caves are used for speleotherapy in Middle and Eastern Europe as follows: Salzgrotten, Saliseum/Vienna (Austria); Slănic, Turda, and Praid (Romania); Wieliczka (Poland); Nakhichevan (Azerbaijan); Chon-Tous (Kirghizstan); Cave Berezniki in Perm (Russia); Solotvino (Ukraine); etc. Halotherapy is a form of speleotherapy, a science aimed to create somewhat similar conditions in a microclimatic environment as in salt caves. In the 1980s, Russia was the pioneer in creating the first salt chambers. These are specially prepared rooms with walls and basements covered with halite, a crystal form of salt (Chervinskaya and Zilber 1995; Hedman et al. 2006; Nica et al. 2007).

8.4 Concluding Remarks

Research in the field of halophilic microbiology attracted huge interest during the last decade, yielding more than 53 new genera in the archaeal family *Halobacteriaceae* and numerous bacterial genera (Table 8.1a and 8.1b). Many of them have their origin in man-made "hypersaline" environments developed for various purposes, e.g., commercial solar salt, salted food, etc. These sources have a connection to marine environments, which apparently are not "hypersaline." We would like to remind the readers of the fact that a relatively small number of the validly published halophilic species of Archaea and Bacteria come from truly hypersaline lakes (like the Dead Sea or the Great Salt Lake). Another fact is that no rigid evidences of the existence of viable haloarchaeal cells in seawater have been presented, except perhaps for halococci (Rodriguez-Valera et al. 1979). Taking into account these facts, future research will be necessary to give an answer to the question of how the halophilic microorganisms originated during the early stages in the evolution of life and how they diversified and were distributed throughout the world, in the past and present (Oren 2004). Although the molecular clock of organisms isolated from inclusion bodies of ancient salt crystals is difficult to be correlated with the geological time of evaporation, the question of longevity of halophilic Archaea and Bacteria continues to be a tremendous fascination to all microbiologists (Grant et al. 1998; Fendrihan et al. 2009a, b; Park et al. 2009; Jaakkola et al. 2016).

Acknowledgment This chapter is dedicated to the 150th anniversary of the Romanian Academy.

References

Anderson H (1954) The reddening of salted hides and fish. Appl Microbiol 2:64–69

Anton J, Meseguer I, Rodriguez-Valera F (1988) Production of an extracellular polysaccharide by *Haloferax mediterranei*. Appl Environ Microbiol 54:2381–2386

Baricz A, Coman C, Andrei AS, Muntean V, Keresztes ZG, Păuşan M, Alexe M, Banciu HL (2014) Spatial and temporal distribution of archaeal diversity in meromictic, hypersaline Ocnei Lake (Transylvanian Basin, Romania). Extremophiles 18:399–413. doi:10.1007/s00792-013-0625-6

Baricz A, Cristea A, Muntean V, Teodosiu G, Andrei AŞ, Molnár I, Alexe M, Rakosy-Tican E, Banciu HL (2015) Culturable diversity of aerobic halophilic archaea (Fam. Halobacteriaceae) from hypersaline, meromictic Transylvanian lakes. Extremophiles 19:525–537. doi:10.1007/s00792-015-0738-1

Borgne SL, Paniagua D, Vazquez-Duhalt R (2008) Biodegradation of organic pollutants by halophilic bacteria and archaea. J Mol Microbiol Biotechnol 15:74–92

Brito-Echeverría J, López-López A, Yarza P, Antón J, Rosselló-Móra R (2009) Occurrence of *Halococcus* spp. in the nostrils salt glands of the seabird *Calonectris diomedea*. Extremophiles 13:557–565. doi:10.1007/s00792-009-0238-2

Brosteanu C (1901) In: Lazareanu GA (ed) Our salt mine. Historic, juridical and economical study towards saline exploitations and salt monopoly on Romans and Romanian (in Romanian), Bucharest

Bulgăreanu VAC (1993) The protection and management of saline lakes of therapeutic value in Romania. Int J Salt Lake Res 2:165–171

Burns DG, Janssen PH, Itoh T, Kamekura M, Echigo A, Dyall-Smith ML (2010) *Halonotius pteroides* gen. nov., sp. nov., an extremely halophilic archaeon recovered from a saltern crystallizer in southern Australia. Int J Syst Evol Microbiol 60:1196–1199

Charlesworth JC, Burns BP (2015) Untapped Resources: Biotechnological potential of peptides and secondary metabolites in Archaea. Archaea 2015:282035. doi:10.1155/2015/282035 eCollection 2015

Chervinskaya AV, Zilber NA (1995) Halotherapy for treatment of respiratory diseases. J Aerosol Med 8:221–232

Cojoc R, Merciu S, Oancea P, Pincu E, Dumitru L, Enache M (2009) Highly thermostable exopolysaccharide produced by the moderately halophilic bacterium isolated from a man-made young salt lake in Romania. Pol J Microbiol 58:289–294

Drăgănescu L (1990) Dates from historic salt exploitation at Slanic-Prahova (in Romanian). Rev Muzeelor 27:68–71

Drăgănescu L, Drăgănescu S (2001) The history of the evolution of salt working methods in Romania, from antiquity to the present. 17th Intl. Mining Congress and Exhibition of Turkey – IMCET, pp 627–633

Duggen S, Hoernle K, van den Bogaard P, Rüpke L, Morgan JP (2003) Deep roots of the Messinian salinity crisis. Nature 422:602–606

Dumitru L, Teodosiu G, Enache E (2002) The tolerance of extremely halophilic archaea to heavy metals. Proc Inst Biol 4:261–267

Dumitru L, Teodosiu-Popescu G, Enache M, Cojoc R, Kleps I, Ignat T (2007) S-layer of *Haloferax* sp. GR 2 (JCM 13922): isolation, characterization and binding to silicon nanostructurated substrates. In: Kleps I, Ion AC, Dascalu D (eds) Progress in nanoscience and nanotechnologies, Seria Vol. 11, Micro and nanoengineering. Ed. Acad Române, Bucharest, pp 146–152

Eichler J (2003) Facing extremes: archaeal surface-layer (glyco) proteins. Microbiology 149:3347–3351

Enache M, Faghi AM (1999) Detection of extracellular halophilic amylase activity from the extreme halophilic genus *Haloferax*. Proc Inst Biol 2:143–146

Enache M, Dumitru L, Faghi AM (1999a) Occurrence of halocins in mixed archaebacteria culture. Proc Inst Biol 2:151–154

Enache M, Teodosiu G, Dumitru L, Faghi AM, Zarnea G (1999b) Diversity of halobacteria in accordance with their membrane lipids composition. Proc Inst Biol 2:147–149

Enache M, Teodosiu G, Faghi AM, Dumitru L (2000a) Identification of halophilic Archaebacteria isolated from some Romanian salts lakes on the basis of lipids composition. Rev Roum Biol Ser Biol Veg 45:93–99

Enache M, Faghi AM, Teodosiu G, Dumitru L, Zarnea G (2000b) Effect of temperature and NaCl concentration of growth media on neutral lipid in *Haloferax volcanii* and *Haloferax* sp. GR1 strain. Proc Inst Biol 3:251–256

Enache E, Faghi AM, Teodosiu G, Dumitru L, Zarnea G (2000c) The halophilic microorganisms response to heavy metals. Rev Roum Biol Ser Biol Veg 45:111–116

Enache M, Faghi AM, Dumitru L, Zarnea G (2001) Halophilic α-amylase from the extremely halophilic archaea *Haloferax* sp. strain GR1. Proc Rom Acad Series B 2:107–109

Enache M, Teodosiu G, Dumitru L, Zarnea G (2004a) The effect of NaCl concentrations on the growth and lipase activity at *Haloferax* sp. GR1. Proc Inst Biol 6:233–236

Enache M, Faghi AM, Dumitru L, Teodosiu G, Zarnea G (2004b) Halocin HF 1 a bacteriocin produced by *Haloferax* sp. GR 1. Proc Rom Acad Series B 1:27–32

Enache M, Itoh T, Kamekura M, Teodosiu G, Dumitru L (2007a) *Haloferax prahovense* sp. nov., an extremely halophilic archaeon isolated from a Romanian salt lake. Int J Syst Evol Microbiol 57:393–397

Enache M, Itoh T, Fukushima UR, Dumitru L, Kamekura M (2007b) Phylogenetic relationship within the family *Halobacteriaceae* inferred from *rpoB'* gene and protein sequences. Int J Syst Evol Microbiol 57:2289–2295

Enache M, Itoh T, Kamekura M, Popescu G, Dumitru L (2008a) Halophilic archaea of *Haloferax* genus isolated from anthropocentric Telega (Palada) salt lake. Proc Rom Acad Series B 10:11–16

Enache M, Itoh T, Kamekura M, Popescu G, Dumitru L (2008b) Halophilic archaea isolated from man-made young (200 years) salt lakes in Slănic, Prahova, Romania. Cent Eur J Biol 3:388–395

Enache M, Popescu G, Dumitru L, Kamekura M (2009) The effect of Na^{+}/Mg^{2+} ratio on the amylase activity of haloarchaea isolated from Techirghiol lake, Romania, a low salt environment. Proc Rom Acad Series B 11:3–7

Enache M, Cojoc R, Kamekura M (2015) Halophilic microorganisms and their biomolecules: Approaching into frame of bio(nano) technologies. In: Maheshwari DK, Saraf M (eds) Halophiles. Sustainable development and biodiversity, vol 6. Springer International Publishing, Switzerland, pp 161–172

Faghi AM, Teodosiu G, Dumitru L (1999) The dynamic of the halophilic microorganisms growth at different values of pH and temperature. I. Moderately halophilic bacteria. Proc Inst Biol 2:155–161

Fathepure BZ (2014) Recent studies in microbial degradation of petroleum hydrocarbons in hypersaline environments. Front Microbiol 5:173. doi:10.3389/fmicb.2014.00173

Fendrihan S, Bérces A, Lammer H, Musso M, Rontó G, Polacsek TK, Holzinger A, Kolb C, Stan-Lotter H (2009a) Investigating the effects of simulated Martian ultraviolet radiation on *Halococcus dombrowskii* and other extremely halophilic archaebacteria. Astrobiology 9: 104–112

Fendrihan S, Musso M, Stan-Lotter H (2009b) Raman spectroscopy as a potential method for the detection of extremely halophilic archaea embedded in halite in terrestrial and possibly extra-terrestrial samples. J Raman Spectrosc 40:1996–2003

Fendrihan S, Dornmayr-Pfaffenhuemer M, Gerbl FW, Holzinger A, Grösbacher M, Briza P, Erler A, Gruber C, Plätzer K, Stan-Lotter H (2012) Spherical particles of halophilic archaea correlate with exposure to low water activity – implications for microbial survival in fluid inclusions of ancient halite. Geobiology 10:424–433

Gâştescu P (1965) On the origin of salt lakes from Romanian Plain (in Romanian). Natura – Seria Geografie–Geologie 2:42–45

Gâştescu P (1971) The lake from Romania (in Romanian). Ed Acad Rep Soc România, Bucharest

Gibbons NE (1974) Family V. Halobacteriaceae fam. nov. In: Buchanan RE, Gibbons NE (eds) Bergey's manual of determinative bacteriology, 8th edn. Williams & Wilkins, Baltimore, pp. 269–273

Gramain A, Chong Díaz GC, Demergasso C, Lowenstein TK, McGenity TJ (2011) Archaeal diversity along a subterranean salt core from the Salar Grande (Chile). Environ Microbiol 13:2105–2121

Grant WD, Gemmell RT, McGenity TJ (1998) Halobacteria: the evidence for longevity. Extremophiles 2:279–287

Grant WD, Kamekura M, McGenity TJ, Ventosa A (2001) The order *Halobacteriales*. In: Boone DR, Castenholz RW (eds) Bergey's manual of systematic bacteriology, vol 1, 2nd edn. Springer, New York, pp. 294–334

Gruber C, Legat A, Pfaffenhuemer M, Radax C, Weidler G, Busse HJ, Stan-Lotter H (2004) *Halobacterium noricense* sp. nov., an archaeal isolate from a bore core of an alpine Permian salt deposit, classification of *Halobacterium* sp. NRC-1 as a strain of *H. salinarum* and emended description of *H. salinarum*. Extremophiles 8:431–439

Har N, Barbu O, Codrea V, Petrescu I (2006) New data on the mineralogy of the salt deposit from Slănic Prahova (Romania). Studia UBB, Geologia 51:29–33

Hedman J, Hagg T, Sandell J, Haahtela T (2006) The effect of salt chamber treatment on bronchial hyperresponsiveness in asthmatics. Allergy 61:605–610

Horikoshi K (1999) Alkaliphiles: some applications of their products for biotechnology. Microbiol Mol Biol Rev 63:735–750

Itoh T, Yamaguchi T, Zhou P, Takashina T (2005) *Natronolimnobius baerhuensis* gen. nov., sp. nov., and *Natronolimnobius innermongolicus* sp. nov., novel haloalkaliphilic archaea isolated from soda lakes in Inner Mongolia, China. Extremophiles 9:111–116

Jaakkola ST, Ravantti JJ, Oksanen HM, Bamford DH (2016) Buried alive: Microbes from ancient halite. Trends Microbiol 24:148–160. doi:10.1016/j.tim.2015.12.002

Javor B (1989a) Chapter 1. Geology and chemistry. In: Hypersaline environments. Microbiology and biogeochemistry. Springer, Berlin/Heidelberg/New York, pp. 5–25

Javor B (1989b) Chapter 15. Gavish Sabkha and other hypersaline marine sabkhas, pools and lagoons. In: Hypersaline environments. Microbiology and biogeochemistry. Springer, Berlin/Heidelberg/New York, pp. 223–235

Juez G (1988) Taxonomy of extremely halophilic archaebacteria. In: Rodriguez-Valera F (ed) Halophilic Bacteria, vol II. CRC Press, Boca Raton, FL, pp. 3–24

Kamekura M, Seno Y (1993) Partial sequence of the gene for a serine protease from a halophilic archaeum *Haloferax mediterranei* R4, and nucleotide sequences of 16S rRNA encoding genes from several halophilic archaea. Experientia 49:503–513

Kamekura M, Dyall-Smith ML, Upasani V, Ventosa A, Kates M (1997) Diversity of alkaliphilic halobacteria: proposal for transfer of *Natronobacterium vacuolatum*, *Natronobacterium magadii*, and *Natronobacterium pharaonis* to *Halorubrum*, *Natrialba* and *Natronomonas* gen. nov., respectively, as *Halorubrum vacuolatum* comb. nov., *Natrialba magadii* comb. nov., and *Natronomonas pharaonis* comb. nov., respectively. Int J Syst Bacteriol 47:853–857

Kates M (1995) Adventures with membrane lipids. Biochem Soc Trans 23:697–709

Kleps I, Ignat T, Miu M, Simion M, Teodosiu-Popescu G, Enache M, Dumitru L (2009) Protein-mesoporous silicon matrix obtained by S-layer technology. Physica Status Solidi C 6:1605–1609

Kocur M, Hodgkiss W (1973) Taxonomic status of the genus *Halococcus* Schoop. Int J Syst Bacteriol 23:151–156

Kunte HJ, Trueper HG, Stan-Lotter H (2002) Halophilic microorganism. In: Horneck G, Baumstark-Khan C (eds) Astrobiology. The quest for the conditions of life. Springer, Berlin/Heidelberg

Kushner DJ (1985) The Halobacteriaceae. In: Woese CR, Wolfe RS (eds) The Bacteria, vol VIII. Academic Press, New York, pp. 171–214

Lequerica JL, O'Connor JE, Such L, Alberola A, Meseguer I, Dolz M, Torreblanca M, Moya A, Colom F, Soria B (2006) A halocin acting on Na^+/H^+ exchanger of Haloarchaea as a new type of inhibitor in NHE of mammals. J Physiol Biochem 62:253–262

Lochhead AG (1934) Bacteriological studies on the red discoloration of salted hides. Can J Res 10:275–286

Lowenstein TK, Brian A, Schubert BA, Timofeeff MN (2011) Microbial communities in fluid inclusions and long-term survival in halite. GSA Today 21:4–9. doi:10.1130/GSATG81A.1

Magrum LJ, Luehrsen KR, Woese CR (1978) Are extreme halophiles actually "bacteria"? J Mol Evol 11:1–8

Mellado E, Ventosa A (2003) Biotechnological potential of moderately and extremely halophilic microorganisms. In: Barredo JL (ed) Microorganisms for health care, food and enzyme production. Research Signpost, Kerala, pp. 233–256

Merciu S, Vacaroiu C, Filimon R, Popescu G, Preda S, Anastasescu C, Zaharescu M, Enache M (2009) Nanotubes biologically active in media with high salt concentration. Biotechnol Biotechnol Eq 23:827–831

Minegishi H, Echigo A, Nagaoka S, Kamekura M, Usami R (2010) *Halarchaeum acidiphilum* gen. nov., sp. nov., a moderately acidophilic haloarchaeon isolated from commercial solar salt. Int J Syst Evol Microbiol 60:2513–2516

Multhauf RP (1978) Neptune's gift. A history of common salt. The Johns Hopkins University Press, Baltimore/London
Nica SA, Meilă AM, Macovei L (2007) Speleotherapy (in Romanian). Rom J Rheumatol 4:269–273
Nieto JJ (1991) The response of halophilic bacteria to heavy metals. In: Rodriguez-Valera F (ed) General and applied aspects of halophilic microorganisms. Plenum Press, New York/London, pp. 173–179
O'Connor EM, Shand RF (2002) Halocins and sulfolobicins: The emerging story of archaeal protein and peptide antibiotics. J Ind Micrbiol Biotechnol 28:23–31
Oncescu T, Oancea P, Enache M, Popescu G, Dumitru L, Kamekura M (2007) Halophilic bacteria are able to decontaminate dichlorvos, a pesticide, from saline environments. Cent Eur J Biol 2:563–573
Oren A (2002) Preface. In: Halophilic microorganisms and their environments. Kluwer Academic Publishers, Dordrecht, pp xv–xviii
Oren A (2004) Prokaryote diversity and taxonomy: current status and future challenges. Philos Trans R Soc Lond B Biol Sci 359:623–638
Oren A (2010) Industrial and environmental applications of halophilic microorganisms. Environ Technol 31:825–834. doi:10.1080/09593330903370026
Park JS, Vreeland RH, Cho BC, Lowenstein TK, Timofeeff MN, Rosenzweig WD (2009) Haloarchaeal diversity in 23, 121 and 419 MYA salts. Geobiology 7:1–9
Popescu G, Dumitru L (2009) Biosorption of some heavy metals from media with high salt concentrations by halophilic Archaea. Biotechnol Biotechnol Eq 23:791–795
Rees CH, Grant DW, Jones EB, Heaphy S (2004) Diversity of Kenyan soda lake alkaliphiles assessed by molecular methods. Extremophiles 8:63–71
Rodriguez-Valera F (1992) Biotechnological potential of halobacteria. Biochem Soc Symp 58:135–147
Rodriguez-Valera F, Ruiz-Berraquero F, Ramos-Cormenzana A (1979) Isolation of extreme halophiles from seawater. Appl Environ Microbiol 38:164–165
Rodriguez-Valera F, Juez G, Kushner DJ (1982) Halocins: salt-dependent bacteriocins produced by extremely halophilic rods. Can J Microbiol 28:151–154
Saiki RK, Gelfand DH, Stoffel S, Scharf SJ, Higuchi R, Horn GT, Mullis KB, Erlich HA (1988) Primer-directed enzymatic amplification of DNA with a thermostable DNA polymerase. Science 239:487–491
Schuster B, Sleytr UB (2000) S-layer-supported lipid membranes. Rev Mol Biotechnol 74:233–254
Schuster B, Györvary E, Pum D, Sleytr UB (2005) Nanotechnology with S-layer protein. In: Vo-Dinh T (ed) Methods in molecular biology, Protein nanotechnology, protocols, instrumentation and applications, vol 300. Humana Press Inc, Totowa, NJ, pp. 101–123
Sencu V (1968) Salt Mount from Slănic Prahova (in Romanian). Ocrot Nat 12:167–179
Shimane Y, Hatada Y, Minegishi H, Mizuki T, Echigo A, Miyazaki M, Ohta Y, Usami R, Grant WD, Horikoshi K (2010) *Natronoarchaeum mannanilyticum* gen. nov., sp. nov., an aerobic, extremely halophilic member of the Archaea isolated from commercial salt made in Niigata, Japan. Int J Syst Evol Microbiol 60:2529–2534
Simyonka YM (1989) Some particular features of infections and inflammatory processes, and immune status in patients with infection-dependent bronchial asthma during speleotherapy in salt-mine microclimate (in Russian). In: Bronchial asthma Leningrad. pp 136–140
Sleytr UB, Sara M (1997) Bacterial and archaeal S-layer proteins: structure-function relationships and their biotechnological applications. Trends Biotechnol 15:20–26
Sleytr UB, Messner P, Pum D, Sara M (1999) Crystalline bacterial cell surface layers (S-layers): from supramolecular cell structure to biomimetics and nanotechnology. Angew Chem Int Ed Engl 38:1034–1054
Sleytr UB, Schuster B, Egelseer EM, Pum D (2014) S-layers: principles and applications. FEMS Microbiol Rev 38:823–864. doi:10.1111/1574-6976.12063
Srivastava P, Kowshik M (2015) Biosynthesis of nanoparticles from halophiles. In: Maheshwari DK, Saraf M (eds) Halophiles, sustainable development and biodiversity. Springer International Publishing, pp 145–159

Srivastava P, Braganca J, Ramanan SR, Kowshik M (2013) Synthesis of silver nanoparticles using haloarchaeal isolate *Halococcus salifodinae* BK3. Extremophiles 17:821–831

Srivastava P, Braganca J, Ramanan SR, Kowshik M (2014) Green synthesis of silver nanoparticles by haloarchaeon *Halococcus salifodinae* BK6. Adv Mater Res 938:236–241

Stan-Lotter H, Radax C, Gruber C, Legat A, Pfaffenhuemer M, Wieland H, Leuko S, Weidler G, Kömle N, Kargle G (2003) Astrobiology with haloarchaea from Permo-Triassic rock salt. Int J Astrobiol 1:271–284

Teodosiu G, Dumitru L, Faghi AM (1999) The dynamic of the halophilic microorganisms growth at different values of pH and temperature. II. Extremely halophilic archaea. Proc Inst Biol 2:163–168

Thongthai C, McGenity TJ, Suntinanalert P, Grant WD (1992) Isolation and characterization of an extremely halophilic archaebacterium from traditionally fermented Thai fish sauce (nam-pla). Lett Appl Microbiol 14:111–114

Tindall BJ, Mills AA, Grant WD (1980) An alkalophilic red halophilic bacterium with low magnesium requirement from a Kenyan soda lake. J Gen Microbiol 116:257–260

Tindall BJ, Ross HNM, Grant WD (1984) *Natronobacterium* gen. nov. and *Natronococcus* gen. nov., two new genera of haloalkaliphilic archaebacteria. Syst Appl Microbiol 5:41–57

Torreblanca M, Rodriguez-Valera F, Juez G, Ventosa A, Kamekura M, Kates M (1986) Classification of non-alkaliphilic halobacteria based on numerical taxonomy and polar lipid composition, and description of *Haloarcula* gen. nov. and *Haloferax* gen. nov. Syst Appl Microbiol 8:89–99

Torreblanca M, Meseguer I, Ventosa A (1994) Production of halocin is a practically universal feature of archaea halophilic rods. Lett Appl Microbiol 19:201–205

Trachtenberg S, Pinnick B, Kessel M (2000) The cell surface glycoprotein layer of the extreme halophile *Halobacterium salinarum* and its relation with *Haloferax volcanii*: cryo-electron tomography of freeze-substituted cells and projection studies of negatively stained envelopes. J Struct Biol 130:10–26

Vellieux F, Madern D, Zaccai G, Ebel C (2007) Molecular adaptation to high salt. In: Gerday C, Glansdorff N (eds) Physiology and biochemistry of extremophiles. ASM Press, Washington, DC, pp. 240–253

Ventosa A (2004) Halophilic microorganisms. Springer, Berlin/Heidelberg

Ventosa A, Nieto JJ, Oren A (1998) Biology of moderately halophilic aerobic bacteria. Microbiol Mol Biol Rev 62:504–544

Ventosa A, Sanchez-Porro C, Martin S, Mellado E (2005) Halophilic archaea and bacteria as a source of extracellular hydrolytic enzymes. In: Gunde-Cimerman N, Oren A, Plemenitas A (eds) Adaptation to life at high salt concentrations in archaea, bacteria, and eukarya. Springer, Dordrecht, pp. 337–354

Ventosa A, Mellado E, Sanchez-Porro C, Marquez MC (2008) Halophilic and halotolerant microorganism from soils. In: Dion P, CS N (eds) Microbiology of extreme soils. Springer, Berlin/Heidelberg, pp. 87–115

Weisburg WG, Barns SM, Pelletier DA, Lane DJ (1991) 16S ribosomal DNA amplification for phylogenetic study. J Bacteriol 173:697–703

Williams WD (1998) Guidelines of lake management, vol. 6. In: Management of inland saline waters. International lake environment committee foundation and the United Nations environment programme, Kusatsu, Japan

Xue Y, Fan H, Ventosa A, Grant WD, Jones BE, Cowan DA, Ma Y (2005) *Halalkalicoccus tibetensis* gen. nov., sp. nov., representing a novel genus of haloalkaliphilic archaea. Int J Syst Evol Microbiol 55:2501–2505

Yamamura A, Ichimura T, Kamekura M, Mizuki T, Usami R, Makino T, Ohtsuka J, Miyazono K, Okai M, Nagata K, Tanokura M (2009) Molecular mechanism of distinct salt-dependent enzyme activity of two halophilic nucleoside diphosphate kinases. Biophys J 96:4692–4700

Yim KJ, Kwon J, Cha IT, Oh KS, Song HS, Lee HW, Rhee JK, Song EJ, Rho JR, Seo ML, Choi JS, Choi HJ, Lee SJ, Nam YD, Roh SW (2015) Occurrence of viable, red-pigmented haloarchaea in the plumage of captive flamingoes. Sci Rep 5:16425. doi:10.1038/srep16425

Acidophiles: Diversity and Mechanisms of Adaptation to Acidic Environments

9

Salvador Mirete, Verónica Morgante, and José Eduardo González-Pastor

Contents

S. Mirete • V. Morgante • J.E. González-Pastor (✉)
Laboratory of Molecular Adaptation, Department of Molecular Evolution, Centro de Astrobiología (CSIC-INTA), Carretera de Ajalvir km 4, Torrejón de Ardoz (28850), Madrid, Spain
e-mail: gonzalezpje@cab.inta-csic.es

H. Stan-Lotter, S. Fendrihan (eds.), *Adaption of Microbial Life to Environmental Extremes*, DOI 10.1007/978-3-319-48327-6_9

9.1 Introduction

Acidic environments are among the most extreme on Earth and include acid mine drainage (AMD) or acid rock drainage (ARD), marine volcanic vents, and acidic sulfur springs. Acidophiles are extremophiles that thrive in these environments under highly acidic conditions (pH 3.0 and below) and they are distributed in all three domains of life: Archaea, Bacteria, and Eukarya (Amaral-Zettler et al. 2002; Johnson and Hallberg 2003). Physiologically, the acidophiles are very diverse: aerobic and facultative anaerobic, chemolithotrophs and different types of heterotrophic prokaryotes, photoautotrophic eukaryotes, predatory protozoa, and others. This chapter summarizes various investigations about acidophiles, mainly from Bacteria and Archaea domains, which inhabit AMD and ARD environments generated as a result of mining and natural weathering of rocks containing an abundance of sulfide minerals. Exposure of metal sulfides to oxygen and water is responsible for the oxidation of metal sulfides (often pyrite, which is iron sulfide – FeS_2) and the production of highly acidic effluents enriched in toxic metals and metalloids released from the minerals. Communities of chemolithoautotrophic bacteria and archaea, which obtain their energy from the oxidation of sulfide and Fe(II), accelerate the process of oxidation of the sulfidic minerals. In fact, it was shown that the oxidation of pyrite was accelerated in the presence of the chemolithoautotrophic bacterium *Acidithiobacillus ferrooxidans* up to 10^6 times (Singer and Stumm 1970).

Acidophiles from AMD or ARD environments are not only exposed to extremely low pH but also to high concentrations of toxic metals and metalloids. Thus, these microorganisms have developed networked cellular adaptations to maintain their intracellular pH to around neutral (Baker-Austin and Dopson 2007; Dopson 2012), and also they have multiple and more efficient metal and metalloid resistance systems than those described in neutrophiles. Different molecular mechanisms have been proposed for maintaining the pH homeostasis and metal/metalloid resistance and are summarized in this chapter.

On the other hand, acidophiles are also an economically important group of microorganisms, since they are being applied in biomining to solubilize metals from ores (Rawlings and Johnson 2007), and some of their enzymes (extremozymes) can be used under harsh industrial conditions (Dopson 2012; Sharma et al. 2012). Furthermore, the discovery of iron minerals on Mars that on Earth are exclusively formed in the presence of water has focused the interest of astrobiology toward the study of AMD/ARD environments and acidophiles, to explore if these or similar microorganisms may exist on Mars.

9.2 Geographical Localizations of AMD Sites and Related Environments

AMD and related environments are distributed all around the world (Table 9.1) and include well-studied examples of acidic environments such as the Tinto River (Huelva, Spain) present in the Iberian Pyrite Belt (IPB) (Fig. 9.1). This river is

Table 9.1 Distribution of representative prokaryotes in different AMD and related environments around the world, pH values of the different sites, and molecular techniques of identification employed. Filled blocks depict the presence of a particular microorganism in a particular site, whereas empty blocks indicate its absence or presence in very low numbers

	Location	Frasassi cave system, Italy (Jones et al. 2012)	Richmond mine at Iron Mountain, California, USA (Tyson et al. 2004)	Copahue–Caviahue system, Argentina (Urbieta et al. 2012)	AMD site DX2, Southeast China (Kuang et al. 2013)	Harz Mountains, Germany (Ziegler et al. 2013)	Cae Coch, acid streamer L3, Wales, UK (Kimura et al. 2011)	Iberian Pyrite Belt pit lake NSC, Spain (Santofimia et al. 2013)	Coal mining area in Svalbard, Norway (García Moyano et al. 2015)	Rio Tinto, Spain (García Moyano et al. 2012)	Fankou lead/zinc mine tailing T6, Guangdong, China (Chen et al. 2013)	Mining effluent sample DBS, Guangdong, China (Chen et al. 2015)	Stream OB, coal mining area in South Island, New Zealand (Lear et al. 2009)	Copper tailing sample S87, Anhui Province, China (Liu et al. 2014)	Nevado Pastoruri glacier, sampling point P1A (Central Andes, Perú)(González-Toril et al. 2015)	Xiang Mountain sulfide mine, China (Hao et al. 2010)	South Shetland Islands, Antarctica (Dold et al. 2013)	Rio Tinto rhizosphere, Spain (Mirete et al. 2007)	Tailing dump in Kristineberg, Sweden (Kock and Schippers 2008)
	Environmental source	Biofilm (snottites)	Biofilm	Plankton	Plankton	Biofilm (snottites)	Plankton	Plankton	Sediment	Plankton	Sediment	Sediment	Biofilm	Sediment	Sediment	Plankton	Plankton	Rhizosphere	Sediment
	pH	0.5	0.83	1	1.9	2.05	2.15	2.2	2.3	2.4	2.4	2.7	2.8	2.9	3	3	3.3	3.5	3.5
	Temperature (°C)	13	42	59	32	–	8.5	–	4	–	–	21.4	16.6	–	9.4	30	−0.2	–	0.7
	Molecular procedures	454 pyroseq., FISH	16S rRNA gene library, FISH	16S rRNA gene library, CARD-FISH	454 pyroseq.	454 pyroseq., CARD-FISH, RFLP analysis	T-RFLPs, FISH and 16S rRNA gene library	16S rRNA gene library	16S rRNA gene library	DGGE, FISH, 16S rRNA gene library,	454 pyroseq.	454 pyroseq.	16S rRNA gene library	454 pyroseq.	454 pyroseq.	16S rRNA gene library	16S rRNA gene library	16S rRNA gene library	Q-PCR, FISH
	Acidophilic microorganism																		
Bacteria	*Leptospirillum* spp. (*Nitrospira*)		■	■	■	■	■	■	■	■	■	■	■	■	■				■
	Acidithiobacillus spp. (*Gammaproteobacteria*)	■		■	■	■	■	■	■	■	■	■		■	■		■		■
	Sulfobacillus spp. (*Firmicutes*)		■		■						■			■					■
	'*Ferrovum myxofaciens*' (*Betaproteobacteria*)				■		■			■						■			
	Ferrimicrobium/*Acidimicrobium* group (*Actinobacteria*, *Acidimicrobiaceae*)	■			■	■	■	■	■	■			■		■		■	■	
	Acidiphilium spp. (*Alphaproteobacteria*)			■	■			■	■	■		■			■	■		■	
	Acidisphaera spp. (*Alphaproteobacteria*)							■	■	■			■		■		■	■	
	Acidobacterium spp.				■				■						■		■	■	
Archaea	*Ferroplasma* spp. (*Thermoplasmata*, *Euryarchaeota*)	■	■	■	■	■		■		■	■	■		■					
	Uncultured *Crenarchaeota*															■		■	

Fig. 9.1 Acid mine drainage (AMD) environment at the origin of the Tinto River, Huelva, Spain (Photos by José Eduardo González-Pastor)

characterized by highly acidic pH values (between 1.5 and 3.1) and high concentrations of dissolved heavy metals (e.g., iron 2.3 g/L, zinc 0.22 g/L, and copper 0.11 g/L) (López-Archilla et al. 2001). The extreme conditions present in the water column and sediments of the Tinto River are caused by the metabolic activity of chemolithoautotrophic acidophiles growing in the complex sulfides of the pyrite. Therefore, most of the microbial diversity detected in this extreme environment belongs to strains of *Leptospirillum ferrooxidans*, *At. ferrooxidans*, and *Acidiphilium* spp., all related to the iron cycle (González-Toril et al. 2003; García-Moyano et al. 2012). Also located in the Iberian Pyrite Belt are diverse pit lakes characterized by their low

Fig. 9.2 Acid rock drainage (ARD) environment in Bahia Almirantazgo, King George Island, Antarctic (Photo courtesy of Angeles Aguilera)

pH and high concentration of toxic metals and sulfate. These acidic lakes host a microbial diversity typical of acid mine drainage environments, including not only iron-oxidizing bacteria but also facultative iron-reducing bacteria and archaea (Santofimia et al. 2013). Pyrite oxidation caused by acidophilic microorganisms has been reported in ARD environments present in cold environments such as Antarctica (Fig. 9.2) where sulfide-rich rocks are more exposed due to increasing glacier melting caused by global warming (Dold et al. 2013) and in the Nevado Pastoruri glacier (Fig. 9.3) (González-Toril et al. 2015). As a result of the biogeochemical activity of acid mine drainage microorganisms such as *At. ferrivorans*, the waters and sediments of these environments can turn acidic ($pH \approx 3$). In addition, the presence of an AMD environment has recently been studied in a coal mining area located in the Arctic where iron-oxidizing bacteria (*At. ferrivorans*, *Acidobacteria*, and *Actinobacteria*) were dominant within the bacterial community (García-Moyano et al. 2015).

On the other hand, the presence of acidophiles has also been documented in the geothermal waters of the Río Agrío, a volcanic acidic river (pH values from 1 to 4) which flows through the Copahue–Caviahue system in the Patagonia, Argentina. In this river the microbial diversity found was typical of other acidic environments and included members of sulfur-oxidizing bacteria such as strains of *Acidithiobacillus* in addition to members of iron-oxidizing bacteria such as *Leptospirillum* and *Ferrimicrobium* and the iron-oxidizing archaea *Ferroplasma* spp. (Urbieta et al. 2012). Subterranean life has been detected within the low-temperature (8–9 °C) and extremely acidic Cae Coch mine ($pH < 2.5$), an abandoned underground pyrite mine in North

Fig. 9.3 Wetland formed by accumulation of acidic water generated by the melting of the Nevado Pastoruri Glacier (Central Andes, Perú). The retracement of this glacier exposes fresh sulfide-rich rocks, which combined to the action of glacier meltwater generates an ARD environment (Photo courtesy of Angeles Aguilera)

Wales with iron-rich waters, whose bacterial diversity was dominated by the presence of the iron-oxidizing acidophiles *At. ferrivorans*, "*Ferrovum myxofaciens*," and, to a lesser extent, *L. ferrooxidans* (Kimura et al. 2011).

Other AMD environments can be found in diverse mining areas such as tailings in Sweden (Kock and Schippers 2008) and China (Hao et al. 2010; Chen et al. 2013, 2015) and acidic streams in the South Island of New Zealand (Lear et al. 2009).

In addition, there are other particular extreme environments where acidophiles can grow and thrive such as the extremely acidic (pH 0–2.5) microbial biofilms known as "snottites" which are found in the walls and ceilings of caves. Snottites have been found in an abandoned pyrite mine (Ziegler et al. 2013) and also in caves enriched with hydrogen sulfide (H_2S) but where iron has not been detected (Jones et al. 2012). Interestingly, the microbial diversity present in these two samples, regardless the presence of iron, was found to be overall similar to that observed in typical acid mine drainage environments, which included members of *Acidithiobacillus* and *Ferroplasma* (Jones et al. 2012; Ziegler et al. 2013).

9.3 Diversity and Distribution of Acidophiles Associated to AMD Sites and Related Environments

Here we summarize recent studies on microbial diversity of relevant acidophiles found in AMD and ARD sites by using culture-independent approaches. The more abundant prokaryotes adapted to acidic environments are those chemolithoautotrophs that can oxidize iron (iron oxidizers) and sulfide minerals (sulfur oxidizers). These prokaryotic microorganisms include members of *Leptospirillum* spp., *Acidithiobacillus* spp., *Sulfobacillus* spp., *Ferrimicrobium* spp., "*Ferrovum myxofaciens*," and *Acidimicrobium* spp. and of the archaeon *Ferroplasma* spp. (Table 9.1). These microorganisms have been detected in a number of acidic sites related to

AMD and ARD environments including mine waters, stalactite-like biofilms (snottites) formed on the walls of caves, subterranean habitats, in the Antarctic, and in glacier and permafrost areas by using diverse molecular techniques such as FISH, 16S rRNA gene libraries, 454 pyrosequencing, T-RFLPs, Q-PCR, and DGGE (Table 9.1). Recent investigations have revealed that the microbial diversity observed in AMD sites are significantly correlated with pH, suggesting that this parameter may be mainly responsible for structuring the whole microbial diversity detected (Lear et al. 2009; Kuang et al. 2013; Liu et al. 2014). From these acidophiles, *L. ferrooxidans* strains are obligate chemolithoautotrophs that are able to oxidize iron (pyrite) aerobically, thus playing an important role in the iron cycle. This species constitutes an abundant fraction of the bacteria detected in the Tinto River water column and is one of the main species involved in the geomicrobiological processes of this ecosystem (González-Toril et al. 2003; García-Moyano et al. 2012). Also, AMD biofilm communities present within the Richmond Mine at Iron Mountain (pH 0.5–1) are dominated by *Leptospirillum* spp. (Tyson et al. 2004; Denef et al. 2010). Other AMD environments where *Leptospirillum* spp. are present include acidic lakes (Hao et al. 2010; Santofimia et al. 2013), mine residues (Kock and Schippers 2008; Chen et al. 2013, 2015; Liu et al. 2014), and acidic streams (Lear et al. 2009; Kuang et al. 2013). These species were also found to be the most abundant bacteria within the microbial community inhabiting the toxic part of snottites formed in an abandoned pyrite mine in the Harz Mountains (Ziegler et al. 2013). *Leptospirillum* species have been documented in other particular environments such as the geothermal Copahue–Caviahue system in Neuquén, Argentina (Urbieta et al. 2012), and the low-temperature environments of the pyrite mine of Cae Coch located in North Wales, characterized by a microbial community underpinned by chemolithotrophy (Kimura et al. 2011), a coal abandoned mine in Svalbard, and a permafrost area in the High Arctic (García-Moyano et al. 2015) and in the ARD present in the Nevado Pastoruri glacier (González-Toril et al. 2015).

In AMD environments, other frequent acidophiles that may appear belong to *Acidithiobacillus* spp., a gammaproteobacterial genus that includes species capable of oxidizing both sulfur (*At. thiooxidans*, *At. caldus*) and iron (*At. ferrooxidans*). In addition, *At. ferrooxidans* is able to reduce iron, and thus, members of this genus are, along with *L. ferrooxidans*, responsible for the constant acidic pH found in the Río Tinto River (González-Toril et al. 2003) and have also been detected in extreme acidic pit lakes, residues, and streams from diverse AMD systems (Chen et al. 2013, 2015; Kuang et al. 2013; Santofimia et al. 2013; Liu et al. 2014). Members of *Acidithiobacillus* spp. represent the most abundant bacteria detected within the snottites present in the sulfide-rich Frasassi cave system (Macalady et al. 2007) and are also considered the main primary producers within this extremely acidic biofilms (Jones et al. 2012; Ziegler et al. 2013). Diverse species belonging to this genus have also been found within microbial communities in the Copahue–Caviahue system (Urbieta et al. 2012). The recently described *At. ferrivorans* (Hallberg et al. 2010), a psychrotolerant and iron-/sulfur-oxidizing bacterium, dominated the microbial community in subterranean samples of pyrite mines (Kimura et al. 2011), and its abundance in acidic mine tailings was attributed to a potential role in the acidification process (Chen et al. 2013). This acidophile has also been found within the prokaryotic population that catalyzed the acid rock drainage (ARD) present in cold environments: the Antarctic (Dold et al. 2013), the High Arctic (García-Moyano et al. 2015), and a glacier area located in the Huascarán National Park, Perú (González-Toril et al. 2015).

Other relevant acidophilic bacteria found in diverse environments include the iron oxidizer/reducer and sulfur oxidizer *Sulfobacillus* spp. (Firmicutes), the iron-oxidizing betaproteobacterium "*Ferrovum myxofaciens*," and the iron oxidizer/reducer *Acidimicrobium–Ferrimicrobium* group (*Acidimicrobiaceae*, *Actinobacteria*). Strains of *Sulfobacillus* spp. (Firmicutes) have been detected in acidic environments associated to AMD processes and residues, with a pH range of 0.83–3.5 (Tyson et al. 2004, Kock and Schippers 2008; Chen et al. 2013; Kuang et al. 2013; Liu et al. 2014). A less acidophilic bacterium is "*Ferrovum myxofaciens*" which has been detected in subterranean samples (Kimura et al. 2011) and in mine waters (Hao et al. 2010; García-Moyano et al. 2012; Kuang et al. 2013) in a pH range of 2–3. In contrast, members of the *Acidimicrobiaceae* genera (*Actinobacteria*) were found over a wide pH range within the microbial communities that formed snottites (Jones et al. 2012; Ziegler et al. 2013), in acidic waters (Lear et al. 2009; García-Moyano et al. 2012; Kuang et al. 2013; Santofimia et al. 2013), as well as in iron-rich and low-temperature environments such as subterranean mines (Kimura et al. 2011), the Antarctic (Dold et al. 2013), the High Arctic permafrost (García-Moyano et al. 2015), and the Nevado Pastoruri glacier (González-Toril et al. 2015).

Heterotrophic acidophiles are also associated with acidic environments and preferentially detected in less acidic sites at a pH of 2.5–3.5. They include bacterial members of the *Acidobacterium* genus and of the *Alphaproteobacteria* such as *Acidiphilium* spp. and *Acidisphaera* spp. Both genera *Acidobacterium* and *Acidiphilium* are involved in the reductive dissolution of minerals by reducing ferric iron to ferrous (Coupland and Johnson 2008) and can also form commensal associations with iron and sulfur oxidizers, removing organic compounds that are toxic for chemolithotrophic bacteria (Johnson and Hallberg 2003). Members of *Acidobacterium* spp. usually occur in a great variety of soils, but they have also been described in natural acidic environments as diverse as acidic streams (Kuang et al. 2013), the rhizosphere of plants growing at the banks of the Tinto River (Mirete et al. 2007), and samples from ARD in Antarctica (Dold et al. 2013), the High Arctic (García-Moyano et al. 2015), and a glacier area in the Huascarán National Park (González-Toril et al. 2015). On the other hand, heterotrophic *Alphaproteobacteria* have been detected in the microbial communities associated with AMD and ARD environments such as acidic rivers (García-Moyano et al. 2012; Urbieta et al. 2012), lake waters (Hao et al. 2010; Santofimia et al. 2013), mining residues (Chen et al. 2015), rhizosphere (Mirete et al. 2007), cold waters (Dold et al. 2013), permafrost (García-Moyano et al. 2015), an acidic stream (Lear et al. 2009), and a glacier sediment (González-Toril et al. 2015).

In addition to these bacterial acidophiles, members belonging to the iron-oxidizing archaeon *Ferroplasma* spp. have been documented in several of the AMD sites reviewed here (Table 9.1). They have been described within the microbial communities found at a pH below 2.7, whereas in less acidic AMD sites (pH from 3 to 3.5), the archaeal fraction either has not been detected or is primarily constituted by members belonging to uncultured Crenarchaeota. (Table 9.1). In addition, in all the sites where *Ferroplasma* was present, they co-occurred with members of the genera *Leptospirillum* and *Acidithiobacillus*. As there is no evidence that *Ferroplasma* spp. can fix nitrogen but its genome contains several amino acid transporters and ammonia permeases, this general co-occurrence supports the notion that this archaeon obtains nitrogen from certain members of *Leptospirillum* and *Acidithiobacillus* since their genomes contain genes that encode nitrogen fixation enzymes (Tyson et al. 2004; Chen et al. 2013).

Here we have focused on bacterial and archaeal microorganisms as most of them are responsible for the main biogeochemical processes found in AMD environments, such as iron oxidation, but there can also be present an outstanding eukaryotic diversity within these extreme acidic environments. For example, the acidic waters of the Tinto River host a relevant microbial diversity which include several eukaryotic lineages that can thrive under such harsh conditions such as algae, which account for 65 % of the biomass (López-Archilla et al. 2001), fungi, diatoms, euglenoids, ciliates, cercomonads, and amoebae (López-Archilla et al. 2001; Amaral-Zettler et al. 2002, 2011; Aguilera et al. 2007). A highly similar AMD system to the Tinto River is the Richmond Mine at Iron Mountain where both fungi and red algae have been detected within microbial communities by using culture-independent approaches (Baker and Banfield 2003; Baker et al. 2004). Other AMD environments where microbial eukaryotes have been detected include the acidic waters of the Nuestra Señora del Carmen (NSC) Pit Lake in the Iberian Pyrite Belt (Santofimia et al. 2013); the biofilms present in a very acidic stream (pH of 2.8) in the South Island, New Zealand (Lear et al. 2009); the snottites found in the Frasassi cave system, Italy (Jones et al. 2012); a coal mining area in Svalbard, Norway (García-Moyano et al. 2015); and samples from mine tailings (pH of 3) from an AMD lake at Xiang Mountain, China (Hao et al. 2010) and within a sulfidic tailing dump in Kristineberg, Sweden (Kock and Schippers 2008). It is worth to note that the eukaryotic microorganisms from these acidic environments seem to be mainly exclusive of each of the sites. This is in contrast to the overlapping prokaryotic diversity observed among different sites and may be attributable to different metabolic roles shared by prokaryotic and eukaryotic microorganisms although this observation would need to be confirmed by further research.

9.4 Mechanisms of Low pH Resistance in Acidophiles

Most microorganisms, including acidophiles, require a circumneutral intracellular pH. Acidophiles are exposed to pH gradients several orders of magnitude greater than neutrophiles. This pH difference (ΔpH) across the membrane is connected to cellular bioenergetics since it is the main contributor to the proton motive force (PMF) in acidophiles. PMF is the force that promotes movement of proton across membranes downhill the electrochemical potential and drives the conversion of

ADP to ATP via membrane-bound ATPases, as well as it is necessary to transport substrates across membranes. Thus, in acidophiles with a near neutral cytoplasm, the net force across the membrane can drive energy-dependent processes. Nevertheless, the influx of protons to produce ATP increases cellular protonation and if it is not regulated could rapidly dissipate the ΔpH. In addition, the excess of intracellular protons would affect the function of nucleic acids and proteins, thus impairing cellular processes such as DNA replication, transcription, protein synthesis, and enzyme activities (Madshus 1988).

The mechanisms of pH homeostasis are not well known in acidophiles. Baker-Austin and Dopson (2007) reviewed various mechanisms, some of which were postulated in the 1980s and early 1990s, but they were not rigorously supported. However, the recent sequencing of several acidophile genomes and functional metagenomic studies have provided further insights to support those early theories and also to suggest other mechanisms.

9.4.1 The Cell Membrane Is Highly Impermeable to Protons

One of the adaptations that contributes to maintain the intracellular pH of acidophiles is the impermeability of the cell membrane, which restricts the influx of protons into the cytoplasm. In archaea, there is a strong association between the presence of tetraether lipids in the cell membrane and tolerance to acid pH. It was shown that liposomes derived from *Picrophilus oshimae* membrane lipids, which contain tetraether lipids, have extremely low proton permeability (van de Vossenberg et al. 1998). An evidence of this mechanism is the presence of tetraether lipids in the cell membranes of different archaea, such as *Thermoplasma acidophilum* (Shimada et al. 2002), *Ferroplasma acidiphilum* Y^T and Y-2 (Pivovarova et al. 2002; Batrakov et al. 2002), "*F. acidarmanus*" (Macalady et al. 2004), and *P. oshimae* (van de Vossenberg et al. 1998). There are no reports showing this mechanism in bacteria. However, it was observed that the genome of *Leptospirillum* group II (*L. ferriphilum*) contains a large number and variety of genes for cell membrane biosynthesis, which suggested the presence of a complex cell wall structure responsible for acid tolerance in this bacterium (Tyson et al. 2004).

9.4.2 Reduced Pore Size in Membrane Channels

Reducing the pore size of membrane channels has been proposed as another mechanism to maintain pH homeostasis to prevent protons from entering the cells. The outer membrane porin Omp40 of *At. ferrooxidans* has a larger external loop that could be responsible for controlling the size of the pores in the cells and also the ion selectivity, and it is highly expressed as a result of a pH shift from pH 3.5 to 1.5 (Amaro et al. 1991). In addition, it was reported that at a pH of 2.5, the external loop would be charged (+2 compared to −4 at neutral pH for a similar porin in *Escherichia coli*), and this charge could control the inflow of protons across the outer membrane (Guiliani and Jerez 2000).

9.4.3 Inhibition of Proton Influx by a Chemiosmotic Gradient Created by a Donnan Potential

Acidophiles are able to generate a transmembrane electric potential ($\Delta\psi$) that is positive inside the cell relative to outside, which is opposite to the inside negative $\Delta\psi$ of neutrophiles (Baker-Austin and Dopson 2007). A Donnan potential is formed when the diffusion of ions through the membrane in one direction equals the flux of ions in the opposite direction, resulting in net zero mass and charge transport. However, the membrane is selectively permeable and does not allow the passage of all the ions. It is proposed that this positive $\Delta\psi$ in acidophiles is generated by a Donnan potential of positively charged molecules and inhibits the influx of protons using a chemiosmotic barrier against the proton gradient. It is suggested that this potential is produced by a greater influx of potassium ions than the outward flux of protons. In fact, numerous potassium-transporting ATPases and other cation transporters have been observed in the genomes of several acidophiles such as *P. torridus* (Fütterer et al. 2004) and *Leptospirillum* group II (*L. ferriphilum*) (Tyson et al. 2004).

9.4.4 Transporters to Pump Out the Excess Protons

One important strategy to maintain pH homeostasis in acidophiles is to remove excess protons from the cytoplasm. For instance, it was reported that bacteria such as *Bacillus acidocaldarius* and *T. acidophilum* have the ability to actively pump protons out of their cytoplasm to maintain pH homeostasis, and this process is associated with the respiratory chain (Michels and Bakker 1985). In fact, genes encoding putative proton efflux systems (H^+ ATPases, antiporters, and symporters) have been identified in all the sequenced genomes from acidophiles. For instance, a predominance of proton efflux systems was reported in the genomes of *Ferroplasma* type II and *Leptospirillum* group II (*L. ferriphilum*) identified in a whole-community sequencing of an AMD biofilm from Iron Mountain (Tyson et al. 2004). The genomes of the archaeon *P. torridus*, which is able to grow at around pH 0 (Fütterer et al. 2004), and of *T. acidophilum* have revealed a predominance of proton-driven secondary transporters against primary transporters (overall ratio of secondary to primary transporters for *P. torridus* and for *T. acidophilum* is 10:1 and 5.6:1, respectively) (Crossman et al. 2004; Fütterer et al. 2004). By using a predominance of secondary transporters, the PMF can be connected for metabolic purposes. In addition the numerous proton-driven secondary transporters might represent a strategy of these microorganisms for the adaptation to the low pH environment.

9.4.5 Cytoplasmic Buffering to Maintain the Intracellular pH

In the event of protons entering in the cytoplasm, one of the strategies of acidophiles is the buffering capacity to sequester and release protons. The cells contain buffer molecules that have basic amino acids such as lysine, histidine, and arginine, and these molecules help in the proton sequestration. It has been proposed that in the acidophilic bacterium *Thiobacillus acidophilus*, the amino acid side chains were responsible for the cytoplasmic buffering (Zychlinsky and Matin 1983). This could be supported by the fact that decarboxylation of amino acids such as arginine and glutamate induced cell buffering in *E. coli* by consuming the protons which replace the carboxyl group of the amino acids. Then, antiporter membrane proteins exchange extracellular arginine or glutamate for the intracellular end product of decarboxylation (Castanie-Cornet et al. 1999; Richard and Foster 2004).

9.4.6 Proton Uncoupling by Organic Acids

Organic acids (for instance, acetic or lactic acid) function as uncouplers of the respiratory chain at low pH by diffusion of the protonated form into the cell followed by dissociation of a proton, a process called cytoplasmic protonation. Therefore, the degradation of these organic acids might be used by heterotrophic acidophiles to prevent their harmful effect. In agreement, different acidophile genomes [*P. torridus* (Angelov and Liebl 2006) and "*F. acidamarnus*"] contain genes encoding organic acid degradation pathways. However, it is unclear if this mechanism is regulated in response to low intracellular pH in these archaea, since they obtain energy from organic acids.

9.4.7 Chaperones to Protect DNA and Proteins from Damage Caused by Low pH

Biomolecules damaged by low pH could require an efficient repair. The genomes of different extreme acidophiles, for instance, *P. torridus* (Crossman et al. 2004), contain a large number of DNA and protein repair genes. On the other hand, chaperones involved in protein refolding were highly expressed in microorganisms from an environmental AMD biofilm (Ram et al. 2005). Chaperones were also highly expressed in "*F. acidamarnus*" in aerobic (Dopson et al. 2005) and anaerobic (Dopson et al. 2006) cultures.

9.4.8 Functional Metagenomics to Explore Novel Genes and Mechanisms Involved in Acid Resistance

Two main problems for the research in obligate acidophiles are (i) the lack of genetic tools (thus it is impossible to construct mutants, which would be necessary to understand the genetic and biochemical basis of the mechanisms of adaptation of these microorganisms in acidic environments) and (ii) the uncultured fraction of acidophiles, which can only be studied using culture-independent techniques, such as metagenomics, metatranscriptomics, and metaproteomics.

An approach to identify genes related to acid resistance from cultured and uncultured microorganisms is the use of functional metagenomics. This approach has recently allowed the recovery of enzymatic activities and physiological mechanisms involved in low pH resistance of microorganisms from an AMD environment. Metagenomic libraries from planktonic and rhizosphere microbial communities of the Tinto River were screened to search for genes involved in acid resistance (Guazzaroni et al. 2013). A total of 15 different genes conferring acid resistance to *E. coli*, the host of the metagenomic libraries, were identified, of which seven confer also acid resistance to *Pseudomonas putida* and six to *Bacillus subtilis*. Some of the recovered genes encoded proteins related to the mechanisms of low pH resistance previously proposed, for instance:

(i) A putative ClpXP protease, which could be involved in a number of cellular processes such as degradation of misfolded proteins and housekeeping removal of dysfunctional proteins (Chandu and Nandi 2004)
(ii) Proteins related to protection of nucleic acid, such as a histone-like protein HU, a Dps protein, and an RNA-binding protein, which could bind to DNA or RNA molecules to protect them (Choi et al. 2000; Oberto et al. 2009)
(iii) A component of the ABC-type nitrate/sulfonate/bicarbonate transporter that could be involved in bicarbonate entry, which has a crucial role in the physiological pH buffering system in eukaryotic organisms, although this system has never been reported in bacteria

Interestingly, this approach also revealed one unknown gene, three hypothetical genes, and other genes encoding known proteins not previously reported to be involved in acid resistance; all of them could represent novel mechanisms of acid resistance (Guazzaroni et al. 2013).

9.5 Metal Resistance Mechanisms in Acidophiles

As a result of the oxidation of minerals containing sulfides by chemolithoautotrophic bacteria, AMD-impacted sites contain elevated concentrations of toxic metals and metalloids such as copper, nickel, cadmium, zinc, arsenic, etc., which are more soluble under acidic conditions. Thus, in addition to acid tolerance, microorganisms that thrive in AMD sites have developed strategies of adaptation to elevated concentrations of metals, which most of them could be similar to those described for neutrophilic microorganisms (Dopson et al. 2003). Those strategies can be summed up in the following categories: (i) exclusion by permeability barrier, (ii) intra- and extracellular sequestration, (iii) enzymatic detoxification, (iv) reduction in the sensitivity of cellular targets to metal ions, and (v) active transport efflux pumps (Nies 1999; Bruins et al. 2000). From these, the active transport through P-type ATPases (pumps) can be considered as the basic strategy employed by microorganisms to cope with elevated concentrations of metal ions (Nies 2003). These efflux systems and other mechanisms of resistance have been identified in a number of acidophiles associated to AMD environments in the last years by using genomic, proteomic, and metagenomic approaches. For example, it has been described in *At. ferrooxidans* a copper resistance mechanism by which metal–phosphate complexes are formed and transported out of the cell (Alvarez and Jerez 2004) and also a number of genes encoding proteins involved in copper homeostasis such as novel chaperones and transport P-type ATPases (Luo et al. 2008; Navarro et al. 2009). In the archaeon "*F. acidarmanus*" F1, a copper efflux system and diverse stress proteins involved in protein folding and DNA repair were induced upon copper exposure (Baker-Austin et al. 2005). Detoxification of arsenite through the ArsB efflux pump has been described in "*F. acidarmanus*" F1 (Baker-Austin et al. 2007) and *L. ferriphilum* (Li et al. 2010). Also, the metal exporter P-type ATPase ZntA was constitutively expressed in the presence of excess zinc in *At. caldus* and *Acidimicrobium ferrooxidans* suggesting that these acidophiles are well adapted to elevated zinc concentrations (Mangold et al. 2013). An efflux system encoded by the nickel resistance determinant *ncrABCY* was responsible for the nickel resistance observed in *L. ferriphilum* UBK03 (Tian et al. 2007; Zhu et al. 2011). Moreover, the screening of a genomic library of *Acidiphilium* sp. PM, an environmental isolate from the acidic and heavy metal-rich waters of the Tinto River (Southwestern Spain) (San Martin-Uriz et al. 2011), revealed a total of seven different genes conferring nickel resistance to *E. coli* (San Martin-Uriz et al. 2014). Two of these genes form an operon encoding the ATP-dependent protease HslVU (ClpQY), which increases resistance to both Ni and Co in *E. coli*, a function not previously described for this family of proteases. Other Ni-resistance determinants include genes involved in the synthesis of branched amino acids and lipopolysaccharide biosynthesis. Interestingly, increased branched amino acid concentrations have been previously reported in *Pseudomonas pseudoalcaligenes* in response to metal stress (Tremaroli et al. 2009). The diversity of molecular functions of the genes recovered in this screening suggested that Ni resistance in *Acidiphilium* sp. PM relies on different molecular mechanisms (San Martin-Uriz et al. 2014). In addition, novel genes conferring nickel and arsenic resistance have been identified by using a functional metagenomics approach applied to the microbial communities from the Tinto River (Huelva, SW Spain), considered a well-known environment impacted by acid mine drainage. To search for nickel resistance genes, metagenomic libraries from the microorganisms of the rhizosphere of the endemic heather *Erica andevalensis*, which grows on the banks of the Tinto River, were screened and 13 clones were detected and analyzed (Mirete et al. 2007; González-Pastor and Mirete 2010). One clone encoded a bacterial serine O-acetyltransferase (SAT), whose expression in *E. coli* allows nickel accumulation in the cell. In plants from the *Thlaspi* genus, expression of the SAT enzyme increases the level of glutathione, which protects against the oxidative damage caused by nickel and allowing the hyperaccumulation of this metal in plant cells (Freeman et al. 2004). Two clones encoded putative ABC transporters belonging to a family not previously related to metal uptake and possibly involved in active efflux pumping outside the cells, since their expression in *E. coli* decreased the accumulation of nickel in the cells. Five clones encoded proteins similar to well-characterized proteins but not previously reported to be related to nickel resistance. An important aspect of using a functional metagenomics approach is the ability to identify genes that have no assigned functions (Mirete et al. 2016), and in fact, the remaining six clones identified in this study encoded unknown and conserved hypothetical proteins (Mirete et al. 2007).

To identify arsenic resistance genes in the microbial communities of the Tinto River, metagenomic libraries from microorganisms of rhizosphere (*Erica andevalensis*) and from planktonic microorganisms at the origin of the river (pH ≈ 1.8) were screened (Morgante et al. 2015). A total of 13 functionally active genes involved in arsenic resistance were identified, and they were assigned to different processes such as transport (a transporter of the major facilitator superfamily – MFS), stress response (a chaperone and an RNA-modifying enzyme), target protection leading to cellular accumulation of arsenic (DNA damage repair), phospholipids biosynthesis, and amino acid biosynthesis. Only two genes, which encode the proteins similar to the MFS transporter and the ClpB chaperone, were previously related to heavy metal resistance but were not previously described in acidophilic microorganisms (Celerin et al. 1998; Yuan et al. 2007; Cai et al. 2009). On the other hand, most of the retrieved genes (11) encode proteins not previously related to heavy metal resistance including one hypothetical and one unknown protein. In addition, the expression in *E. coli* of the genes encoding the ClpB chaperone and the two tRNA-modifying enzymes increased the cell survival under different stress conditions (heat shock and UV radiation) (Morgante et al. 2015).

In addition, metal and metalloid resistance in some isolated acidophilic bacterial species could be due to a combination of genes that are present in plasmids or transposons (Butcher et al. 2000; Barreto et al. 2003; Rawlings 2005). These genes may be recruited by horizontal gene transfer by acquisition of plasmids or transposons containing metal resistance genes. For instance, a strain of the acidophilic bacterium *At. caldus* isolated from a 10-year continuous-flow tank of biooxidation of arsenopyrite process contained the transposon Tn *AtcArs*, a Tn*21*-like transposon (de Groot et al. 2003), whose sequence revealed the presence of a set of *ars* genes previously described in *E. coli* cells (Carlin et al. 1995) or plasmid R773 (Chen et al. 1986). However, the *ars* resistance genes were arranged in an unusual manner with the *arsA* (ATPase) and *arsD* (regulator protein) being duplicated. The DNA sequence of the transposon Tn *AtcArs* indicated that the closest relative of the *ars* genes is present in the heterotrophic bacterium *Alcaligenes faecalis* (Tuffin et al. 2005). The arsenic originated primarily from the biooxidation of arsenopyrite may be present in high concentrations in AMD environments. The major difficulty in the initial stages of biooxidation of the arsenopyrite is that the microorganisms responsible for this process are sensitive to the high concentrations of arsenic released [levels of arsenic frequently reach saturation, (Dopson et al. 2003)]. Interestingly, the operation of continuous-flow tanks over many years promoted a selection for arsenic resistance resulting in highly arsenic-resistant microorganisms.

9.6 Biotechnological Applications of Acidophiles

In terms of "extremophile biotechnology," the acidophiles are an ecologically and an economically important group of microorganisms. They possess networked cellular adaptations to maintain their intracellular pH to around neutral (Dopson 2012). The mechanisms of pH homeostasis and resistance to toxic metals and metalloids used by acidophiles have been reviewed in the previous sections. Different "omics" approaches such as genomics, proteomics, and metagenomics have been applied to the study of acidophiles as they represent useful tools for studying whole microbial communities within their natural environments. In addition, these procedures provide a rich but largely unexploited hunting ground for genes and predicted proteins that might have useful functions in biotechnological applications. Considering such promising properties, both extremophile-derived enzymes and the whole microorganism may be exploitable. Industrial process conditions can be extremely harsh and continuously demand biocatalysts that can withstand those conditions (e.g., temperature, pH, and ionic strength). The majority of the enzymes used to date originate from mesophilic organisms and, despite their many advantages, mesophilic enzymes usually lack stability during severe industrial processes. For this reason, they represent only a small fraction of the potential industrial market (Gomes and Steiner 2004). Consequently, the discovery of microorganisms that are able to thrive in extreme environments and their enzymes has had a great impact on the field of biocatalysis.

This section focuses on providing information about the use of several acidophiles (mainly thriving in AMD environments) or acidophilic enzymes as significant biocatalysts for industry processes and bioremediation strategies. However, since many acidophiles are also multiextremophiles (i.e., microorganisms adapted to more than one type of extreme environment, such as the archaeon *Sulfolobus acidocaldarius*, which grows at pH 3 and 80 °C or *At. ferrivorans* SS3, a psychrotolerant acidophilic bacterium able to oxidize ferrous iron at 5 °C), special attention should be drawn to those publications and reviews where the missing information could be found (Demirjian et al. 2001; Liljeqvist et al. 2011; Dopson 2012; Sharma et al. 2012; Taylor et al. 2012; Zheng et al. 2012).

9.6.1 Whole-Cell Biocatalyst Exploitation

The most direct application of extremophiles in biotechnological processes involves the organisms themselves. Among their most established uses, bioleaching refers to the ability of some microorganisms to transform solid compounds into extractable entities. Although this term is commonly applied to biomining (i.e., solubilization of metals, principally copper, from ores) (Rawlings and Johnson 2007), the bioleaching process has also been proposed to recover heavy metal values from industrial solid waste (Asghari et al. 2013), spent refinery waste (Pradhan et al. 2010), or accidental petroleum hydrocarbons released in arctic environments (Brakstad and Bonaunet 2006; Head et al. 2006).

While bioleaching (or biomining) occurs in heaps of crushed ores, a related process, termed biooxidation, takes place in stirred reactors and is used principally for the recovery of gold (Rawlings and Johnson 2007). Both bioleaching and biooxidation can use many similar microorganisms. These processes involve iron- or sulfur-oxidizing acidophilic microorganisms adapted to different temperature ranges, from mesophiles (bacteria such as *Acidithiobacillus*, *Leptospirillum* and the archaeon *Ferroplasma*) to thermophiles (archaea from the genera *Sulfolobus*, *Metallosphaera*, and *Acidianus*) (Johnson and Hallberg 2003; Hallberg 2010). When the activity of such extreme acidophiles is not controlled in biomining operations, this can lead to acid mine drainage (AMD), which is one of the most pernicious forms of pollution in the world (Johnson and Hallberg 2003; Hallberg 2010). In comparison with conventional technologies such as mechanical or chemical methods (Johnson 1995; Mayes et al. 2009), the biotechnological leaching processes mediated by acidophiles cells may not seem competitive since a significantly longer leaching time is required to achieve the same extraction efficiencies when using traditional procedures. Nevertheless, bioleaching can offer to biohydrometallurgy industries other attractive features: the processes are more cost-efficient, simpler, and more environmentally friendly than their chemical counterparts (Johnson 1995; Beolchini et al. 2010; Dopson 2012).

The microorganisms responsible for bioleaching and AMD production or bioremediation are inherently adapted to extremely low pH, often pH values from 2 to 3, and elevated concentrations of metals (e.g., iron, aluminum, manganese, copper, cadmium, etc.) and metalloids (e.g., arsenic) depending on the mineralogy of the host rock. Thus, the physiology and genetic adaptability of biomining microbes dynamically and continuously adapt to the particular conditions of the environment. For instance, during the biooxidation process in gold-bearing arsenopyrite bioreactors, a noticeably increased resistance to arsenic was reported due to the arsenic resistance genes have been horizontally transferred via a transposon from an unidentified bacterium to *At. caldus* and *L. ferriphilum* (Rawlings 2005; Tuffin et al. 2005). For commercial bioleaching operations, this natural mechanism can contribute to the selection of more robust and efficient strains with reduced sensitivity to metal toxicity (Rawlings 2005; Podar and Reysenbach 2006).

9.6.2 The Industrial Applications of Extremozymes

The term extremozyme refers to enzymes from extremophilic microorganisms. Extremozymes from various acidophilic as well as acidothermophilic microbes such as amylases, proteases, ligases, cellulases, xylanases, α-glucosidases, endoglucanases, and esterases were shown to be stable at low pH and high temperature and functionally active under harsh industrial conditions (Bertoldo et al. 2004; Schepers et al. 2006; Kocabiyik and Özel 2007; Maurelli et al. 2008; Jaramillo et al. 2012; Sharma et al. 2012). Many of them have been used or have been proposed for use in biotechnological applications such as biofuel production, degradation of carbon polymers, pharmaceutical as well as food and textile industry, for laundry detergents and paper manufacturing, and also for the production of ultra-purified enzymes and substrates used in molecular biology (Dopson 2012; Sharma et al. 2012).

Besides the above mentioned biocatalysts, other biomolecules such as an electron transfer protein called rusticyanin (Yamada et al. 2004), a maltose-binding protein associated to bacterial transport and chemotaxis systems (Schäfer et al. 2004), and plasmids (Keeling et al. 1996; Yamashiro et al. 2006) have also been reported from acidophiles. Zheng and coworkers (Zheng et al. 2012) reported the first application of an antibiotic selectable marker in genetic study for a hyperthermophilic acidophile and in the crenarchaeal lineage. The novel selectable marker of *Sulfolobus islandicus* is a cassette composed of the *sac7d* promoter and the *hmg* gene coding for the 3-hydroxy-3-methylglutaryl coenzyme A (HMG-CoA) reductase, which confers simvastatin resistance to this crenarchaeon (Zheng et al. 2012). Antibiotic selection has successfully been employed for a few archaeal species, and interestingly, simvastatin antibiotic is a suitable selectable marker to confer effective genetic selection for hyperthermophilic acidophiles, including *Sulfolobus* as well as *Acidianus* and *Thermoplasma* species. Specially, species of *Sulfolobus* genus serve as model organisms for research on metabolic pathways, transcription, translation, and replication, all of them important insights into the molecular mechanisms for the third domain of life. *Sulfolobus* is also an important model in geomicrobiological studies.

Another remarkable advantage of the biotechnological application of extremozymes is that several extracellular enzymes from acidophiles are known to be functional at much lower pH than the cytoplasmic pH. For instance, the genes of several intracellular or cell-bound enzymes from *Ferroplasma acidiphilum* were cloned and expressed in *E. coli* and the products were purified and characterized (Golyshina et al. 2006). The characterized enzymes were found to be stable and functionally active in the acidic pH range of 1.7–4.0, suggesting undetected cellular compartmentalization providing cytoplasmic pH patchiness and low pH environments for the analyzed enzymes.

Finally, the possibility of improving extremozymes by genetic engineering and directed evolution will further enhance their industrial applications. Since the cultivation of extremophiles is associated with many potential difficulties, several researchers consider that only genetic engineering of the desired extremozymes into mesophilic hosts will allow large-scale production of them. On the other hand, to increase the activity of a microorganism for some specific process or boost the production of a substance within a cell by metabolic engineering (i.e., manipulating the genetic and regulatory processes) has been exploited in some thermophilic bacteria (Shaw et al. 2008), but not in any extreme acidophile (Cárdenas et al. 2010).

9.7 Relevance of Acidophiles in Astrobiology

Since 2004, observational and remote sensing data from different region of Mars confirmed the presence of iron minerals that can only be formed in the presence of water (Moore 2004; Morris et al. 2004; Squyres et al. 2004). Such remarkable findings led the scientific community to emphasize the study of analogous environments to Mars on Earth.

Chemolithoautotrophic acidophiles provide a link to early Earth when oxygen was unavailable and ecosystems were sustained by inorganic redox reactions (González-Toril 2005; Canfield et al. 2006). Environments which are inhabited by acidophilic microorganisms, such as Tinto River (Spain) and Eagle plains (Canada), are also considered to be important terrestrial analogs of Mars (Fairén et al. 2010).

Therefore, acid-stable enzymes might provide useful information about early earth enzymology.

To date, a promising device of biotechnology that would be used to detect organic molecules on Mars was performed based on two acidophile–chemolithotrophic bacterial strains *L. ferrooxidans* and *At. ferrooxidans* (Parro et al. 2005). Both are native and abundant microorganisms in the Tinto River ecosystem and also inhabiting AMD environments and have demonstrated ways of life for which it is feasible to identify a past or present hypothetical niche on Mars (López-Archilla et al. 2001; González-Toril et al. 2003). By integrated approaches, V. Parro and coworkers developed an antibody microarray-based biosensor based on *L. ferrooxidans* and *At. ferrooxidans* (Parro et al. 2005). Due to its novelty, this autonomous and remote operation instrument called SOLID (Signs Of LIfe Detector) is proven for searching biosignatures on acid and metal-rich environments and was also demonstrated to process a variety of samples for the detection of specific biomarkers as individual molecules or families of compounds fundamental in the chemistry of life (Parro et al. 2011a, b).

Acknowledgments This work was funded by the Spanish Ministry of Science and Innovation (CGL2012-39627-C03-02). Photos included in Fig. 9.2 and Fig. 9.3 were kindly provided by Angeles Aguilera [Centro de Astrobiología (CSIC-INTA), Madrid, Spain].

References

Aguilera A, Zettler E, Gómez F, Amaral-Zettler L, Rodríguez N, Amils R (2007) Distribution and seasonal variability in the benthic eukaryotic community of Río Tinto (SW, Spain), an acidic, high metal extreme environment. Syst Appl Microbiol 30:531–546

Alvarez S, Jerez CA (2004) Copper ions stimulate polyphosphate degradation and phosphate efflux in *Acidithiobacillus ferrooxidans*. Appl Environ Microbiol 70:5177–5182

Amaral-Zettler LA, Gomez F, Zettler E, Keenan BG, Amils R, Sogin ML (2002) Microbiology: eukaryotic diversity in Spain's River of Fire. Nature 417:137

Amaral-Zettler LA, Zettler ER, Theroux SM, Palacios C, Aguilera A, Amils R (2011) Microbial community structure across the tree of life in the extreme Rio Tinto. ISME J 5:42–50

Amaro AM, Chamorro D, Seeger M, Arredondo R, Peirano I, Jerez CA (1991) Effect of external pH perturbations on in vivo protein synthesis by the acidophilic bacterium *Thiobacillus ferrooxidans*. J Bacteriol 173:910–915

Angelov A, Liebl W (2006) Insights into extreme thermoacidophily based on genome analysis of *Picrophilus torridus* and other thermoacidophilic archaea. J Biotechnol 126:3–10

Asghari I, Mousavi SM, Amiri F, Tavassoli S (2013) Bioleaching of spent refinery catalysts: a review. J Ind Eng Chem 19:1069–1081

Baker BJ, Banfield JF (2003) Microbial communities in acid mine drainage. FEMS Microbiol Ecol 44:139–152

Baker BJ, Lutz MA, Dawson SC, Bond PL, Banfield JF (2004) Metabolically active eukaryotic communities in extremely acidic mine drainage. Appl Environ Microbiol 70:6264–6271

Baker-Austin C, Dopson M (2007) Life in acid: pH homeostasis in acidophiles. Trends Microbiol 15:165–171

Baker-Austin C, Dopson M, Wexler M, Sawers RG, Bond PL (2005) Molecular insight into extreme copper resistance in the extremophilic archaeon '*Ferroplasma acidarmanus*' Fer1. Microbiology 151:2637–2646

Baker-Austin C, Dopson M, Wexler M, Sawers RG, Stemmler A, Rosen BP, Bond PL (2007) Extreme arsenic resistance by the acidophilic archaeon '*Ferroplasma acidarmanus*' Fer1. Extremophiles 11:425–434

Barreto M, Quatrini R, Bueno S, Arriagada C, Valdes J, Silver S, Jedlicki E, Holmes DS (2003) Aspects of the predicted physiology of *Acidithiobacillus ferrooxidans* deduced from an analysis of its partial genome sequence. Hydrometallurgy 71:97–105

Batrakov SG, Pivovarova TA, Esipov SE, Sheichenko VI, Karavaiko GI (2002) β-D-Glucopyranosyl caldarchaetidylglycerol is the main lipid of the acidophilic, mesophilic, ferrous iron-oxidising archaeon *Ferroplasma acidiphilum*. Biochim Biophys Acta 1581:29–35

Beolchini F, Fonti V, Ferella F, Vegliò F (2010) Metal recovery from spent refinery catalysts by means of biotechnological strategies. J Hazard Mater 178:529–534

Bertoldo C, Dock C, Antranikian G (2004) Thermoacidophilic microorganisms and their novel biocatalysts. Eng Life Sci 4:521–532

Brakstad OG, Bonaunet K (2006) Biodegradation of petroleum hydrocarbons in seawater at low temperatures (0–5 C) and bacterial communities associated with degradation. Biodegradation 17:71–82

Bruins MR, Magning JP, Oehme FW (2000) Microbial resistance to metals in the environment. Ecotoxicol Environ Saf 45:198–207

Butcher BG, Deane SM, Rawlings DE (2000) The chromosomal arsenic resistance genes of *Thiobacillus ferrooxidans* have an unusual arrangement and confer increased arsenic and antimony resistance to *Escherichia coli*. Appl Environ Microbiol 66:1826–1833

Cai L, Rensing C, Li X, Wang G (2009) Novel gene clusters involved in arsenite oxidation and resistance in two arsenite oxidizers: *Achromobacter* sp. SY8 and *Pseudomonas* sp. TS44. Appl Microbiol Biotechnol 83:715–725

Canfield DE, Rosing MT, Bjerrum C (2006) Early anaerobic metabolisms. Philos Trans R Soc Lond B Biol Sci 361:1819–1836

Cárdenas JP, Valdés J, Quatrini R, Duarte F, Holmes DS (2010) Lessons from the genomes of extremely acidophilic bacteria and archaea with special emphasis on bioleaching microorganisms. Appl Microbiol Biotechnol 88:605–620

Carlin A, Shi W, Dey S, Rosen BP (1995) The *ars* operon of *Escherichia coli* confers arsenical and antimonial resistance. J Bacteriol 177:981–986

Castanie-Cornet M-P, Penfound TA, Smith D, Elliott JF, Foster JW (1999) Control of acid resistance in *Escherichia coli*. J Bacteriol 181:3525–3535

Celerin M, Gilpin AA, Schisler NJ, Ivanov AG, Miskiewicz E, Krol M, Laudenbach DE (1998) ClpB in a cyanobacterium: predicted structure, phylogenetic relationships, and regulation by light and temperature. J Bacteriol 180:5173–5182

Chandu D, Nandi D (2004) Comparative genomics and functional roles of the ATP-dependent proteases Lon and Clp during cytosolic protein degradation. Res Microbiol 155:710–719

Chen L-X, Hu M, Huang L-N, Hua Z-S, Kuang J-L, Li S-J, Shu W-S (2015) Comparative metagenomic and metatranscriptomic analyses of microbial communities in acid mine drainage. ISME J 9:1579–1592

Chen LX, Li JT, Chen YT, Huang LN, Hua ZS, Hu M, Shu WS (2013) Shifts in microbial community composition and function in the acidification of a lead/zinc mine tailings. Environ Microbiol 15:2431–2444

Chen CM, Misra TK, Silver S, Rosen BP (1986) Nucleotide sequence of the structural genes for an anion pump. The plasmid-encoded arsenical resistance operon. J Biol Chem 261:15030–15038

Choi SH, Baumler DJ, Kaspar CW (2000) Contribution of *dps* to acid stress tolerance and oxidative stress tolerance in *Escherichia coli* O157:H7. Appl Environ Microbiol 66:3911–3916

Coupland K, Johnson DB (2008) Evidence that the potential for dissimilatory ferric iron reduction is widespread among acidophilic heterotrophic bacteria. FEMS Microbiol Lett 279:30–35

Crossman L, Holden M, Pain A, Parkhill J (2004) Genomes beyond compare. Nat Rev Microbiol 2:616–617

de Groot P, Deane SM, Rawlings DE (2003) A transposon-located arsenic resistance mechanism from a strain of *Acidithiobacillus caldus* isolated from commercial, arsenopyrite biooxidation tanks. Hydrometallurgy 71:115–123

Demirjian DC, Moŕis-Varas F, Cassidy CS (2001) Enzymes from extremophiles. Curr Opin Chem Biol 5:144–151

Denef VJ, Mueller RS, Banfield JF (2010) AMD biofilms: using model communities to study microbial evolution and ecological complexity in nature. ISME J 4:599–610

Dold B, González-Toril E, Aguilera A, López-Pamo E, Cisternas ME, Bucchi F, Amils R (2013) Acid rock drainage and rock weathering in Antarctica: important sources for iron cycling in the Southern Ocean. Environ Sci Technol 47:6129–6136

Dopson M (2012) Physiological adaptations and biotechnological applications of acidophiles. In: Anitori RP (ed) Extremophiles: microbiology and biotechnology. Horizon Scientific Press, Norwich, pp. 265–294

Dopson M, Baker-Austin C, Bond PL (2005) Analysis of differential protein expression during growth states of *Ferroplasma* strains and insights into electron transport for iron oxidation. Microbiology 151:4127–4137

Dopson M, Baker-Austin C, Bond P (2006) Towards determining details of anaerobic growth coupled to ferric iron reduction by the acidophilic archaeon '*Ferroplasma acidarmanus*' Fer1. Extremophiles 11:159–168

Dopson M, Baker-Austin C, Koppineedi PR, Bond PL (2003) Growth in sulfidic mineral environments: metal resistance mechanisms in acidophilic micro-organisms. Microbiology 149: 1959–1970

Fairén AG, Davila AF, Lim D et al (2010) Astrobiology through the ages of Mars: the study of terrestrial analogues to understand the habitability of Mars. Astrobiology 10:821–843

Freeman JL, Persans MW, Nieman K, Albrecht C, Peer W, Pickering IJ, Salt DE (2004) Increased glutathione biosynthesis plays a role in nickel tolerance in *Thlaspi* nickel hyperaccumulators. Plant Cell 16:2176–2191

Fütterer O, Angelov A, Liesegang H, Gottschalk G, Schleper C, Schepers B, Dock C, Antranikian G, Liebl W (2004) Genome sequence of *Picrophilus torridus* and its implications for life around pH 0. Proc Natl Acad Sci U S A 101:9091–9096

García-Moyano A, Austnes AE, Lanzén A, González-Toril E, Aguilera Á, Øvreås L (2015) Novel and unexpected microbial diversity in acid mine drainage in Svalbard (78° N), revealed by culture-independent approaches. Microorganisms 3:667–694

García-Moyano A, González-Toril E, Aguilera A, Amils R (2012) Comparative microbial ecology study of the sediments and the water column of the Rio Tinto, an extreme acidic environment. FEMS Microbiol Ecol 81:303–314

Golyshina OV, Golyshin PN, Timmis KN, Ferrer M (2006) The 'pH optimum anomaly' of intracellular enzymes of *Ferroplasma acidiphilum*. Environ Microbiol 8:416–425

Gomes J, Steiner W (2004) The biocatalytic potential of extremophiles and extremozymes. Food Technol Biotechnol 42:223–235

González-Pastor JE, Mirete S (2010) Novel metal resistance genes from microorganisms: a functional metagenomic approach. Methods Mol Biol 668:273–285. doi:10.1007/978-1-60761-823-2_19

González-Toril E (2005) Iron meteorites can support the growth of acidophilic chemolithoautotrophic microorganisms. Astrobiology 5:406–414

González-Toril E, Llobet-Brossa E, Casamayor EO, Amann R, Amils R (2003) Microbial ecology of an extreme acidic environment, the Tinto River. Appl Environ Microbiol 69:4853–4865

González-Toril E, Santofimia E, Blanco Y, López-Pamo E, Gómez MJ, Bobadilla M, Cruz R, Palomino EJ, Aguilera Á (2015) Pyrosequencing-based assessment of the microbial community structure of Pastoruri glacier area (Huascarán National Park, Perú), a natural extreme acidic environment. Microb Ecol 70:936–947

Guazzaroni M-E, Morgante V, Mirete S, González-Pastor JE (2013) Novel acid resistance genes from the metagenome of the Tinto River, an extremely acidic environment. Environ Microbiol 15:1088–1102

Guiliani N, Jerez CA (2000) Molecular cloning, sequencing, and expression of *omp-40*, the gene coding for the major outer membrane protein from the acidophilic bacterium *Thiobacillus ferrooxidans*. Appl Environ Microbiol 66:2318–2324

Hallberg KB (2010) New perspectives in acid mine drainage microbiology. Hydrometallurgy 104:448–453

Hallberg KB, Gonzalez-Toril E, Johnson DB (2010) *Acidithiobacillus ferrivorans*, sp. nov.; facultatively anaerobic, psychrotolerant iron-, and sulfur-oxidizing acidophiles isolated from metal mine-impacted environments. Extremophiles 14:9–19

Hao C, Wang L, Gao Y, Zhang L, Dong H (2010) Microbial diversity in acid mine drainage of Xiang Mountain sulfide mine, Anhui Province, China. Extremophiles 14:465–474

Head IM, Jones DM, Roling WFM (2006) Marine microorganisms make a meal of oil. Nat Rev Microbiol 4:173–182

Jaramillo ML, Abanto M, Quispe RL, Calderón J, del Valle LJ, Talledo M, Ramírez P (2012) Cloning, expression and bioinformatics analysis of ATP sulfurylase from *Acidithiobacillus ferrooxidans* ATCC 23270 in *Escherichia coli*. Bioinformation 8:695–704

Johnson DB (1995) Acidophilic microbial communities: candidates for bioremediation of acidic mine effluents. Int Biodeter Biodegr 35:41–58

Johnson DB, Hallberg KB (2003) The microbiology of acidic mine waters. Res Microbiol 154:466–473

Jones DS, Albrecht HL, Dawson KS, Schaperdoth I, Freeman KH, Pi Y, Pearson A, Macalady JL (2012) Community genomic analysis of an extremely acidophilic sulfur-oxidizing biofilm. ISME J 6:158–170

Keeling PJ, Klenk H-P, Singh RK, Feeley O, Schleper C, Zillig W, Doolittle WF, Sensen CW (1996) Complete nucleotide sequence of the *Sulfolobus islandicus* multicopy plasmid pRN1. Plasmid 35:141–144

Kimura S, Bryan CG, Hallberg KB, Johnson DB (2011) Biodiversity and geochemistry of an extremely acidic, low-temperature subterranean environment sustained by chemolithotrophy. Environ Microbiol 13:2092–2104

Kocabiyik S, Özel H (2007) An extracellular—Pepstatin insensitive acid protease produced by *Thermoplasma volcanium*. Bioresour Technol 98:112–117

Kock D, Schippers A (2008) Quantitative microbial community analysis of three different sulfidic mine tailing dumps generating acid mine drainage. Appl Environ Microbiol 74:5211–5219

Kuang J-L, Huang L-N, Chen L-X, Hua Z-S, Li S-J, Hu M, Li J-T, Shu W-S (2013) Contemporary environmental variation determines microbial diversity patterns in acid mine drainage. ISME J 7:1038–1050

Lear G, Niyogi D, Harding J, Dong Y, Lewis G (2009) Biofilm bacterial community structure in streams affected by acid mine drainage. Appl Environ Microbiol 75:3455–3460

Li B, Lin J, Mi S (2010) Arsenic resistance operon structure in *Leptospirillum ferriphilum* and proteomic response to arsenic stress. Bioresour Technol 101:9811–9814

Liljeqvist M, Sundkvist J-E, Saleh A, Dopson M (2011) Low temperature removal of inorganic sulfur compounds from mining process waters. Biotechnol Bioeng 108:1251–1259

Liu J, Hua Z-S, Chen L-X, Kuang J-L, Li S-J, Shu W-S, Huang L-N (2014) Correlating microbial diversity patterns with geochemistry in an extreme and heterogeneous environment of mine tailings. Appl Environ Microbiol 80:3677–3686

López-Archilla AI, Marin I, Amils R (2001) Microbial community composition and ecology of an acidic aquatic environment: the Tinto River, Spain. Microb Ecol 41:20–35

Luo Y, Liu Y, Zhang C, Luo H, Guan H, Liao H, Qiu G, Liu X (2008) Insights into two high homogenous genes involved in copper homeostasis in *Acidithiobacillus ferrooxidans*. Curr Microbiol 57:274–280

Macalady JL, Jones DS, Lyon EH (2007) Extremely acidic, pendulous cave wall biofilms from the Frasassi cave system, Italy. Environ Microbiol 9:1402–1414

Macalady JL, Vestling MM, Baumler D, Boekelheide N, Kaspar CW, Banfield JF (2004) Tetraether-linked membrane monolayers in *Ferroplasma* spp: a key to survival in acid. Extremophiles 8:411–419

Madshus IH (1988) Regulation of intracellular pH in eukaryotic cells. Biochem J 250:1–8

Mangold S, Potrykus J, Bjorn E, Lovgren L, Dopson M (2013) Extreme zinc tolerance in acidophilic microorganisms from the bacterial and archaeal domains. Extremophiles 17: 75–85

Maurelli L, Giovane A, Esposito A, Moracci M, Fiume I, Rossi M, Morana A (2008) Evidence that the xylanase activity from *Sulfolobus solfataricus* Oα is encoded by the endoglucanase precursor gene (*sso1354*) and characterization of the associated cellulase activity. Extremophiles 12:689–700

Mayes WM, Batty LC, Younger PL, Jarvis AP, Kõiv M, Vohla C, Mander U (2009) Wetland treatment at extremes of pH: a review. Sci Total Environ 407:3944–3957

Michels M, Bakker EP (1985) Generation of a large, protonophore-sensitive proton motive force and pH difference in the acidophilic bacteria *Thermoplasma acidophilum* and *Bacillus acidocaldarius*. J Bacteriol 161:231–237

Mirete S, de Figueras CG, Gonzalez-Pastor JE (2007) Novel nickel resistance genes from the rhizosphere metagenome of plants adapted to acid mine drainage. Appl Environ Microbiol 73:6001–6011

Mirete S, Morgante V, González-Pastor JE (2016) Functional metagenomics of extreme environments. Curr Opin Biotechnol 38:143–149

Moore JM (2004) Mars: blueberry fields for ever. Nature 428:711–712

Morgante V, Mirete S, de Figueras CG, Postigo Cacho M, Gonzalez-Pastor JE (2015) Exploring the diversity of arsenic resistance genes from acid mine drainage microorganisms. Environ Microbiol 17:1910–1925

Morris RV, Klingelhöfer G, Bernhardt B et al (2004) Mineralogy at Gusev Crater from the Mössbauer spectrometer on the Spirit Rover. Science 305:833–836

Navarro CA, Orellana LH, Mauriaca C, Jerez CA (2009) Transcriptional and functional studies of *Acidithiobacillus ferrooxidans* genes related to survival in the presence of copper. Appl Environ Microbiol 75:6102–6109

Nies DH (1999) Microbial heavy-metal resistance. Appl Microbiol Biotechnol 51:730–750

Nies DH (2003) Efflux-mediated heavy metal resistance in prokaryotes. FEMS Microbiol Rev 27:313–339

Oberto J, Nabti S, Jooste V, Mignot H, Rouviere-Yaniv J (2009) The HU regulon is composed of genes responding to anaerobiosis, acid stress, high osmolarity and SOS induction. PLoS One 4:e4367

Parro V, de Diego-Castilla G, Rodríguez-Manfredi JA et al (2011b) SOLID3: A Multiplex antibody microarray-based optical sensor instrument for in situ life detection in planetary exploration. Astrobiology 11:15–28

Parro V, Fernández-Remolar D, Rodríguez-Manfredi JA et al (2011a) Classification of modern and old Río Tinto sedimentary deposits through the biomolecular record using a life marker biochip: implications for detecting life on Mars. Astrobiology 11:29–44

Parro V, Rodríguez-Manfredi JA, Briones C et al (2005) Instrument development to search for biomarkers on mars: terrestrial acidophile, iron-powered chemolithoautotrophic communities as model systems. Planet Space Sci 53:729–737

Pivovarova TA, Kondrat'eva TF, Batrakov SG, Esipov SE, Sheichenko VI, Bykova SA, Lysenko AM, Karavaiko GI (2002) Phenotypic features of *Ferroplasma acidiphilum* strains YT and Y-2. Microbiology 71:698–706

Podar M, Reysenbach A-L (2006) New opportunities revealed by biotechnological explorations of extremophiles. Curr Opin Biotechnol 17:250–255

Pradhan D, Mishra D, Kim DJ, Ahn JG, Chaudhury GR, Lee SW (2010) Bioleaching kinetics and multivariate analysis of spent petroleum catalyst dissolution using two acidophiles. J Hazard Mater 175:267–273

Ram RJ, VerBerkmoes NC, Thelen MP, Tyson GW, Baker BJ, Blake RC, Shah M, Hettich RL, Banfield JF (2005) Community proteomics of a natural microbial biofilm. Science 308: 1915–1920

Rawlings DE (2005) Characteristics and adaptability of iron- and sulfur-oxidizing microorganisms used for the recovery of metals from minerals and their concentrates. Microb Cell Fact 4:1–15

Rawlings DE, Johnson DB (2007) The microbiology of biomining: development and optimization of mineral-oxidizing microbial consortia. Microbiology 153:315–324

Richard H, Foster JW (2004) *Escherichia coli* glutamate- and arginine-dependent acid resistance systems increase internal pH and reverse transmembrane potential. J Bacteriol 186:6032–6041

San Martin-Uriz P, Gomez MJ, Arcas A, Bargiela R, Amils R (2011) Draft genome sequence of the electricigen *Acidiphilium* sp. strain PM (DSM 24941). J Bacteriol 193:5585–5586

San Martin-Uriz P, Mirete S, Alcolea PJ, Gomez MJ, Amils R, Gonzalez-Pastor JE (2014) Nickel-resistance determinants in *Acidiphilium* sp. PM dentified by genome-wide functional screening. PLoS One 9:e95041

Santofimia E, Gonzalez-Toril E, Lopez-Pamo E, Gomariz M, Amils R, Aguilera A (2013) Microbial diversity and its relationship to physicochemical characteristics of the water in two extreme acidic pit lakes from the Iberian Pyrite Belt (SW Spain). PLoS One 8:e66746

Schäfer K, Magnusson U, Scheffel F, Schiefner A, Sandgren MOJ, Diederichs K, Welte W, Hülsmann A, Schneider E, Mowbray SL (2004) X-ray structures of the maltose–maltodextrin-binding protein of the thermoacidophilic bacterium *Alicyclobacillus acidocaldarius* provide insight into acid stability of proteins. J Mol Biol 335:261–274

Schepers B, Thiemann V, Antranikian G (2006) Characterization of a novel glucoamylase from the thermoacidophilic archaeon *Picrophilus torridus* heterologously expressed in *E. coli*. Eng Life Sci 6:311–317

Sharma A, Kawarabayasi Y, Satyanarayana T (2012) Acidophilic bacteria and archaea: acid stable biocatalysts and their potential applications. Extremophiles 16:1–19

Shaw AJ, Podkaminer KK, Desai SG, Bardsley JS, Rogers SR, Thorne PG, Hogsett DA, Lynd LR (2008) Metabolic engineering of a thermophilic bacterium to produce ethanol at high yield. Proc Natl Acad Sci U S A 105:13769–13774

Shimada H, Nemoto N, Shida Y, Oshima T, Yamagishi A (2002) Complete polar lipid composition of *Thermoplasma acidophilum* HO-62 determined by high-performance liquid chromatography with evaporative light-scattering detection. J Bacteriol 184:556–563

Singer PC, Stumm W (1970) Acidic mine drainage: the rate-determining step. Science 167: 1121–1123

Squyres SW, Grotzinger JP, Arvidson RE et al (2004) In situ evidence for an ancient aqueous environment at Meridiani Planum, Mars. Science 306:1709–1714

Taylor MP, van Zyl L, Tuffin M, Cowan D (2012) Extremophiles and biotechnology: how far have we come? In: Anitori RP (ed) Extremophiles: microbiol biotechnol, vol. 1. Horizon Scientific Press, Norwich, pp. 1–24

Tian J, Wu N, Li J, Liu Y, Guo J, Yao B, Fan Y (2007) Nickel-resistant determinant from *Leptospirillum ferriphilum*. Appl Environ Microbiol 73:2364–2368

Tremaroli V, Workentine ML, Weljie AM et al (2009) Metabolomic investigation of the bacterial response to a metal challenge. Appl Environ Microbiol 75:719–728

Tuffin IM, de Groot P, Deane SM, Rawlings DE (2005) An unusual Tn21-like transposon containing an *ars* operon is present in highly arsenic-resistant strains of the biomining bacterium *Acidithiobacillus caldus*. Microbiology 151:3027–3039

Tyson GW, Chapman J, Hugenholtz P, Allen EE, Ram RJ, Richardson PM, Solovyev VV, Rubin EM, Rokhsar DS, Banfield JF (2004) Community structure and metabolism through reconstruction of microbial genomes from the environment. Nature 428:37–43

Urbieta MS, Gonzalez Toril E, Aguilera A, Giaveno MA, Donati E (2012) First prokaryotic biodiversity assessment using molecular techniques of an acidic river in Neuquen, Argentina. Microb Ecol 64:91–104

van de Vossenberg MJLC, Driessen MAJ, Zillig W, Konings NW (1998) Bioenergetics and cytoplasmic membrane stability of the extremely acidophilic, thermophilic archaeon *Picrophilus oshimae*. Extremophiles 2:67–74

Yamada T, Hiraoka Y, Gupta TKD, Chakrabarty AM (2004) Rusticyanin, a bacterial electron transfer protein, causes G1 arrest and apoptosis in human cancer cells. Cell Cycle 3:1180–1185

Yamashiro K, S-i Y, Oshima T, Yamagishi A (2006) Structural analysis of the plasmid pTA1 isolated from the thermoacidophilic archaeon *Thermoplasma acidophilum*. Extremophiles 10:327–335

Yuan L, Rodrigues PH, Bélanger M, Dunn W, Progulske-Fox A (2007) The *Porphyromonas gingivalis clpB* gene is involved in cellular invasion in vitro and virulence in vivo. FEMS Immunol Med Microbiol 51:388–398

Zheng T, Huang Q, Zhang C, Ni J, She Q, Shen Y (2012) Development of a simvastatin selection marker for a hyperthermophilic acidophile, *Sulfolobus islandicus*. Appl Environ Microbiol 78:568–574

Zhu T, Tian J, Zhang S, Wu N, Fan Y (2011) Identification of the transcriptional regulator NcrB in the nickel resistance determinant of *Leptospirillum ferriphilum* UBK03. PLoS One 6:e17367

Ziegler S, Dolch K, Geiger K et al (2013) Oxygen-dependent niche formation of a pyrite-dependent acidophilic consortium built by archaea and bacteria. ISME J 7:1725–1737

Zychlinsky E, Matin A (1983) Cytoplasmic pH homeostasis in an acidophilic bacterium, *Thiobacillus acidophilus*. J Bacteriol 156:1352–1355

Extremophiles in Spacecraft Assembly Cleanrooms

10

Christine Moissl-Eichinger

Contents

C. Moissl-Eichinger
Medical University Graz, Department of Internal Medicine, Section of Infectious Diseases and Tropical Medicine, Auenbruggerplatz 15, 8036 Graz, Austria
e-mail: christine.moissl-eichinger@medunigraz.at

H. Stan-Lotter, S. Fendrihan (eds.), *Adaption of Microbial Life to Environmental Extremes*, DOI 10.1007/978-3-319-48327-6_10

10.1 Introduction

10.1.1 Planetary Protection

The major goal of planetary protection is to avoid any compromise of scientific exploration of celestial bodies by (biological) contamination. This includes (a) the protection of extraterrestrial ecosystems from terrestrial biomolecules and life forms and (b) the protection of the integrity of missions to avoid false-positive signals from, e.g., life-detection instruments. Besides forward contamination of celestial bodies, a reverse contamination of Earth by extraterrestrial material is also a fundamental concern: "States parties shall pursue studies of outer space, including the Moon and other celestial bodies, and conduct explorations of them so as to avoid their harmful contamination and also adverse changes in the environment of the Earth resulting from the introduction of extraterrestrial matter and, when necessary, adopt appropriate measures for this purpose" (UN Outer Space Treaty 1967; Conley 2011). This scope of the planetary protection policy emphasizes that forward contamination by terrestrial life and even by biomolecules needs to be avoided in order to preserve extraterrestrial bodies and to prevent confounding of future life-detection experiments on other planets (Kminek and Rummel 2015). According to the COSPAR (Committee of Space Research) recommendations, space missions are divided in five categories, considering the scientific interest and also the probability of possible contamination of other planets and, in case of return missions, also Earth (Table 10.1).

Landing missions on Mars are generally assigned to category IV with subcategories a, b, and c. Although Mars appears hostile for life, some so-called special regions could support Earth or own indigenous Mars life, should it exist (Rummel et al. 2014; Kminek and Rummel 2015; Rettberg et al. 2016). The parameters for Martian special regions are currently defined as follows: water activity between 0.5 and 1.0 and 25 °C as the lower limit for temperature (Kminek and Rummel 2015). However, those special regions remain currently untackled by lander missions, due to extremely strict protection regulations.

There is great fear that in particular the search for life could be affected by the contamination of landing spacecraft and their sensitive biosensors. False positives could possibly mask present signatures of Martian life and therefore inhibit the successful search for extraterrestrial life forms. Additionally, although not very likely, Earth organisms could possibly proliferate and therefore contaminate Mars' biotopes, competing with potential indigenous life.

Table 10.1 Planetary protection mission categories (Kminek and Rummel 2015)

Category	Mission type	Possible target
I	Missions to a target body without direct interest for understanding the process of chemical evolution or the origin of life. Since no protection of these bodies is warranted, no planetary protection requirements are necessary	Venus, undifferentiated asteroids
II	Missions to target bodies with significant interest relative to the process of chemical evolution and the origin of life, but in which there is only a remote chance of contamination. Planetary protection requirements include mainly simple documentation and passive contamination control (cleanroom assembly)	Jupiter, Saturn, comets, Uranus
III	Flyby and orbiter missions, targeting a body of significant interest for chemical evolution and/or origin of life, with high risk of contamination that could jeopardize future search for life missions. COSPAR requirements are a documentation that include also a possible bioburden reduction if necessary. Furthermore, an inventory of the microbial community present is required if an impact is very probable	Mars, Europa, Enceladus
IV	Mostly probe and lander missions, targeting bodies of high interest concerning chemical evolutions and/or origin of life, with a significant chance of contamination. Category IV lander missions are separated into three subcategories (a, b, c) with different requirements based on the location of the landing site and the objectives of that mission. IVb and c have the strictest bioburden limits and require detailed documentations, bioassays for bioburden measurements, (partial) sterilization of hardware, aggressive cleaning, protection from recontamination, and aseptic assembly	Mars, Europa, Enceladus
V	All Earth return missions, distinguishing unrestricted and restricted Earth return, depending on the probability of the presence of indigenous life forms on the visited solar body. Restricted Earth return missions require strict containment of samples	Unrestricted Earth return: Moon; restricted Earth return: Mars, Europa

Thus, complete sterility of a spacecraft is a desirable goal. However, nowadays sensitive instruments and detectors onboard do not allow to heat-sterilize the entire spacecraft as done for the Viking lander in the 1970s (111.7 °C ± 1.7 °C, 23–30 h; Puleo et al. 1977). Instead, all assembly procedures are performed in microbiologically controlled cleanrooms, including the integration of pre-cleaned and presterilized spacecraft hardware.

10.1.2 Cleanrooms

In order not to affect or even to confound future life-detection missions on celestial bodies, which are of interest for their chemical and biological evolution, spacecraft are constructed in so-called cleanrooms and are subject to severe cleaning processes and microbiological controls before launch (Crawford 2005). Cleanrooms are certified according to ISO14644-1. For instance, the cleanroom class ISO 5 corresponds to the former cleanroom class 100 (US FED STD 209E), allowing a maximum of 3.5 particles with a maximum 0.5 μm diameter per liter air.

During assembly, test, and launch operations (ATLO) of, e.g., Mars landers, appropriate cleanliness and sterility levels have to be guaranteed: The proper maintenance of the cleanroom includes a repeated cleaning with antimicrobial agents, particulates are filtered from the air (HEPA filtering), and even staff, working in the cleanroom, must take appropriate actions to minimize the particulate and microbial contamination. Cleanroom personnel have to follow specific access procedures (air locks, tacky mats) to minimize the influx of particulate matter. Staff has to wear special suits, use sterile tools, observe possible biocontamination risks, and even undergo frequent health checks.

Spacecraft assembly cleanrooms are quite similar to pharmaceutical or hospital cleanrooms. In the pharmaceutical industry, for aseptic production, cleanrooms are required, and monitoring microbial and also particle counts are part of good manufacturing practices (Nagarkar et al. 2001).

10.1.3 Contamination Control and Examinations

To date, space missions in preparation have to follow an implementation plan describing all actions necessary to reduce and measure bioburden. This plan also includes the requirement of (daily) sampling of the spacecraft and hardware using swabs and wipes. The recommended sampling size for swabs is 25 cm^2 only, whereas polyester wipes are used for the sampling of larger surfaces. The bacterial spore count is then assessed by culturing a heat-shocked sample according to a standard protocol and aims to reflect the most resistant component of the aerobic, heterotrophic, and mesophilic microbial community present ("bioburden," Administration NASA, Technical Handbook 2010). The current NASA standard is based on the methods that were originally developed for the Viking missions in the 1970s. In brief, the surface of a spacecraft is either swabbed (cotton swabs) or wiped (for larger surfaces). The sampling tools are extracted in liquid, by a combination of vortexing/shaking and sonication. After a heat shock (80 °C, 15 min), the suspension is plated on TSA plates and incubated at 32 °C. After a final count (72 h), the resulting plate count is used as a basis for the calculation of the overall microbial cleanliness of the spacecraft surface.

ESA's new standard methodology is based on the usage of the nylon-flocked swab instead of cotton swabs. Additionally, for improved cultivation of low-nutrient adapted cleanroom microorganisms, the cultivation medium used is R2A instead of TSA, as given in the new ESA standard protocol (2008). If space hardware surfaces are contaminated above the accepted levels, biocleaning is necessary: alcohols (IPA), disinfectants, and UV exposure are some methods applied to reduce the present contaminants. Furthermore, bioshields can be used to enclose certain (clean) hardware or the entire spacecraft to avoid contamination (Debus 2006).

In the case of Mars, limits on bioburden are based on requirements first imposed on the Viking missions. Based on these data, the acceptable microbial contamination for a category IVa mission is limited to 3×10^5 bacterial spores per Mars landing spacecraft or 300 spores per m^2 surface (Viking presterilization biological burden levels; Pillinger et al. 2006). For instance, for Beagle 2, an ESA IVa mission, the overall surface bioburden was estimated to be 2.3×10^4 spores, the total bioburden was 1.01×10^5 spores, and bioburden density was approximately 20.6 spores/m^2 and therefore within the acceptable range (Pillinger et al. 2006).

10.1.4 Cleanroom Microbiology

Examination protocols for spacecraft assembly cleanrooms focus on the detection and enumeration of cultivable, mesophilic, heterotrophic organisms. Nevertheless, cleanrooms are unique environments for microbes: due to low nutrient levels, dry and clean conditions, and constant control of humidity and temperature, these environments are inhospitable to microbial life and even considered "extreme" (Venkateswaran et al. 2001).

Several procedures keep contamination from the outside as low as possible, but these conditions are also highly selective for indigenous extremotolerant microbial communities (Crawford 2005). For space missions, it is crucial to generally control the contaminating bioburden as much as possible. But on the other hand, for the development of novel cleaning/sterilization methods, it is also important to identify and characterize (understand) the microbial community of spacecraft cleanrooms.

The low biomass is generally problematic for microbiological surveys, since the sampling procedure itself is biased and characterized by significant losses during the procedure (Probst et al. 2010a). Furthermore, it is estimated that only 0.1–1 % of all microbes present in one biotope can be cultivated using standard cultivation techniques, increasing the unseen microbial portion in cleanrooms significantly (Amann et al. 1995).

An increasing body of literature about the microbial diversity in spacecraft-associated cleanrooms and on spacecraft hardware is available, including extensive cultivation workup to studies based on next-generation sequencing.

The first article about the microbial analyses of the two Viking spacecraft reported that about 7000 samples were taken from the spacecraft surface during prelaunch activities in order to determine the cultivable microbial load (Puleo et al. 1977). Besides human-associated bacteria (including opportunistic pathogens), which were predominant among the microbes isolated from these samples, aerobic spore-forming microorganisms (*Bacillus*) were found frequently on spacecraft and within the facilities. The predominance of human-associated microorganisms and sporeformers has been confirmed in subsequent publications (e.g., Moissl et al. 2007; Stieglmeier et al. 2009), whereas the portion of *Bacillus* and, e.g., *Micrococcus* was reported to be substantial. In general, 85 % of all isolated microorganisms of the NASA JPL group until 2005 were identified as Gram-positive bacteria

(Newcombe et al. 2005). All of these cultivated microorganisms were obtained from isolation attempts on heterotrophic full media. However, chances of survival of space flights are higher for organisms that can thrive under more restrictive and extreme conditions.

In preparation of missions that intend to land on Mars (such as ESA's ExoMars) or other celestial bodies but also for future lander and Mars Sample Return (MSR) missions, the knowledge about the biological contamination in spacecraft assembly, integration, testing, and launch facilities is an important prerequisite.

In a recent, separate study, two European and one South American spacecraft assembly cleanrooms were analyzed concerning their microbial diversity, using standard procedures, new cultivation approaches, and molecular methods, with the aim to shed light onto the presence of planetary protection-relevant microorganisms within their facilities. This study served as a preparation study for the current ExoMars mission and included the microbial analysis of the Herschel Space Observatory (launched in May 2009) and its housing cleanrooms during ATLO activities at three different locations. Although Herschel did not demand planetary protection requirements, all cleanrooms were under full operation when sampled. The following table gives details about the sampling locations and their characteristics (Table 10.2).

The cultivation procedures and media are summarized in Table 10.3 (see also: Stieglmeier et al. 2009; Moissl-Eichinger et al. 2012).

Table 10.2 Sampling locations and specifics

	Friedrichshafen 1 (FR1)	Friedrichshafen 2 (FR2)	ESTEC (ES)	Kourou (KO)
Location	Friedrichshafen, EADS (European Aeronautic Defense and Space Company)	Friedrichshafen, EADS (European Aeronautic Defense and Space Company)	ESTEC (European Space Research and Technology Centre), Noordwijk	Centre Spatial Guyanais (CSG), Kourou
Sampling date	Apr-07	Nov-07	Mar-08	Apr-09
Cleanroom facility	Hall 6, room 6101–04	Hall 6, room 6101–04	Hydra	BAF
Cleanroom specifics	ISO 5[a]	ISO 5	ISO 8	ISO 8
Sampled surfaces	Various cleanroom surfaces, e.g., floor, stairs, door knobs; spacecraft	Various cleanroom surfaces, e.g., floor, stairs; spacecraft	Various cleanroom surfaces, mainly floor; spacecraft	Various cleanroom surfaces, mainly floor; floor of Ariane5 container

Further details are given in Stieglmeier et al. (2009); Moissl-Eichinger (2011)

[a]Cleanroom was nominally operated at ISO 5 but opened to ISO 8 section just before sampling

Table 10.3 Assortment of media used for the enrichment of cleanroom microbes.

Target microbes	Basic medium	Supplemented with/modification/conditions	Gas phase	"Extreme conditions"
Oligotrophs	R2A	Diluted 1:10, 1:100	Ae	1:100
Alkaliphiles	R2A	pH 9, 11	Ae	pH 11
Acidophiles	R2A	pH 5, 3	Ae	pH 3
Autotrophs	*MM* (methanogenic archaea medium)[a]		H_2/CO_2	Carbon present as CO_2 only
	AHM (autotrophic homoacetogen medium)[a]		H_2/CO_2	
	ASR (autotrophic sulfate- reducer medium)[a]		H_2/CO_2	
	AAM (autotrophic all-rounder medium)[a]		N_2/CO_2	
Nitrogen fix	*N2 fix* (Hino and Wilson N2-free medium)[a]		N_2, N_2/O_2	N_2 as nitrogen source
Anaerobes	TGA (thioglycolate agar)[a]		N_2	Anaerobic medium
	TSA (trypticase soy agar)[a]			
	SRA (sulfate-reducer agar)[a]			
	TS (trypticase soy medium)[a]			
	TG (thioglycolate medium)[a]			
Thermophiles	R2A	Incubated at 50/60 °C	Ae	10 °C
Psychrophiles	R2A	Incubated at 10/4 °C	Ae	4 °C
Halophiles	R2A	Addition of 3.5 % and 10 % (w/v) NaCl	Ae	10 % NaCl
Additional media:				
Heterotrophs	R2A		ma	
Archaea	*MM* (methanogenic Archaea medium)[a]	Sodium acetate, methanol	N_2/CO_2	
	ASM (Archaea supporting medium)[a]	Antibiotics mixture; NH_4Cl or yeast extract	N_2, ma, ae	

Liquid media are given in italics. Abbreviations: *Ae* aerobic, *ma* microaerophilic (<3 % O_2). The last column indicates the extreme conditions applied that are discussed in this article. If not given otherwise, the incubation temperature was 32 °C

[a]Medium recipe and preparation given in Stieglmeier et al. (2009)

It shall be mentioned that most of the isolates obtained from planetary protection-relevant cleanrooms and spacecraft are collected in public culture collections. One, the ESA collection, is maintained by the German Collection of Microorganisms and Cell Cultures DSMZ (Moissl-Eichinger et al. 2012); the other one, for isolates obtained during the NASA Phoenix mission specifically, is maintained at the US Department of Agriculture's Agricultural Research Service Culture Collection in Peoria, Illinois (Venkateswaran et al. 2014).

In the following chapters, results from cultivation attempts performed during the Herschel ATLO activities, focusing on the extremophilic microbial community in spacecraft assembly cleanrooms, will be displayed comprehensively and compared to obtained results from US studies.

10.2 Extremophiles and Extremotolerants: Definition

Generally, extremophiles are microorganisms that require extreme conditions for growth. For instance, psychrophiles are adapted to low-temperature environments and require temperatures lower than 15 °C for optimal proliferation. A cleanroom itself is an extreme environment but hosts mainly extremotolerant microorganisms, accepting the extreme circumstances but preferring moderate conditions for growth. In the following, the author will (for simplifying the terminology) not distinguish between real extremophiles and extremotolerants: For instance, the term alkaliphiles concerning cleanroom isolates includes also microorganisms that tolerate but don't require alkaline conditions.

10.3 Spore-Forming Microorganisms

10.3.1 Background

Spores are resting states of certain bacteria and are usually formed when the organism recognizes a lack in nutrients, such as the carbon or nitrogen source. Endospores of *Bacillus subtilis* are highly resistant to inactivation by environmental stresses, like biocidal agents, toxic chemicals, desiccation, pressure, temperature extremes, higher doses of UV, and ionizing radiation (Nicholson et al. 2000, 2005). Spores possess thick layers of coating proteins, and even their DNA is protected by small proteins (SASPs, Moeller et al. 2008). The gel-like core of a desiccated spore contains only 10–25 % of the water available in a vegetative cell. Enzymes and therefore the metabolism of a spore are more or less inactive. Spores can survive hundreds or maybe even million years, when kept dry and protected against mechanical forces and lethal doses of radiation (Cano and Borucki 1995).

The germination generally needs an activator, e.g., moderate heat. In culture, amino acids like alanine seem to support the germination process. In total, almost

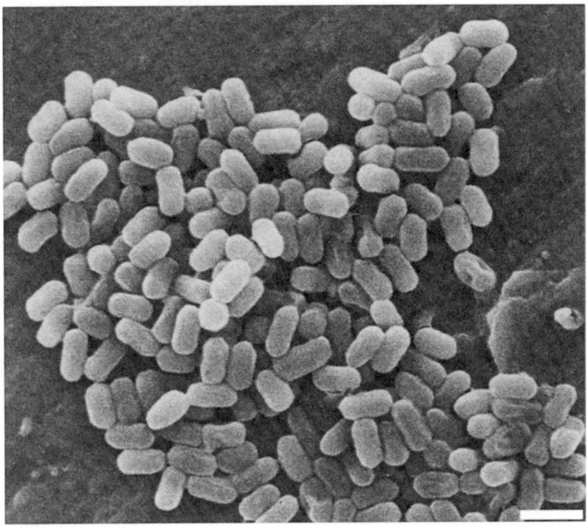

Fig. 10.1 *Bacillus* spores on a spacecraft relevant surface (Probst et al. 2010a), scanning electron micrograph. Bar: 2 μm

20 bacterial genera are able to form spores, but only *Bacillus* and *Clostridium* spores have been subjected to deeper characterization studies (Fig. 10.1).

The multiple resistance properties of such spore-forming microorganisms make them ideal candidates for the survival of a space flight. Additionally, commonly applied sterilization conditions of dry heat or chemical disinfectants that do not harm the spacecraft and its hardware are not able to kill most bacterial spores (Crawford 2005). Since the microbial analysis of the Viking mission has proven the presence of a broad diversity of spore-forming microorganisms on spacecraft surfaces, the main focus of attention has been on them for the past decades.

Although 99.9 % of all *B. subtilis* spores were killed when exposed to a few minutes of Mars simulated surface conditions (in terms of UV irradiation, pressure, gas composition, and temperature), it has also been shown that dried spores were resistant to UV inactivation when mixed with Mars surrogate soil (Schuerger et al. 2003; Crawford et al. 2003; Osman et al. 2008). They were even resistant to sterilizing UV, as long as protected by a shallow layer of sand (Crawford et al. 2003). It therefore can be assumed that highly resistant spores, delivered to Mars, could survive the travel to and the stay on Mars without further damage when located on lander parts not fully exposed to radiation or covered by a thin layer of dust (Osman et al. 2008).

For the selective enrichment of spore-forming microbes, a heat shock at 80 °C for 15 min is one important step within the bioburden-level determination procedure. Besides the effect that most vegetative cells are killed at 80 °C, this heat step is also helpful in stimulation of *Bacillus* spores to germinate. Newcombe et al. (2005) reported that members of the genus *Bacillus* were the predominant microbes among the heat shock survivors; nevertheless the isolation of heat shock-resistant *Staphylococcus*, *Planococcus*, and *Micrococcus* also has been reported (Venkateswaran et al. 2001; Moissl-Eichinger et al. 2013).

For the sake of completeness, it shall also be mentioned that some vegetative microbial cells can resist very harsh conditions such as extreme doses of (UV and ionizing) radiation and desiccation (e.g., *Deinococcus radiodurans*, *Halobacterium* sp. NRC-1; Cox and Battista 2005; DeVeaux et al. 2007). Nevertheless, the information about vegetative, resistant bacteria from spacecraft assembly cleanrooms is very limited.

For sure, the space agency's standard procedures are not able to cover the broadest diversity of tolerant microbes but give a number and a proxy estimation to work with.

10.3.2 Results

Bacillus is a typical spore-forming contaminant in spacecraft assembly cleanrooms. Already Puleo et al. (1977) reported the detection of more than 14 different *Bacillus* strains on the Viking spacecraft.

Newer studies from spacecraft assembly cleanrooms confirm the presence or even predominance of spore-forming bacteria in cultivation assays based on full, heterotrophic media. Six different *Bacillus* strains have been detected on the Mars Odyssey spacecraft (La Duc et al. 2003), some of them revealing resistances against 0.5 Mrad γ-radiation, 5 % H_2O_2 (60 min), or higher doses of UV. In another study, further spore-forming microorganisms, like *Sporosarcina*, *Paenibacillus*, *Actinomycetes*, and *Aureobasidium*, have been detected (La Duc et al. 2004). In newer studies, in particular of the microbial analysis of the Phoenix spacecraft, the presence of (extremophilic) microbial sporeformers has been confirmed (Ghosh et al. 2010), and some new species have been described, such as *Bacillus horneckiae* (Vaishampayan et al. 2010). Another spore-forming bacterium from a genus besides *Bacillus*, *Paenibacillus purispatii*, was isolated from a spacecraft assembly cleanroom at ESA ESTEC (European Space Research and Technology Centre) (Behrendt et al. 2010).

In the here presented study of three cleanrooms, 32 different culture media were used to target a wide range of different microorganisms (see Table 10.3). With this approach, the presence of a broad variety of spore-forming microorganisms in spacecraft assembly cleanrooms was obtained. *Bacillus* and *Paenibacillus* were found in every facility. Overall 13 different *Bacillus* strains, 11 different paenibacilli, *Brevibacillus*, *Clostridium*, *Desulfotomaculum*, *Geobacillus*, *Micromonospora*, *Sporosarcina*, and two *Streptomyces* species were isolated. In general, spore-forming microorganisms accounted for about 25 % of all microbes, but this portion was highly depending on the cultivation strategy (Moissl-Eichinger et al. 2013). The lowest percentage of sporeformers was found during the second Friedrichshafen sampling. During that, the cleanroom was operated at ISO 5, resulting in a higher percentage of human-associated microorganisms and a lower percentage of sporeformers. Most of the spore-forming bacteria observed are associated with environmental biotopes (like soil) and therefore most likely introduced on items moved into the cleanrooms or attached to humans and clothes. It can be concluded that the higher the operational cleanliness of a facility, the less spore-forming microorganisms could be expected.

B. pumilus SAFR-032, an isolate originally obtained from a class 100 K (ISO 8) cleanroom at the Jet Propulsion Laboratory Spacecraft Assembly Facility (JPL-SAF), was described to form spores with extraordinary UV resistance outcompeting even a standard dosimetry strain of *B. subtilis* (Newcombe et al. 2005). A larger percentage of SAFR-032 spores was found to be even able to survive exposure to dark space conditions (EuTEF, European Technology Exposure Facility) outside of the international space station for 18 months (Vaishampayan et al. 2012).

Different strains of *B. pumilus* have very frequently been isolated from US spacecraft assembly cleanrooms, and many of them were described to possess amazing resistances against H_2O_2 (Kempf et al. 2005) or UV (Newcombe et al. 2005; Link et al. 2004). Generally, these cleanroom isolates revealed a higher resistance to UV irradiation than the type strain of *B. pumilus* (Newcombe et al. 2005).

Strains of *B. pumilus* have been isolated also from the first Friedrichshafen and the Kourou sampling, but these strains were underrepresented (0.6 and 1.4 %, respectively) among all isolates obtained. The most frequent *Bacillus* strains obtained were *B. megaterium* or bacilli affiliated to the *B. thuringiensis/cereus* group. As reported by Newcombe et al. (2005), from 125 aerobic strains isolated from US spacecraft assembly facilities, 65 % were resistant against the heat shock implemented by the NASA standard protocol. Among 15 different *Bacillus* identified, *B. licheniformis* (25 %) and *B. pumilus* (15 %) were the most prevalent species.

10.4 Oligotrophic Microorganisms

10.4.1 Background

Oligotrophs (or oligophilic microorganisms) are microbes that are adapted to low nutrient conditions. Standard laboratory media are usually fully heterotrophic media providing a broad variety of carbon and other nutrient sources. Nevertheless, most of the microbes thriving in natural biotopes have to deal with nutrient restriction and competition with other organisms. Similarly, cleanrooms are characterized by a significant lack of nutrients. Frequent cleaning and air filtering procedures remove particles that could provide nutritive substances, so that the microorganisms present either have to retreat into a resting state (like spores) or have to adapt their metabolism to the extreme circumstances. To date, NASA's standard procedures recommend the usage of TSA medium (Trypticase soy broth) for the cultivation of microbial contaminants in spacecraft assembly facilities. Nevertheless, a pharmaceutical cleanroom study revealed that the portion of cultivables from the cleanroom production unit could be increased by two orders of magnitude when a low nutrient medium was applied instead of a full medium (Nagarkar et al. 2001). Additionally, if looking for possible hitchhikers to Mars, the search for microbes adapted to low nutrient conditions is even more reasonable: So far, no complex organic molecules have been detected on the Martian surface or in its atmosphere.

10.4.2 Results

Until now, data on oligophilic microorganisms from spacecraft assembly cleanrooms are rather sparse, and the mentioned study from a pharmaceutical cleanroom has not delivered data about the microbial strains detected (Nagarkar et al. 2001).

Until now, no data have been published with respect to oligotrophic microorganisms from spacecraft assembly cleanrooms, and the aforementioned study from a pharmaceutical cleanroom has not delivered data about the microbial strains which were detected (Nagarkar et al. 2001).

In our recent and ongoing studies, R2A medium was used for various cultivation attempts. This medium was originally developed to study microorganisms inhabiting potable water (Reasoner and Geldreich 1985); it is a low nutrient medium that could stimulate the growth of stressed and slow growing microbes. For the detection of oligotrophs in this here discussed study, R2A medium was applied even in a 1:10 and 1:100 dilution, respectively. Interestingly, a broad variety of bacteria was cultured on R2A 1:100, including *Acinetobacter*, *Balneimonas*, *Brevundimonas*, *Citrobacter*, *Kocuria*, *Microbacterium*, *Micrococcus*, *Moraxella*, *Paenibacillus*, *Sanguibacter*, *Staphylococcus*, *Stenotrophomonas*, and *Streptomyces*, whereas *Paenibacillus* and *Streptomyces* are capable of forming spores.

Interestingly, isolates from the first Friedrichshafen sampling that were grown on 1:10 and 1:100 R2A outperformed the number of cultivables on other (nutrient-rich) media (2.8×10^4 and 4.0×10^4 oligotrophic cultivables per m^2 cleanroom surface). Comparably, the samples from the other cleanrooms revealed a high number of oligotrophs present. Since our approach was the first in this field searching for oligotrophs, further studies in other cleanrooms will be necessary and are highly recommended.

Based on own observations and results from other studies, the new ESA standard for biocontamination measurement relies on R2A instead of TSA for the cultivation of spacecraft assembly-related microorganisms.

10.5 Alkaliphiles and Acidophiles

10.5.1 Background

Although the pH of Mars' regolith was estimated to be rather neutral (7.2± 0.1, Plumb et al. 1993), the presence of alkaliphiles and acidophiles in spacecraft assembly cleanrooms can deliver valuable information for planetary protection considerations and in particular for cleanroom maintenance: Most of the disinfectants and detergents used for (bio)cleaning in such cleanrooms are either pH neutral or alkaline. It is unclear if the consequent treatment with, e.g., alkaline detergents could result in a positive selection effect. A preference of alkaline or acidic media by the microbial diversity in cleanrooms was analyzed in two independent studies.

10.5.2 Results

Samples from diverse spacecraft assembly cleanrooms were plated on R2A with pH 9, 10.6, and 3 (La Duc et al. 2007), pH 11 and 3 (Ghosh et al. 2010), or pH 9, 11, 5, and 3 (this project; Moissl-Eichinger et al. 2013). A broad variety of bacteria tolerating pH 10.6 were reported to be present in US cleanrooms (e.g., *Bacillus*, *Staphylococcus*, *Sphingomonas*, and many others; La Duc et al. 2007), and also our studies observed various alkaliphiles (pH 9, pH 11; Table 10.4). Interestingly, *Bacillus* and *Brevundimonas* were detected in all studies, accepting a medium pH of 10.6 and 11, respectively. Alkaliphiles were observed in every facility looked at, with numbers ranging from 1.6×10^2 (ISO 6, JSC-GCL) to 2.0×10^6 per m^2 (ISO 8, LMA-MTF). During the Herschel campaign, 4.6×10^2 (second sampling Friedrichshafen, ISO 5) up to 8.4×10^3 (ESTEC, ISO 8) were measured, whereas the number of alkaliphiles in Kourou samples could not be determined due to overgrowth of the agar plates. Similar levels were obtained from the Phoenix assembly cleanroom (up to 68×10^2 CFU per m^2 (Ghosh et al. 2010).

Nevertheless, the detection of acidophiles was much more difficult. The colony counts on agar plates with pH 5 were very low, no isolate was obtained during the Herschel campaign tolerating pH 3, whereas an acid-tolerant colony count was reported for US LMA-MTF and KSC-PHSF facility only (La Duc et al. 2007). All other samplings were negative for acidophiles. Nevertheless, none of the isolates obtained during the US study was analyzed further or revealed multiresistant properties (La Duc et al. 2007).

A substantial preference of alkaline media was found in all facilities analyzed so far. One isolate, *B. canaveralius*, was obtained recently and was found to be salt and alkali tolerant (Newcombe et al. 2009). To date, the reasons for the shifts toward alkaliphily are unclear, but a positive selection by the usage of (alkaline) cleaning detergents seems probable and could result in an outperformance of acidophiles or non-alkali tolerants. If so, the selectivity via the pH of cleaning agents could be circumvented by using detergents with alternating pH.

10.6 Autotrophic and Nitrogen-Fixing Microorganisms

10.6.1 Background

The capability to fix nitrogen from the gaseous atmosphere or to grow autotrophically on CO_2 is an important property of primary producers. The activities of these microbes are the prerequisites for other microorganisms to colonize new nutrient-poor environments (Thomas et al. 2006). Since cleanrooms and also the Martian environment are depleted in organic materials, primary producers could pave the way for secondary settlers.

Nevertheless, most studies have not reported the attempt to grow primary producers. The majority of studies looking for cultivables from spacecraft and associated cleanrooms have used heterotrophic, solid full media. The assortment of special media used in our studies also included media selectively for chemolithoautotrophs with CO_2 as the only carbon source. Media providing nitrogen only in the gas phase were also applied.

10.6.2 Results

Seven isolated bacterial genera were able to fix CO_2; eight were able to fix N_2. Further details about the isolated bacterial genera are given in Table 10.4.

An interesting observation was that solely species of the bacterial genus *Paenibacillus*, *Micrococcus*, and *Sanguibacter* were able to perform both reactions, whereas *Sanguibacter marinus* was the only species that was isolated on an autotrophic and N_2 fixer medium in parallel (Kourou sampling). The type strain of this species was originally isolated from coastal sediment (Fujian province of China), and none of the observed properties has been reported in the original strain description (Huang et al. 2005). It can be imagined that the cleanroom isolate shows distinct properties due to the adaptation to the extreme biotope or the type strain has not been tested concerning these metabolic capabilities. *Paenibacillus* and *Streptomyces* were the only spore-forming microorganisms that were able to fix N_2 (both) or CO_2 (*Paenibacillus*).

10.7 Anaerobes

10.7.1 Background

The atmospheres of most planets within the reach of space missions contain only traces of oxygen, most likely not enough to support aerobic life as we know it from terrestrial biotopes (Thomas et al. 2006). Since the Martian surface is exposed to radiation and the soil is very oxidizing, the Martian subsurface could be an anaerobic biotope for possible life (Boston et al. 1992; Schulze-Makuch and Grinspoon 2005). On Earth, (facultative) anaerobes are widespread in different environments and can be detected in, e.g., oxic soil, aerobic desert soils, or other biotopes such as the human body (Küsel et al. 1999; Peters and Conrad 1995; Tally et al. 1975). The latter makes them potential contaminants of spacecraft assembly facilities through staff, having close contact to flight hardware.

Generally, there are different types of anaerobic organisms. Facultative anaerobes always prefer aerobic conditions but are able to grow under conditions with or without oxygen; aerotolerant anaerobes do not require oxygen for their growth and show no preference. Strict anaerobes (e.g., methanogens) never require oxygen for

Table 10.4 Extremophilic isolates from global spacecraft assembly cleanrooms

	Oligotrohs [1]	Psychrophiles [2]	Alkaliphiles [3]	Anaerobes	Thermophiles [4]	Halophiles [5]	CO2 fix	N2 fix	Species (location/extremophily)
Acinetobacter	o	oo	o						*A. sp.*, *A. ursingii* (FR2), *A. johnsonii* (KSC/alk,psy)
Actinotalea			o						*A. fermentans* (KO)
Aerococcus				o		o			*A. urinaeequi* (KO)
Agrococcus			o						*A. jenensis* (KSC)
Arsenicicoccus				o			♦		*A. bolidensis* (FR2)
Arthrobacter		o	o	o				♦	*A. sp.* (KSC/alk, KO/an, N_2), *A. polychromogenes* (KSC/psy)
Bacillus		o	oo oo oo oo oo	oo oo	oo o	o			*B. thermoamylovorans* (ES/alk, FR1/an), *B. gibsonii* (FR2/alk), *B. licheniformis* (FR1/alk, ES/an,therm), *B. pumilus* (KO/an, JSC/alk, KSC/alk), *B.* tc (FR2, ES,KO/an), *B. badius* (KO/therm), *B. coagulans* (ES,KSC/therm), *B. megaterium* (KO/halo), *B. sp.* (JSC/alk; KSC/alk,psy), *B. oshimensis* (KSC/alk), *B. pseudoalcaliphilus* (KSC/alk), *B. thuringiensis* (KSC/alk)
Balneimonas	o		o						*B. sp.* (KO)
Blastococcus		o							*B. sp.* (KSC)
Brachybacterium			o						*B. paraconglomeratum* (LMA)
Brevibacillus		oo	o		o				*B. agri* (FR1), *B. invocatus* (KSC/psy)
Brevibacterium			o						*B. frigoritolerans* (ES)
Brevundimonas	o		oo						*B. nasdae* (FR2, KO), *B. diminuta* (KSC/alk)
Cellulomonas				oo			♦		*C. hominis* (FR1, KO/an, KO/aut)
Cellulosimicrobium			o						*C. funkei* (KSC)
Citrobacter	o								*C. werkmanii* (KO)
Clostridium				o					*C. perfringens* (ES)
Corynebacterium				o					*C. pseudogenitalium* (ES)
Cupriavidus				o					*C. gilardii* (KO)
Dermabacter				o					*D. hominis* (FR2)
Desemzia						o			*D. incerta* (KO)
Desulfotomaculum				o					*D. guttoideum* (KO)
Dietzia		o	o						*D. maris* (KSC)
Enterococcus				oo			♦		*E. casseliflavus* (KO/aut), *E. faecalis* (ES/an), *E. faecium* (ES/an)
Facklamia			o	o					*F. sp.* (KO)
Geobacillus					oo oo				*G. caldoxylosilyticus* (ES, KSC), *G. stearothermophilus* (JPL), *G. kaustophilus* (JPL), *G. thermodenitrificans* (JSC)
Georgenia		o		o					*G. muralis* (KO), *G. sp.* (KSC)
Herbaspirillum		o							*H. sp.* (KSC)
Janibacter			o						*J. terrae* (KSC)
Kocuria	o		o						*K. rhizophila* (ES), *K. rosea* (KSC)
Labedella		o							*L. kawkjii* (KSC)
Lysobacter			o						*L. sp.* (KO)
Massilia		o							*M. brevitalaea* (KO)
Microbacterium	o		oo o					♦	*M. oleivorans* (KO/oligo), *M. paraoxydans* (KO/oligo, N_2), *M. schleiferi* (LMA), *M. aurum* (JSC), *M. arborescens* (KSC), *M. testaceum* (KSC)
Micrococcus	o		oo oo			o	♦		*M.* sp. (KO/oligo, alk), *M. flavus* (KO/alk), *M. indicus* (ES/aut) *M. luteus* (KO,FR2,FR1/alk; KO/halo), *M. mucilaginosus* (KSC/alk)
Moraxella	o		o	o					*M. osloensis* (FR2/oligo; FR1/alk,an)
Nocardioides			o						*N. oleivorans* (KSC)
Oceanobacillus			oo o						*O.* sp. (JPL/alk), *O. iheyensis* (KSC/alk), *O. profundus* (KSC/alk)
Paenibacillus	o		oo o	oo o	o		♦ ♦	♦	*P. pasadenensis* (FR1/oligo,alk, N_2; ES/alk), *P. telluris* (ES/alk), *P.* sp. (KO/alk), *P. amylolyticus* (FR1/alk), *P. glucanolyticus* (FR1/alk), *P.* sp. (ES/an), *P. ginsengisoli* (ES/an,auto), *P. barengoltzii* (FR1/an), *P. cookii* (FR1/therm), *P. wynii* (LMA/an)
Paracoccus								♦	*P. yeeii* (ES)
Planctibacter		o	o						*P. flavus* (KSC)
Propionibacterium				oo o					*P. acnes* (FR1,FR2, ES), *P. avidum* (ES)
Pseudomonas		oo	oo					♦	*P. luteola* (FR2/ N_2), *P. xanthomarina* (KO/alk,psy), *P. stutzeri* (KSC/alk), *P. oryzihabitans* (KSC/psy)
Rhodococcus		o	o						*R. fascians* (KSC/psy), *R. globerulus* (KSC/alk), *R. kroppenstedtii* (KSC/alk), *R. sp* (KSC/psy)
Roseomonas		o	o						*R. aquatica* (KO)
Sanguibacter	o	o		o			♦	♦	*S. marinus* (KO), *S. sp.* (KSC)
Sphingomonas		o	oo						*S. oligophenolica* (JSC), *S. trueperi* (JSC), *S. dokdonensis* (KSC), *S. yunnanensis* (KSC), *S. sp* (KSC/psy)
Spirosoma		o							*S. aquatica* (KSC)
Staphylococcus	o o		oo oo	oo oo		oo oo	♦ ♦		*S. haemolyticus* (ES, FR1/oligo, FR1, FR2, ES/alk, ES,KO/an, FR1, FR2, ES/halo,auto), *S. warneri* (KO/an), *S. pasteuri* (FR2/an, KO,ES/halo), *S. lugdunensis* (FR2/an), *S. hominis* (FR1/halo), *S. epidermidis* (JSC/an)

Stenotrophomonas	○		○○	○				◆	*S. maltophilia* (FR2), *S. rhizophila* (KSC/alk)
Streptomyces	○		○					◆	*S. luteogriseus* (ES)
Tessaracoccus				○					*T. flavescens* (KO)
Variovorax		○							*V. paradoxux* (KSC)

Circles stand for each location, where detected. Black: isolates from this study. Abbreviations of locations according to Table 10.2. Red: for comparison, isolates from US studies are listed, too (La Duc et al. 2007; Ghosh et al. 2010); locations: KSC (Kennedy Space Center), JSC (Johnson Space Center), LMA (Lockheed Martin Aeronautics), JPL (Jet Propulsion Laboratory). Further abbreviations: an (anaerobic), alk (alkaliphilic), oligo (oligotrophic), therm (thermophilic), halo (halophilic), aut (autotrophic), N2 (N2 fixing), psy (psychrophilic), B. tc (*Bacillus thuringiensis/cereus* group)

[a]Oligotrophs were grown on R2A 1:100.

[b]Psychrophiles were grown on R2A at 4 °C

[c]Alkaliphiles were grown on R2A with pH 10.6 and 11, respectively

[d]Thermophiles were grown at 50, 60 and 65 °C

[e]Halophiles were grown on R2A containing 10 % NaCl

their reproduction and metabolism and can even be growth inhibited or killed by oxygen.

Until now, not much is known about the presence of anaerobically growing microorganisms in spacecraft cleanrooms. The presence of anaerobic microorganisms (enriched using the BD GasPaK system) in surface samples from US cleanrooms has rarely been reported (Ghosh et al. 2010; La Duc et al. 2007). Overall 25 facultatively anaerobic strains were retrieved from the Phoenix housing facility, whereas members of the facultatively anaerobic genera *Paenibacillus* and *Staphylococcus* have been isolated in the course of a study about extremotolerant microorganisms (La Duc et al. 2007).

A proper anaerobic cultivation necessitates the application of the Hungate technique (Hungate 1969). Although this method has undergone a few simplifications during past decades, it still requires specialized equipment and practical experience. During our research, samples from the Herschel campaign were – for the first time – subjected to growth experiments performed with the Hungate technology, and a broad variety of microbes capable of anaerobic growth were isolated (Stieglmeier et al. 2009).

10.7.2 Results

A variety of anaerobic microorganisms was successfully isolated from all four cleanroom samplings. Overall, 30 strains were isolated on anaerobic media. The greatest number and diversity of bacteria were obtained from the Kourou sampling (13 species). The following chart shows the oxygen requirements of isolates obtained from our campaign (see also Table 10.4 and Fig. 10.2).

In most cases, anaerobically enriched species were identified to be facultative anaerobes, comprising 16–78 % of the total counts (numbers were calculated based on own observations and literature review for microbes grown on aerobic plates only; for comparison see Stieglmeier et al. 2009). Colony counts obtained on anaerobic full media showed the presence of up to 5.8×10^2 anaerobes per m^2 cleanroom surface (Kourou sampling).

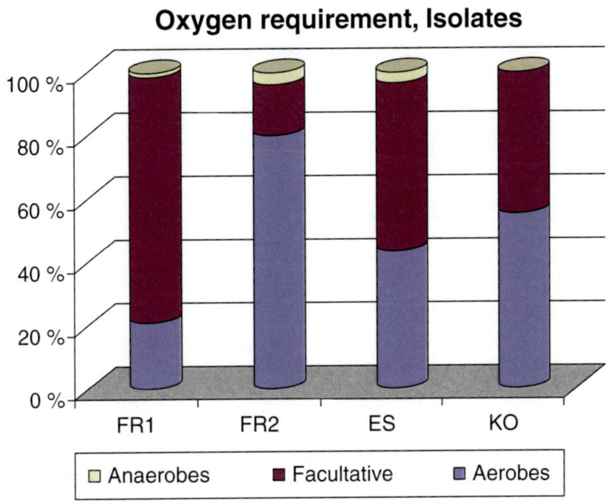

Fig. 10.2 Oxygen requirements of isolates. Physiological capabilities are either based on own experiments or published data (for isolates grown on aerobic media only)

Only a comparatively low percentage of microbes grew strictly anaerobic (*Propionibacterium*, *Corynebacterium*, *Desulfotomaculum*, *Clostridium*; 0.7–4 % of all isolates at each location). *Propionibacterium acnes*, typically found on human skin, was isolated from each European cleanroom and therefore the most prominent strict anaerobe detected. All strict anaerobes except *Desulfotomaculum* were opportunistic pathogens and isolated from full, heterotrophic media. Interestingly, the *Corynebacterium* isolate (*C. pseudogenitalium*) could not be grown under aerobic conditions, although its type strain was described to be facultatively anaerobic (Stieglmeier et al. 2009).

Desulfotomaculum guttoideum was the only strict anaerobe isolated from the Kourou sampling. It was grown on sulfate-reducer specific medium, but without producing black colonies that would indicate a sulfate-reducing activity. As it was clarified in a previous publication, *Desulfotomaculum guttoideum* was misclassified and is actually affiliated to *Clostridium* cluster XIVa (Stackebrandt et al. 1997). This strain is therefore closely related to *Clostridium sphenoides*, a fermentative, saccharolytic, sulfite and thiosulfate (but not sulfate) reducing sporeformer. Other spore-forming microorganisms capable of anaerobic growth were *Bacillus*, *Paenibacillus*, and *Clostridium (Desulfotomaculum)*. Further details are given in Table 10.4.

These data were confirmed by an analysis of the anaerobic microbial diversity in NASA's cleanrooms at the Jet Propulsion Laboratory. This study was based on a microbial enrichment of cleanroom samples under anaerobic conditions, which was then analyzed via cultivation, 16S rRNA gene sequence analysis, and microarray (Probst et al. 2010b). *Clostridium* and *Propionibacterium* were the only strictly anaerobic microbes isolated, whereas additionally *Oerskovia*, *Dermabacter*, *Bacillus*, *Granulicatella*, *Sarcina*, *Leuconostoc*, *Paenibacillus*, *Staphylococcus*, and *Streptococcus* were detected during the molecular approach (Probst et al. 2010b).

Our results indicate that the facultatively and strictly anaerobic microbial community is quite diverse and maybe even dominant in spacecraft assembly cleanrooms.

10.8 Thermophiles and Psychrophiles

10.8.1 Background

The Martian surface is very cold. Although the temperatures can reach up to 20 °C in certain areas in the summer, the average temperatures lie around and way below 0 °C. Actually, also Earth's biosphere is quite cold – more than 70 % of its water occurs as ice, and the world's oceans reveal temperatures below 5 °C (National Research Council 2006). A typical terrestrial biotope used for comparative studies is the permafrost environment exhibiting a lively, highly diverse microbial community. It is assumed that Earth's psychrophiles could survive under certain circumstances in the Martian environment but would grow very slowly (National Research Council 2006).

Although (hyper)thermophiles would probably not be able to proliferate on Mars, this group of microorganisms is often employed in studies concerning the origin and the evolution of life. Hot conditions prevailed on early Earth, and many thermophiles have "basic" metabolic capabilities (for instance, chemolithoautotrophy). Thermophiles are also generally considered more resistant toward environmental stresses than moderate or cold-loving microorganisms.

For these reasons, experiments were carried out searching for microorganisms in cleanroom environments that could grow under significantly higher or lower temperatures than the standard incubation temperature for all other experiments (32 °C).

10.8.2 Results

The study of US spacecraft assembly cleanrooms reported no growth of microorganisms on R2A medium when incubated at 4 °C for 10 days (La Duc et al. 2007). Differently, psychrophiles were reported to be present in the Phoenix assembly area, where 30 strains were identified that were capable of growth at 4 °C (Ghosh et al. 2010). These strains included *Brevibacillus*, *Bacillus*, *Acinetobacter*, *Pseudomonas*, *Sphingomonas*, *Variovorax*, *Herbaspirillum*, *Arthrobacter*, *Curtobacterium*, *Dietzia*, *Labedella*, *Plantibacter*, *Rhodococcus*, *Blastococcus*, *Sanguibacter*, *Rhodococcus*, and *Spirosoma* species.

Our study was selective for microorganisms capable of growing at 10 and 4 °C. The incubation duration was prolonged up to 3 months. *Acinetobacter*, *Massilia*, *Pseudomonas*, and *Roseomonas* were isolated at 4 °C; additionally, *Bacillus*, *Brevundimonas*, *Micrococcus*, *Moraxella*, *Paenibacillus*, *Sanguibacter*, *Sphingomonas*, *Sporosarcina*, *Staphylococcus*, and *Stenotrophomonas* were observed at 10 °C. *Sporosarcina globispora*, a spore-forming bacterium, even was the only isolate that was not able to grow at 32 °C, the standard cultivation temperature.

Nevertheless, most of the "psychrophilic" isolates also had been obtained by using other cultivation methods at higher temperatures (32 °C), too. It can be assumed that many (most?) of the present microorganisms in cleanrooms are capable of growing at lower (very low) temperatures but that cell proliferation takes significantly longer than under higher, optimal temperature conditions.

The selective enrichment of thermophiles on R2A at 65 °C (La Duc et al. 2007; Ghosh et al. 2010) or 60 and 50 °C (Herschel study) allowed the isolation of three *Geobacillus* and one *Bacillus* strain (La Duc et al. 2007) and *Bacillus*, *Brevibacillus*, *Paenibacillus*, and *Geobacillus*, respectively. Also Ghosh et al. (2010) reported the presence of thermophilic microorganisms before spacecraft had been transferred into the cleanroom.

Interestingly, *Bacillus coagulans* and *Geobacillus caldoxylosilyticus* were found in ESA- and NASA-related cleanrooms, whereas the latter was described as an obligate thermophile ("Saccharococcus caldoxylosilyticus," Ahmad et al. 2000), which is congruent to the observations made in this study (no growth was observed at 32 °C). It is unclear how *Geobacillus* (spores) entered the cleanroom (although not capable to proliferate under the thermal conditions of a cleanroom) and why this organism was detected in two independent studies in cleanrooms on different continents. In general, *Geobacillus* spores have been described to be very resistant to environmental (thermal) stress (Head et al. 2008). Because of this high resistance, it possibly survived strict cleanliness control conditions, after being carried into the facilities by human and item traffic.

10.9 Halophiles

10.9.1 Background

Halophiles have been discussed as possible survivors on Mars, since the Martian liquid water is suspected to contain high concentrations of different salts (Landis 2001). Additionally, a high resistance of salt-crystal-associated halophiles against UV radiation has been reported, making them potential survivors on the Martian surface (Fendrihan et al. 2009). Prevention of potential contamination of the Martian surface with halophiles is therefore highly important. Nevertheless, hardly any studies have been carried out thus far to investigate the potential presence of halotolerants and halophiles in spacecraft assembly cleanrooms. In order to obtain insights into the distribution of these organisms in cleanrooms, samples from two studies were plated on R2A containing different NaCl concentrations.

10.9.2 Results

Samples from US cleanrooms were plated onto R2A containing 25 % (w/v) NaCl, but no growth was observed (La Duc et al. 2007; Ghosh et al. 2010). In contrast, samples obtained from the Herschel study were plated on R2A containing 3.5 and

10 % (w/v) NaCl. R2A containing 3.5 % NaCl revealed that most of the organisms isolated via other cultivation attempts were also capable of tolerating this comparatively low concentration of NaCl. The plates containing 10 % NaCl revealed much lower cell counts: *Aerococcus*, *Bacillus*, *Desemzia*, *Micrococcus*, and *Staphylococcus* were observed on this medium (Table 10.4). The most prevalent species accepting higher concentrations of NaCl were staphylococci mainly originating from human skin, where they are exposed to higher levels of salt. Some staphylococci from this campaign were transferable to salt concentration up to 16 % (w/v). Despite obvious presence of halophilic/halotolerant microbes, *Bacillus megaterium* was the only spore-forming isolate that was detected on salty agar plates.

10.10 Other Extremotolerant Bacteria and Eukarya

To complete the data presented here, other resistances of spacecraft assembly microbes shall be mentioned, although their presence has not been tested for the Herschel campaign. The microbial resistance against radiation (UV and gamma) as well as hydrogen peroxide has been in the focus of interest, since these are techniques usually applied for the sterilization of spacecraft components (besides dry heat).

La Duc et al. (2007) and Ghosh et al. (2010) reported the presence of UV-C (1,000 J m^{-2})-resistant and 5 % hydrogen peroxide-resistant microorganisms in the cleanrooms. These resistances were found for spore-forming microbes mainly (*Bacillus*, *Nocardioides*, *Paenibacillus*), whereas *B. pumilus* was resistant against both conditions.

The detection of H_2O_2-resistant *B. pumilus* has been reported even earlier (Kempf et al. 2005) and in particular multiresistant *B. pumilus* SAFR-032 was studied extensively at the NASA Jet Propulsion Laboratory. The whole genome has meanwhile been sequenced and annotated (Gioia et al. 2007). Although the sequence revealed differences and additional, unknown genes to closely related *B. subtilis* or *B. licheniformis*, the *B. pumilus* genome seems to lack genes functioning in UV or H_2O_2 resistance found in other *Bacillus* strains. Further studies will certainly be necessary to understand the molecular basis of extremotolerance in bacteria.

Interestingly, Eukarya have been detected only sparsely during the Herschel campaign. Solely one representative of *Coprinopsis* (fungi) and a few yeast strains have been isolated on R2A. Nevertheless, the entire study focused on microorganisms, and the conditions certainly would have to be adapted for the cultivation of Eukarya. Although fungi can produce spores, their resistance properties, possible extremotolerance, and resulting impact on planetary protection considerations have not been studied yet and are highly recommended. Nevertheless, the detection of *Aureobasidium pullulans* from spacecraft assembly facilities was reported, surviving 1 Mrad gamma radiation for 5.5 h (Bruckner et al. 2008).

For the sake of completeness, and also the discovery of *Tersicoccus phoenicis*, a member of the bacterial *Micrococcaceae* should be mentioned. Although it was not found to be extraordinarily resistant to one specific stress, this microbe, which represents a novel genus, was found solely in spacecraft assembly cleanrooms so far, namely, at the Kennedy Space Center (Phoenix cleanroom) and at the Centre Spatial Guyanais in Kourou, French Guiana, during the Herschel campaign (Vaishampayan et al. 2013a). Due to the unusual distribution and the general resistance against overall cleanroom conditions, this microbe was declared as one of the "Top 10 New Species of 2014" by the International Institute for Species Exploration.

10.11 Archaea

10.11.1 Background

Archaea, the third domain of life, have meanwhile been detected in almost any "normal" biotope, like marine and freshwater or soil. For more than 20 years, they were considered extremophiles that were ecologically restricted and highly adapted to specific and often hostile biotopes. Many of the extremophilic archaea have very interesting properties. In the eyes of many researchers, they are "primitive" in their metabolism (which actually means they can act as primary producers), which could be an advantage when settling in new biotopes. Detailed resistance experiments with vegetative (hyper) thermophilic archaea have unexpectedly revealed tolerances against desiccation, vacuum, and UV or gamma radiation (Beblo et al. 2009, 2011). It is unclear, however, whether these organisms could withstand the extremely harsh conditions during space travel, lack of nutrients, or low temperatures in extraterrestrial environments.

The major procedures to detect microorganisms in cleanrooms are still targeting Bacteria, but Archaea are more and more considered a possible microbial contaminant on spacecraft. Although the possibility that Archaea, such as methanogens or halophiles, might be able to survive a space flight and survive or even to thrive on Mars (Taubner et al. 2015; Kendrick and Kral 2006; Landis 2001), the existence of Archaea in human-controlled and rigorously cleaned environments has not been assessed until a few years ago, so that it was unclear if Archaea even could be found in spacecraft-associated cleanrooms.

The vast majority of mesophilic and psychrophilic Archaea still resist cultivation, and also the attempt to cultivate Archaea from spacecraft assembly facilities failed (Moissl et al. 2008; Moissl-Eichinger 2011). For this reason, the detection of Archaea can be solely based on molecular studies, which are presented in the following chapter.

10.11.2 Results

In 2008, we reported the detection of archaeal 16S rRNA gene signatures in two US spacecraft assembly cleanrooms (Moissl et al. 2008; Tally et al. 1975). Using a very sensitive PCR approach, 30 different cren- and euryarchaeal sequences were derived from NASA facilities. Meanwhile, the relevant branch of the Crenarchaeota has been reclassified as Thaumarchaeota (Brochier-Armanet et al. 2008).

The omnipresence of Archaea in global cleanrooms was confirmed in a follow-up study (Moissl-Eichinger 2011) in which Archaea were detected in all clean-rooms analyzed (second sampling Friedrichshafen, ESTEC and Kourou). As already reported in 2008, most of the gene sequences obtained clustered within the Thaumarchaeota group 1b. The closest cultivated neighbor *Candidatus* Nitrososphaera gargensis showed more than 4 % difference in the 16S rRNA gene sequence; most of the relatives from various natural biotopes are still uncultivated. Nevertheless, many representatives of this group were described to possess genes for ammonia oxidation (Pester et al. 2011).

The detection of these Thaumarchaeota in different spacecraft assembly clean-rooms was substantial and led to the hypothesis that these archaeal signatures could be linked to human presence (Moissl-Eichinger 2011), similar to the majority of the cleanroom bacteria that turned out to be human-associated or even opportunistic pathogens (e.g., *Staphylococcus;* Moissl et al. 2007; Stieglmeier et al. 2009). Meanwhile, this assumption could be proven and these Thaumarchaeota were clearly linked to the human microbiome, and the source of these archaeal signatures, the human skin, could be identified (Probst et al. 2013). This finding again confirmed the staff being the major contamination source for microbes in cleanrooms.

Quantitative PCR revealed that the average number of Archaea per m^2 cleanroom surface is around two to three logs lower than the estimated total number of Bacteria. Although the number of Archaea appears quite low, their presence is as consistent as typical bacterial cleanroom contaminants – and they were found to be potentially alive at the time of sampling (Mahnert et al. 2015). In the latter study, the microbiome of a US cleanroom was deeply sequenced, revealing the presence of intact *Cand.* Nitrososphaera cells (Thaumarchaeota) and *Haloferax* (Euryarchaeota). These assumptions were congruent to earlier studies, detecting methanogens and other halophiles additionally (Moissl et al. 2008; Moissl-Eichinger 2011). Due to the potential of some archaea to survive in particular under harsh conditions, and their detection only inside, not outside the cleanroom, they were suggested as indicators of cleanroom environments (Mahnert et al. 2015).

Meanwhile, it is well accepted that also Archaea are part of the cleanroom microbiome, and due to the finding that they are intact and thus potentially alive, a potential threat with regard to planetary protection aspects has to be discussed. However, further studies will be necessary, and additional attempts for cultivation are recommended.

10.12 Lessons Learned from Cleanroom Microbiomes: Extremophiles Are Everywhere

A broad variety of extremotolerant bacteria was successfully isolated from each facility analyzed so far. Microorganisms that were able to grow under extremely oligotrophic, cold, alkalic, anaerobic, warm, and high-salt conditions were detected (Table 10.4). Besides that, we have shown that many microbes thriving or surviving in cleanrooms are able to fix nitrogen and/or carbon dioxide and could therefore serve as primary producers. The following list summarized the main findings of recent studies:

- Sporeformers were present in each facility analyzed, but the total number did not exceed 25 % of all isolates. The lowest percentage was seen in FR2, the "cleanest" environment sampled.
- The highest cell counts were obtained on media with lower nutrient levels and higher pH, hinting toward an influence of the environmental selective forces.
- Primary producers were found in an unexpected diversity: autotrophs and N_2-fixing microbes were successfully isolated.
- Strictly anaerobic bacteria were successfully isolated but are present in a low number only (up to 4 %), although facultative microbes were found to add up to 78 %.
- Thermophiles, psychrophiles, and halophiles were found in the cleanrooms.
- Archaea were detected in each facility and their presence is significant. The properties of the (thaum-)archaeal cleanroom community are unclear, but Archaea are able to persist and intact cells seem to be present.
- Cleanliness level of a cleanroom definitely influences the microbial diversity. The broadest diversity of cultivables was seen in Kourou samples and ESTEC (both cleanrooms were operated at ISO 8). It is assumed that also the environmental conditions have influence on the microbial diversity within the cleanrooms, since Kourou is located in a very humid environment.
- Most of the microbes detected in the overall study were human associated, but the most resistant strains seem to be environmental organisms.

Furthermore, we have shown that besides *Bacillus*, also other (spore-forming) bacteria can play a significant role in cleanroom environments. For instance, several *Paenibacillus* strains have been detected; at least three of them were identified to be novel isolates. Many bacilli are dependent on complex organic compounds for their metabolism. Interestingly, although closely related, some paenibacilli seem to have the ability of nitrogen and carbon dioxide fixation. First resistance experiments with our isolates revealed a proper capability to survive desiccation, vacuum, heating, or Mars-cycle simulations (Fig. 10.3).

One novel isolate (ES_MS17, candidatus *Paenibacillus purispatii* sp. nov.) was shown to have profound metabolic capabilities in nitrogen conversion processes (Behrendt et al. 2010).

Although paenibacilli have not been explicitly reported to be heat shock survivors, their capabilities hint toward multiresistance, combined with a high metabolic versatility. The study of paenibacilli from cleanrooms could be beneficial for planetary protection considerations, and further research on this fascinating group of microbes is highly recommended.

In sum it can be stated that extremophilic and extremotolerant microorganisms are present in all spacecraft assembly facilities. Many of them reveal multiresistances and primary producer capabilities. Nevertheless, the information about

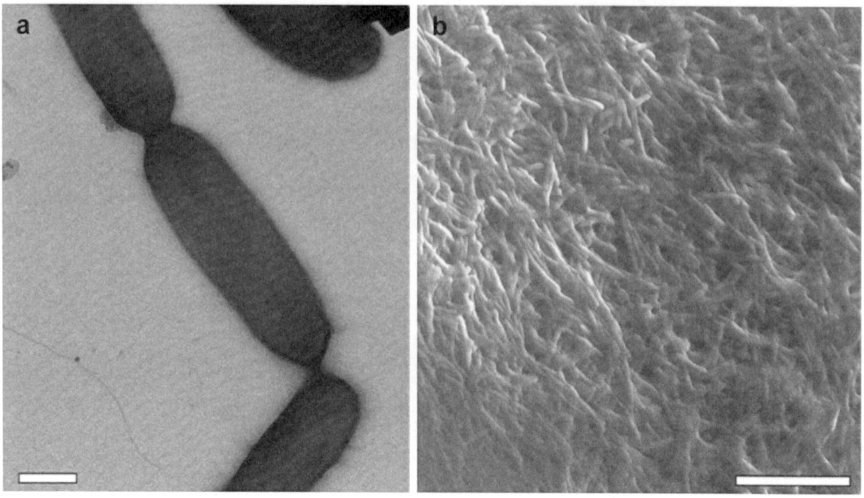

Fig. 10.3 *Paenibacillus cookii* FR1_23. (a) Electron micrograph of a dividing cell; bar, 600 nm. (b) Scanning electron micrograph of a colony; bar, 10 μm

indigenous microbial communities is still very limited: "How many are there?" and "What are they capable of?" are questions that will have to be answered. In order to preserve the integrity of future space travel, the research in the field of planetary protection needs to be enforced further in order to understand the microbial communities in the spacecraft assembly facilities as much as possible.

10.13 The Bacterial Diversity Beyond Cultivation or Cultivation Versus Molecular Analyses

Cultivation as a sole procedure currently does not allow assessing the overall microbial diversity. Previous publications expect a very low percentage (0.1–1 %) of all microbes to be cultivable via standard laboratory techniques (Amann et al. 1995). Nevertheless, our own studies based on the usage of 32 different media and conditions lead to the cultivation of approximately 0.3–5 %, when compared to the quantitative PCR results obtained.

Current molecular microbial diversity methodologies are mainly based on DNA extraction and analysis (mostly depending on the 16S rRNA gene), whereas LAL and ATP measurements have been reported as an acceptable method to obtain insights into the Gram-negative microbial diversity and the ATP content of cleanroom samples. LAL (*Limulus* amoebocyte lysate) analysis is used for the estimation of the Gram-negative, endotoxin-producing bacterial population and measures the presence of lipopolysaccharides. The ATP-based bioluminescence assay can help to obtain insights into the presence and the quantity of viable but non-cultivable cells (Bruckner et al. 2008; La Duc et al. 2007). However, since cells do not contain the same ATP amount (depending on the growth status or the size, Bruckner et al. 2008), quantities can only be estimated.

A strong bias has also been reported for DNA extraction methods and subsequent procedures. Up to now, no extraction method is able to fully extract DNA from spores, without disrupting the nucleic acid. For this reason, DNA-based molecular studies of cleanroom environments detect much more Gram negatives than Gram positives, which are generally harder to lyse or are sporeformers. It can be concluded that many microbes in spacecraft assembly cleanrooms are present as spores, which are not detected by molecular methods but supported by cultivation attempts.

The bias of PCR, still often used to amplify, e.g., the 16S rRNA gene, has also been discussed in several publications, and it is widely accepted that quantitative answers based on standard PCR and subsequent sequencing procedures are rather limited. Furthermore, the selected primers are not universal for the entire microbial group in focus; mispairings can lead to lower PCR efficiency or even to a non-binding of the primers to the target gene. This primer issue is also true for quantitative PCR approaches. Nevertheless, qPCR usually focuses on a very specific microbial group, and the entire methodology is designed for a very effective (up to 100 %) amplification of the target gene. Since measurements are independent other subsequent steps, qPCR allows at least quantitative, comparative predictions.

During the last years, next-generation sequencing methods and also the usage of microarrays ("PhyloChips," La Duc et al. 2009) have been implemented also for cleanroom and other, planetary protection-related studies. Nowadays, molecular, DNA-based methods can even discriminate intact and dead cells and thus allow a more detailed discussion of the results. However, the majority of all detectable microorganisms (99 %) in a cleanroom were found to be dead at the time of sampling (Mahnert et al. 2015; Vaishampayan et al. 2013b).

Molecular high-throughput techniques allow obtaining much more data in a much shorter time and are much more sensitive than cultivation assays. However, even based on metagenomes, we are not able to answer the question whether a microbe or a microbial community is extremophilic or not or could withstand extreme conditions during space flight and thus pose a risk for planetary protection. This information however is crucial for future efforts. It is therefore advisable to put still much effort into novel cultivation strategies (different from standard procedures) in order to increase the percentage of cultivable microorganisms from spacecraft assembly cleanrooms and until the shortcomings of molecular methods are solved.

Acknowledgments I thank the European Space Agency (ESA) for funding our projects. Furthermore, I thank Michaela Stieglmeier and Petra Schwendner for providing data, Alexander Probst and Ruth Henneberger for critically reading the manuscript, and Gerhard Kminek (ESA) for discussions and valuable input. The preparation of graphical illustrations by Petra Schwendner and Alexander Probst is gratefully acknowledged.

References

Administration NASA (2010) Handbook for the microbial examination of space hardware. NASA technical handbook, Washington DC

Ahmad S, Scopes RK, Rees GN, Patel BK (2000) *Saccharococcus caldoxylosilyticus* sp. nov., an obligately thermophilic, xylose-utilizing, endospore-forming bacterium. Int J Syst Evol Microbiol 50:517–523

Amann R, Ludwig W, Schleifer KH (1995) Phylogenetic identification and *in situ* detection of individual microbial cells without cultivation. Microbiol Rev 59:143–169

Beblo K, Rabbow E, Rachel R, Huber H, Rettberg P (2009) Tolerance of thermophilic and hyperthermophilic microorganisms to desiccation. Extremophiles 13:521–531

Beblo K, Douki T, Schmalz G, Rachel R, Wirth R, Huber H, Reitz G, Rettberg P (2011) Survival of thermophilic and hyperthermophilic microorganisms after exposure to UV-C, ionizing radiation and desiccation. Arch Microbiol 193:797–809

Behrendt U, Schumann P, Stieglmeier M, Pukall R, Augustin J, Spröer C, Schwendner P, Moissl-Eichinger C, Ulrich A (2010) Characterization of heterotrophic nitrifying bacteria with respiratory ammonification and denitrification activity – Description of *Paenibacillus uliginis* sp. nov., an inhabitant of fen peat soil and *Paenibacillus purispatii* sp. nov., isolated from a spacecraft assembly clean room. Syst Appl Microbiol 33:328–336

Boston P, Ivanov MV, McKay CP (1992) On the possibility of chemosynthetic ecosystems in subsurface habitats on Mars. Icarus 95:300–308

Brochier-Armanet C, Boussau B, Gribaldo S, Forterre P (2008) Mesophilic Crenarchaeota: proposal for a third archaeal phylum, the Thaumarchaeota. Nat Rev Microbiol 6:245–252

Bruckner JC, Osman S, Venkateswaran K, Conley C (2008) Space microbiology: planetary protection, burden, diversity and significance of spacecraft associated microbes. In: Schaechter M (ed) Encyclopedia of microbiology. Elsevier, Oxford, pp. 52–66

Cano RJ, Borucki MK (1995) Revival and identification of bacterial spores in 25- to 40-million-year-old Dominican amber. Science 268:1060–1064

Conley CA (2011) Outer space treaty. In: Gargaud M (ed) Encyclopedia of astrobiology. Springer Verlag Berlin, Heidelberg, p. 1192

Cox MM, Battista JR (2005) *Deinococcus radiodurans* – the consummate survivor. Nat Rev Microbiol 3:882–892

Crawford RL (2005) Microbial diversity and its relationship to planetary protection. Appl Environ Microbiol 71:4163–4168

Crawford RL, Paszczynski A, Allenbach L (2003) Potassium ferrate [Fe(VI)] does not mediate self-sterilization of a surrogate Mars soil. BMC Microbiol 3:4. doi:10.1186/1471-2180-3-4

Debus A (2006) The European standard on planetary protection requirements. Res Microbiol 157:13–18

DeVeaux LC, Müller JA, Smith J, Petrisko J, Wells DP, DasSarma S (2007) Extremely radiation-resistant mutants of a halophilic archaeon with increased single-stranded DNA-binding protein (RPA) gene expression. Radiat Res 168:507–514

ESA (2008) Microbial examination of flight hardware and clean rooms. ECSS-Q-ST-70-55C. European Cooperation for Space Standardization, ESA-ESTEC, The Netherlands

Fendrihan S, Berces A, Lammer H, Musso M, Ronto G, Polacsek TK, Holzinger A, Kolb C, Stan-Lotter H (2009) Investigating the effects of simulated Martian ultraviolet radiation on *Halococcus dombrowskii* and other extremely halophilic archaebacteria. Astrobiology 9:104–112

Ghosh S, Osman S, Vaishampayan P, Venkateswaran K (2010) Recurrent isolation of extremotolerant bacteria from the clean room where Phoenix spacecraft components were assembled. Astrobiology 10:325–335

Gioia J, Yerrapragada S, Qin X, Jiang H, Igboeli OC, Muzny D, Dugan-Rocha S, Ding Y, Hawes A, Liu W, Perez L, Kovar C, Dinh H, Lee S, Nazareth L, Blyth P, Holder M, Buhay C, Tirumalai MR, Liu Y, Dasgupta I, Bokhetache L, Fujita M, Karouia F, Eswara Moorthy P, Siefert J, Uzman A, Buzumbo P, Verma A, Zwiya H, McWilliams BD, Olowu A, Clinkenbeard KD, Newcombe D, Golebiewski L, Petrosino JF, Nicholson WL, Fox GE, Venkateswaran K, Highlander SK, Weinstock GM (2007) Paradoxical DNA repair and peroxide resistance gene conservation in *Bacillus pumilus* SAFR-032. PLoS One 2(9):e928

Head DS, Cenkowski S, Holley R, Blank G (2008) Effects of superheated steam on *Geobacillus stearothermophilus* spore viability. J Appl Microbiol 104:1213–1220

Huang Y, Dai X, He L, Wang YN, Wang BJ, Liu Z, Liu SJ (2005) *Sanguibacter marinus* sp. nov., isolated from coastal sediment. Int J Syst Evol Microbiol 55:1755–1758

Hungate RE (1969) A roll tube method for cultivation of strict anaerobes. In: Norris JR, Ribbons DW (eds) Methods in microbiology. Academic Press, New York, pp. 117–132

Kempf MJ, Chen F, Kern R, Venkateswaran K (2005) Recurrent isolation of hydrogen peroxide-resistant spores of *Bacillus pumilus* from a spacecraft assembly facility. Astrobiology 5:391–405

Kendrick MG, Kral TA (2006) Survival of methanogens during desiccation: implications for life on Mars. Astrobiology 6:546–551

Kminek G, Rummel JD (2015) COSPAR's Planetary Protection Policy. Space Res Today 193:7–18

Küsel K, Wagner C, Drake HL (1999) Enumeration and metabolic product profiles of the anaerobic microflora in the mineral soil and litter of a beech forest. FEMS Microbiol Ecol 29:91–103

La Duc MT, Nicholson W, Kern R, Venkateswaran K (2003) Microbial characterization of the Mars Odyssey spacecraft and its encapsulation facility. Environ Microbiol 5:977–985

La Duc MT, Kern R, Venkateswaran K (2004) Microbial monitoring of spacecraft and associated environments. Microb Ecol 47:150–158

La Duc MT, Dekas A, Osman S, Moissl C, Newcombe D, Venkateswaran K (2007) Isolation and characterization of bacteria capable of tolerating the extreme. Appl Environ Microbiol 73:2600–2611

La Duc MT, Osman S, Vaishampayan P, Piceno Y, Andersen G, Spry JA, Venkateswaran K (2009) Comprehensive census of bacteria in clean rooms by using DNA microarray and cloning methods. Appl Environ Microbiol 75:6559–6567

Landis GA (2001) Martian water: are there extant halobacteria on Mars? Astrobiology 1:161–164

Link L, Sawyer J, Venkateswaran K, Nicholson W (2004) Extreme spore UV resistance of *Bacillus pumilus* isolates obtained from an ultraclean spacecraft assembly facility. Microb Ecol 47:159–163

Mahnert A, Vaishampayan P, Probst AJ, Auerbach A, Moissl-Eichinger C, Venkateswaran K, Berg G (2015) Cleanroom maintenance significantly reduces abundance but not diversity of indoor microbiomes. PLoS One 10(8):e0134848. doi:10.1371/journal.pone.0134848

Moeller R, Setlow P, Horneck G, Berger T, Reitz G, Rettberg P, Doherty AJ, Okayasu R, Nicholson WL (2008) Roles of the major, small, acid-soluble spore proteins and spore-specific and universal DNA repair mechanisms in resistance of *Bacillus subtilis* spores to ionizing radiation from X-rays and high-energy charged (HZE) particle bombardment. J Bacteriol 190:1134–1140

Moissl C, Osman S, La Duc MT, Dekas A, Brodie E, DeSantis T, Venkateswaran K (2007) Molecular bacterial community analysis of clean rooms where spacecraft are assembled. FEMS Microbiol Ecol 61:509–521

Moissl C, Bruckner JC, Venkateswaran K (2008) Archaeal diversity analysis of spacecraft assembly clean rooms. ISME J 2:115–119

Moissl-Eichinger C (2011) Archaea in artificial environments: their presence in global spacecraft clean rooms and impact on planetary protection. ISME J 5:209–219

Moissl-Eichinger C, Rettberg P, Pukall R (2012) The first collection of spacecraft-associated microorganisms: a public source for extremotolerant microorganisms from spacecraft assembly clean rooms. Astrobiology 12:1024–1034

Moissl-Eichinger C, Pukall R, Probst AJ, Stieglmeier M, Schwendner P, Mora M, Barczyk S, Bohmeier M, Rettberg P (2013) Lessons learned from the microbial analysis of the Herschel spacecraft during assembly, integration, and test operations. Astrobiology 13:1125–1139. doi:10.1089/ast.2013.1024

Nagarkar PP, Ravetkar SD, Watve MG (2001) Oligophilic bacteria as tools to monitor aseptic pharmaceutical production units. Appl Environ Microbiol 67:1371–1374

National Research Council (2006) Expanding our knowledge of the limits of life on Earth. In: Preventing the forward contamination of Mars. National Academies Press, Washington, pp. 69–90

Newcombe DA, Schuerger AC, Benardini JN, Dickinson D, Tanner R, Venkateswaran K (2005) Survival of spacecraft-associated microorganisms under simulated martian UV irradiation. Appl Environ Microbiol 71:8147–8156

Newcombe D, Dekas A, Mayilraj S, Venkateswaran K (2009) *Bacillus canaveralius* sp. nov., an alkali-tolerant bacterium isolated from a spacecraft assembly facility. Int J Syst Evol Microbiol 59:2015–2019

Nicholson WL, Munakata N, Horneck G, Melosh HJ, Setlow P (2000) Resistance of *Bacillus* endospores to extreme terrestrial and extraterrestrial environments. Microbiol Mol Biol Rev 64:548–572

Nicholson WL, Schuerger AC, Setlow P (2005) The solar UV environment and bacterial spore UV resistance: considerations for Earth-to-Mars transport by natural processes and human spaceflight. Mutat Res 571:249–264

Osman S, Peeters Z, La Duc MT, Mancinelli R, Ehrenfreund P, Venkateswaran K (2008) Effect of shadowing on survival of bacteria under conditions simulating the Martian atmosphere and UV radiation. Appl Environ Microbiol 74:959–970

Pester M, Schleper C, Wagner M (2011) The Thaumarchaeota: an emerging view of their phylogeny and ecophysiology. Curr Opin Microbiol 14:300–306

Peters V, Conrad R (1995) Methanogenic and other strictly anaerobic bacteria in desert soil and other oxic soils. Appl Environ Microbiol 61:1673–1676

Pillinger JM, Pillinger CT, Sancisi-Frey S, Spry JA (2006) The microbiology of spacecraft hardware: lessons learned from the planetary protection activities on the Beagle 2 spacecraft. Res Microbiol 157:19–24

Plumb RC, Bishop JL, Edwards JO (1993) The pH of Mars. In: Lunar and Planetary Inst., Mars: past, present, and future. Results from the MSATT program, part 1, pp 40–41

Probst A, Facius R, Wirth R, Moissl-Eichinger C (2010a) Validation of a nylon-flocked-swab protocol for efficient recovery of bacterial spores from smooth and rough surfaces. Appl Environ Microbiol 76:5148–5158

Probst A, Vaishampayan P, Osman S, Moissl-Eichinger C, Andersen GL, Venkateswaran K (2010b) Diversity of anaerobic microbes in spacecraft assembly clean rooms. Appl Environ Microbiol 76:2837–2845

Probst AJ, Auerbach AK, Moissl-Eichinger C (2013) Archaea on human skin. PLoS One 8(6):e65388

Puleo JR, Fields ND, Bergstrom SL, Oxborrow GS, Stabekis PD, Koukol R (1977) Microbiological profiles of the Viking spacecraft. Appl Environ Microbiol 33:379–384

Reasoner DJ, Geldreich EE (1985) A new medium for the enumeration and subculture of bacteria from potable water. Appl Environ Microbiol 49:1–7

Rettberg P, Anesio AM, Baker VR, Baross JA, Cady SL, Detsis E, Foreman CM, Hauber E, Ori GG, Pearce DA (2016) Planetary protection and Mars Special Regions – a suggestion for updating the definition. Astrobiology 16:119–125. doi:10.1089/ast.2016.1472

Rummel JD, Beaty DW, Jones MA, Bakermans C, Barlow NG, Boston PJ, Chevrier VF, Clark BC, de Vera J-PP, Gough RV (2014) A new analysis of Mars "special regions": findings of the second MEPAG Special Regions Science Analysis Group (SR-SAG2). Astrobiology 14:887–968

Schuerger AC, Mancinelli RL, Kern RG, Rothschild LJ, McKay CP (2003) Survival of endospores of *Bacillus subtilis* on spacecraft surfaces under simulated martian environments: implications for the forward contamination of Mars. Icarus 165:253–276

Schulze-Makuch D, Grinspoon DH (2005) Biologically enhanced energy and carbon cycling on Titan? Astrobiology 5:560–567

Stackebrandt E, Sproer C, Rainey FA, Burghardt J, Pauker O, Hippe H (1997) Phylogenetic analysis of the genus *Desulfotomaculum*: evidence for the misclassification of *Desulfotomaculum guttoideum* and description of *Desulfotomaculum orientis* as *Desulfosporosinus orientis* gen. nov., comb. nov. Int J Syst Bacteriol 47:1134–1139

Stieglmeier M, Wirth R, Kminek G, Moissl-Eichinger C (2009) Cultivation of anaerobic and facultatively anaerobic bacteria from spacecraft-associated clean rooms. Appl Environ Microbiol 75:3484–3491

Tally FP, Stewart PR, Sutter VL, Rosenblatt JE (1975) Oxygen tolerance of fresh clinical anaerobic bacteria. J Clin Microbiol 1:161–164

Taubner RS, Schleper C, Firneis MG, Rittmann SK (2015) Assessing the ecophysiology of methanogens in the context of recent astrobiological and planetological studies. Life 5:1652–1686

Thomas DJ, Boling J, Boston PJ, Campbell KA, McSpadden T, McWilliams L, Todd P (2006) Extremophiles or ecopoiesis: desirable traits for and survivability of pioneer Martian organisms. Grav Space Biol 19:91–104

Treaty on Principles Governing the Activities of States in the Exploration and Use of Outer Space, including the Moon and Other Celestial Bodies". United Nations Office for Disarmament Affairs. Retrieved 2013-04–18 http://disarmament.un.org/treaties/t/outer_space/text

Vaishampayan P, Probst A, Krishnamurthi S, Ghosh S, Osman S, McDowall A, Ruckmani A, Mayilraj S, Venkateswaran K (2010) *Bacillus horneckiae* sp. nov., isolated from a spacecraft-assembly clean room. Int J Syst Evol Microbiol 60:1031–1037

Vaishampayan PA, Rabbow E, Horneck G, Venkateswaran KJ (2012) Survival of *Bacillus pumilus* spores for a prolonged period of time in real space conditions. Astrobiology 12:487–497

Vaishampayan P, Moissl-Eichinger C, Pukall R, Schumann P, Spröer C, Augustus A, Roberts AH, Namba G, Cisneros J, Salmassi T (2013a) Description of *Tersicoccus phoenicis* gen. nov., sp. nov. isolated from spacecraft assembly clean room environments. Int J Syst Evol Microbiol 63:2463–2471

Vaishampayan P, Probst AJ, La Duc MT, Bargoma E, Benardini JN, Andersen GL, Venkateswaran K (2013b) New perspectives on viable microbial communities in low-biomass cleanroom environments. ISME J 7:312–324

Venkateswaran K, Satomi M, Chung S, Kern R, Koukol R, Basic C, White D (2001) Molecular microbial diversity of a spacecraft assembly facility. Syst Appl Microbiol 24:311–320

Venkateswaran K, Vaishampayan P, Benardini JN III, Rooney AP, Spry JA (2014) Deposition of extreme-tolerant bacterial strains isolated during different phases of Phoenix spacecraft assembly in a public culture collection. Astrobiology 14:24–26

The Extreme Biology of Meteorites: Their Role in Understanding the Origin and Distribution of Life on Earth and in the Universe

11

Natuschka N. Lee, Jörg Fritz, Marc D. Fries, Jose F. Gil, Andreas Beck, Asta Pellinen-Wannberg, Birger Schmitz, Andrew Steele, and Beda A. Hofmann

N.N. Lee (✉)
Department of Ecology and Environmental Science, Umeå University, SE-901 87 Umeå, Sweden

Biochemical Imaging Centre and the Department of Medical Biochemistry and Biophysics, Umeå University, SE-901 87 Umeå, Sweden
e-mail: natuschka.lee@umu.se

J. Fritz
Saalbau Weltraum Projekt, Liebigstrasse 6, 64646 Heppenheim, Germany

M.D. Fries
NASA Johnson Space Center, Mail Code XI2, Houston, TX, USA

J.F. Gil
Department of Plant Biology, Linnean Centre for Plant Biology, Uppsala BioCentre, SE-75007 Uppsala, Sweden

A. Beck
Botanische Staatssammlung München, D-80638 Munich, Germany

A. Pellinen-Wannberg
Department of Physics, Umeå University, S-901 87 Umeå, Sweden

Swedish Institute of Space Physics, SE-981 28 Kiruna, Sweden

B. Schmitz
Department of Physics, Lund University, SE-22100 Lund, Sweden

A. Steele
Geophysical Laboratory, Carnegie Institution of Washington, Washington, DC 20015, USA

B.A. Hofmann
Natural History Museum Bern, Bernastrasse 15, CH-3005 Bern, Switzerland

Institute of Geological Sciences, University of Bern, CH-3012 Bern, Switzerland

H. Stan-Lotter, S. Fendrihan (eds.), *Adaption of Microbial Life to Environmental Extremes*, DOI 10.1007/978-3-319-48327-6_11

Contents

11.1 Introduction

Meteorites have captured our fascination since our early history – they have evoked awe, fear, an irresistible curiosity, and numerous lively debates. Former historians have indicated that many of the ancient cultures and civilizations in Europe, Africa, Asia, the Inuit, and the native Indians in America regarded both the meteorite and the location of their fall as sacred. Thus, they used the meteorites as religious objects, or for craft design like jewelry, weapons, or even practical things like tools and horse shoes (e.g., Buchner et al. 2012; Chung 1976; Comelli et al. 2016; Johnson et al. 2013; Rehren et al. 2013). Today, meteorites continue to capture our fascination through popular cultural formats such as science fiction, but in particular also as a scientific window that reveals the secrets of the Solar System formation. Within academia, meteorites have always fomented keen scientific debate. It was not until the early 19th century that the cosmic origin of meteorites, i.e., being truly not tellurian, was approved by the scientific community after the late 18 th century work of Ernst F. F. Chladni (1794). This implied for the first time that there are other smaller bodies in the sky besides the Moon (e.g., McCall et al. 2006). After this, several other lively debates followed on controversial findings and hypotheses around the role of meteorites in the Universe and for the evolutionary course of life on Earth, often in connection with the profound difficulties to approach this subject in an adequate scientific way. Principally, the different types of meteorites and their parent bodies (asteroids, Mars, Moon) as well as their precursor materials (solar nebular, molecular clouds) can be viewed as a most extreme or exotic substrate, habitat, and transport mode of chemicals and possibly even of cell-based life forms for several reasons:

(i) They have experienced a remarkable history since their origin as condensates from the Solar Nebula, more or less metamorphosed or molten fragments of asteroids, or rocks from Mars or our Moon.
(ii) The meteorites have been exposed to multiple extreme conditions ranging from milliseconds to billions of years duration when traveling through the interplanetary space, until they fell down on an astronomical body like Earth.

(iii) Once on Earth, the meteorites get exposed to different weathering conditions, which often makes it a challenge to retrieve their former history in an unambiguous way.

Here, we aim to describe the current state of knowledge and methodology to explore the origin, role, fate, and biological colonization of meteorites on Earth, how this can expand our understanding of the evolution of life on Earth, and how this affects our current hypotheses on the habitability of terrestrial and extraterrestrial life in the Universe in general. After all, in a planetary system, the same type of meteorites can accumulate on other astronomical bodies, such as on Mars. Thus, generally seen, meteorites may therefore serve as a familiar habitat on different astronomical bodies.

11.2 The Role of Meteorites in the History of the Universe and Our Solar System

The current understanding is that the Universe is 13.9 billion years old, and that the Solar System was formed 4.5 billion years ago (e.g., Horneck et al. 2016). The study of meteorites and their parent bodies by both petrographic and astronomic observations have significantly aided in our understanding of the history of the Universe, the Solar System, and more specifically of our habitable Earth as part of the planetary system (e.g., Fritz et al. 2014). The first condensates from the Solar Nebula bear testimony of the planetary system formation – a process that can now with increasing capabilities of astronomic observation be studied in distant regions of the Milky Way and other galaxies (Fig. 11.1). In addition, some of the particles preserved in the most primitive meteorites are in fact even older than our Solar System. These so-called pre-solar grains, together with the overall elemental and isotopic composition of our Solar System, thus record the galactic nucleosynthetic processes of a large number of stars that preceded our Sun and formed the atoms that we are made of (Davis 2011; Wang et al. 2009).

Atoms, molecules, and dust particles were exposed to the radiation environment, leading to interactions between hard radiation and matter, and formation of cosmogenic nuclides. The presence of unstable (long-lived radioactive) isotopes in certain minerals indicates that the first planetary bodies in our Solar System formed ~4.57 billion years ago (Bouvier and Wadhwa 2010; Horneck et al. 2016, and references therein). Furthermore, the presence of former, short-lived radionuclides with half-life times of less than 10 million years (which can be traced by isotopic anomalies of the daughter isotopes in CAIs, chondrules, and iron meteorites) enable us to formulate hypotheses about former developmental processes in our Solar System. For example, that these isotopes were generated by nuclear fusion and expelled in a stellar explosion very close in time and space to where and when our Solar System formed and that the first solid planets like Mars were formed within 10 million years (Krot et al. 2005). These isotopic boundary conditions preserved in meteorites thus indicate that our Solar System formed along with other stars and planetary systems during the gravitational collapse of a giant interstellar molecular cloud. Giant

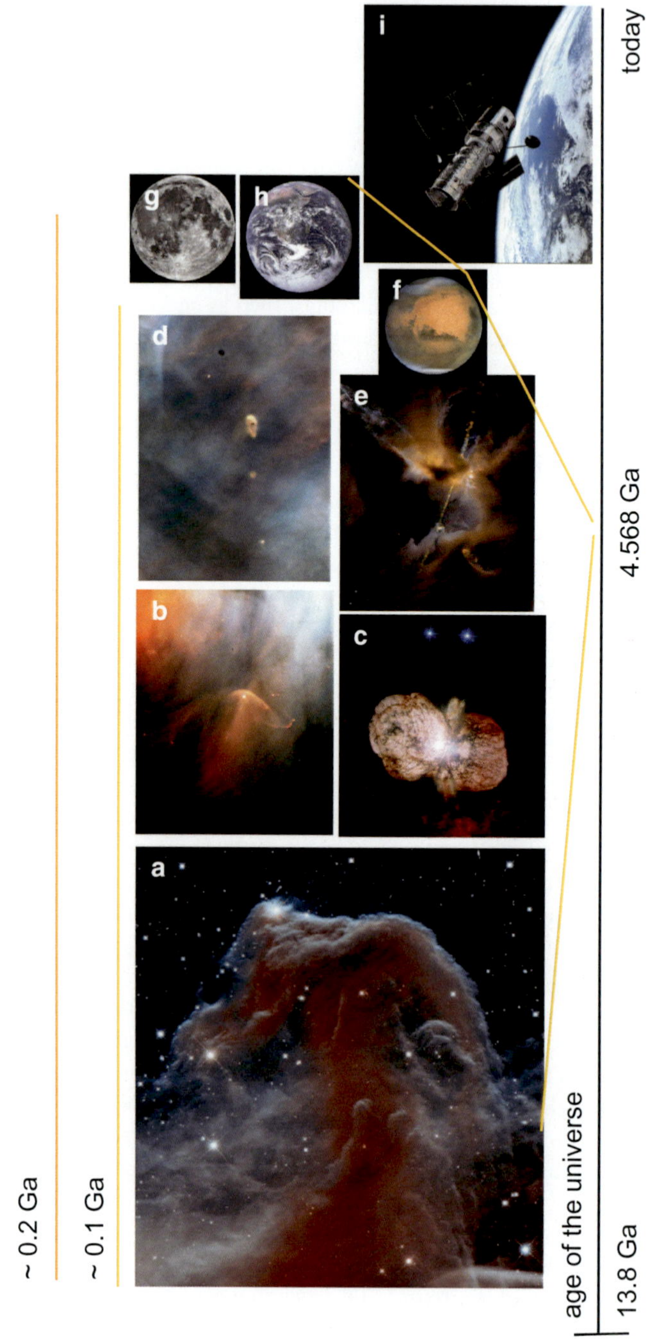

today
4.568 Ga
13.8 Ga
age of the universe
~ 0.1 Ga
~ 0.2 Ga
a
b
c
d
e
f
g
h
i

Fig. 11.1 Hubble Space Telescope images showing some selected examples on stages of planetary system formation along a time arrow. The 200 million years are inferred from the radiometric ages obtained on planetary materials and astronomical observations (see text in Sect. 11.2). (**a**) Infrared light image of the Horsehead Nebula in the Orion constellation is a part of a giant molecular cloud with high stellar formation rates. (**b**) Image showing a bow shock around a very young star in that nebular. (**c**) Composite ultraviolet and visible light images showing the Homunculus Nebula, which represents huge amounts of material ejected from the supernovae imposter system Eta Carinae likely enriched with short-lived radionuclides. (**d**) Five young stars in a small portion of the Orion Nebula. (**e**) Jets of energized gas, ejected from the poles of a young star. (**f**) Planet Mars. (**g**) Our Moon. (**h**) Earth. (**i**) The NASA/ESA Hubble Space Telescope orbiting today's Earth. Credits: NASA, ESA, and the Hubble Heritage Team (AURA/STScI); C.R. O'Dell/Rice University, Padgett (GSFC), T. Megeath (University of Toledo), B. Reipurth (University of Hawaii), James Bell (Cornell Univ.), Michael Wolff (Space Science Inst.); Lick Observatory; NASA/Apollo 17 crew picture taken by either Harrison Schmitt or Ron Evans

molecular clouds, such as the iconic Eagle Nebula or the Orion molecular cloud complex imaged by the Hubble Space Telescope, are regions in the Galaxy with high stellar formation rates. They are, however, transient phenomena as their populations of stars are not older than 10 of million years. As the most massive stars (>10 solar masses) burn their fuel in 10s of million years, these massive stars can potentially enrich the local environment with freshly formed radionuclides, from which subsequently planetary systems may form with elevated abundances of short-lived radionuclides. These temporal constrains show that the formation of giant molecular clouds, their gravitational collapse to stars containing protoplanetary disks, and the formation of planetary bodies occurred in less than 100 million years, which is short compared to the age of the Solar System or the age of the Universe. These processes are recorded in the most primitive meteorites.

Today, our Solar System consists of different types of objects like terrestrial planets, gas planets, ice planets, dwarf planets, smaller objects in the asteroid and Kuiper belt (comets, size > 100 m to < 100 km, or asteroids, size 1 m–1000 km large), or the Oort cloud. Some of these objects remain fixed in Keplerian orbits for billions of years. Breakup and collisional fragmentation might reduce their sizes to meteoroids (size 10 μm to 1 m) and cosmic dust (micrometeoroids below 10 μm). The orbits of objects like asteroids smaller than a few kilometers in diameter are significantly influenced by non-gravitational forces during less than a billion years. During this time, these objects may slowly leak into regions where the gravitational interaction with larger planets (mainly Jupiter) quickly takes them on erratic orbits. By the interplay between non-gravitational and gravitational forces, astronomical objects like dust to kilometer-sized objects may eventually acquire planet-crossing orbits, get scavenged by the Sun, or expelled from the planetary system. During this journey, they are constantly exposed to extreme conditions like space vacuum, solar and cosmic radiation, and extreme cold and occasionally hot temperatures for different time periods, until they may eventually land on another astronomical body like our planet Earth.

While rock fragments with diameters larger than 10–100 m (depending on the material) might strike the Earth surface with cosmic velocity, smaller meteoroids and cosmic dust are efficiently decelerated in the Earth's atmosphere. The frictional heating results in vaporization and ionization of the outer layers of the meteorite and ionization of the atmosphere. These small objects become optically luminous at altitudes of 50 to 100 km, where they will then form a brilliant shooting star or a fireball – a so-called meteor. During deceleration, a significant amount of the meteoritic mass is ablated and efficiently removed in seconds. Because the duration of frictional heating is so short, the inner part of the meteorite remains at or near the temperature it attained by solar radiation in near Earth space. The remaining mass of the cold meteoroid will arrive after minutes of free fall on the Earth's surface as a meteorite. Thus, a meteorite is simply a meteoroid that eventually arrived at the Earth's surface, although reduced in mass and with surficial alteration in the form of fusion crust (see, e.g., Ceplecha et al. 1998; Cockell 2015; Hofmann 2010; Rubin and Grossman 2010).

In summary, meteorites represent a valuable spectrum of stony to stony-iron and iron fragments from different types of parent bodies, such as asteroids, the planet Mars, and our Moon. These meteorites therefore represent a unique source of information about our early Solar System and its evolution over time. Such processes include condensation and agglutination of the first solids in the Solar Nebula (chondrites), thermal and aqueous alteration on the parent body, magmatic differentiation, development of a surficial regolith by space weathering and impact gardening on asteroids or the Moon, and alteration phenomena on Mars (e.g., Agee et al. 2013; Hofmann 2010; Cockell 2015; Forterre and Gribaldo 2007; Horneck et al. 2016). In addition of serving as a fundamentally important source about the history of the Solar System, the material represented by the different types of meteorites have also played at least four fundamental roles for the evolution of Earth (see Fritz et al. 2014 and various references within).

(i) For planetary accretion, meteorites represent samples of some of the building blocks from which the growing planets acquired the main total mass and chemical makeup including volatiles. The duration of the main accretion phase is sometimes constrained using the chemical separation of the lithophile element hafnium (Hf) from the siderophile tungsten (W) during core formation. The short-lived $^{182}Hf/^{182}W$ chronometer indicates that the iron core of Mars and Earth is 10 and ~100 Ma younger than the oldest solids in our Solar System, respectively.

(ii) A "late veneer" of meteoritic material is assumed to explain the appreciable amounts of platinum group elements (PGE) with chondritic (Solar System like) inter-element ratios in the mantle of Earth, Mars, and our Moon. It is assumed that each of these planets accreted between ~0.1 and 2 % of their mass during a so-called late veneer. The PGE-rich "late veneer" occurred after core formation during which the initial PGE inventory of the Earth was efficiently removed from the silicate mantle. In addition to noble metals, the "late veneer" could have delivered volatile elements, such as H, C, N, or prebiotic and organic molecules. For Earth, this "late veneer" followed the hypothetical giant impact that formed the Earth–Moon system and reset the $^{182}Hf/^{182}W$ chronometer. The "late veneer" likely includes material in unstable orbits that were left over from planetary accretion or from heliocentric debris disks produced during giant impact collisions, such as the one forming the Earth–Moon system, or other types of materials that were revolving around the Sun during that distant time on unstable orbits.

(iii) An intense "bombardment" with asteroids during the first 1.5 billion years is, according to some researchers, recorded by ~40 large impact basins (craters >300 km diameter) and order of magnitude more abundant smaller impact craters in the lunar crust (Fernandes et al. 2013). This Heavy Bombardment Eon apparently extended into terrestrial Archean and severely affected the environmental conditions of an increasingly habitable Earth including the delivery of prebiotic and organic compounds, and some have conjectured even extraterrestrial life (although the latter is not scientifically proven; see discussion in Sect. 11.7). Altogether, this bombardment may also have facilitated the evolution of life on Earth (e.g., Alexander et al. 2012; Caleb et al. 2013; Kleine 2011; Martin et al. 2016; Pasek and Lauretta 2008, see also Sects. 11.4 and 11.5).

(iv) As initiators, or at least as contributors to some of Earths' species extinction or diversification events, e.g., the mid-Ordovician event (440–450 Ma, see also Sect. 11.3 about fossil meteorites), the Cretaceous–Paleogene extinction event, which eliminated the dinosaurs approximately 66 million years ago, or the late Eocene event (33.9 Ma, e.g., Cockell 2015).

11.3 Meteorites Found on Earth

The amount of fragmented astronomical objects like asteroids, comets, meteoroids, and cosmic dust (micrometeoroids) traveling in our Solar System is immense – thus, most astronomical bodies like planets and moons are constantly hit by large amounts of rocky material. During Earth's early history, the meteorite fall rate was considerably higher than today (e.g., Fernandes et al. 2013; Fassett and Minton 2013; Cockell 2015) – however, the amounts of meteorites and cosmic dust that reach today's Earth are still impressive. Several attempts have been made to estimate the amount of meteorites that have reached Earth, e.g., vast measurements in marine sediments, but so far it has unfortunately not yet been possible to determine the precise total amount and size of the meteors and meteorites that have reached Earth's atmosphere or ground throughout time (Halliday 2001, Peucker-Ehrenbrink et al. 2016). All methods employed so far are based on different kinds of estimations, which produce different error ranges depending on the size of the meteors and the areas considered. The amounts usually are within the range from 37,000 to 78,000 tons/year. The majority stems from smaller particles (cosmic dust), while rock-sized meteorites only contribute a minor fraction, except for very rare large events. Bland et al. (1996) and Bland (2001) conducted a survey of the meteorite falls during the last 50,000 years and estimated that the total mass flux of meteorites to the Earth's surface over the 10 g to 1 kg interval was 2900–7300 kg/year.

Although billions of tons of meteorites have fallen on Earth since its early history, it is relatively rare to find intact, large meteorites today. So far, the Meteoritical

Society has catalogued at least 63,008 meteorites (data from 1 June 2016, http://www.lpi.usra.edu/meteor/metbull.php); however, only 1294 of these were retrieved from observed falls. Fortunately, the observation, search, and documentation of astronomical objects like meteorites and cosmic dust have improved during the last decade (e.g., Evatt et al. 2016; Fries et al. 2014; Hofmann et al. 2004; Jenniskens et al. 2012; Pellinen-Wannberg et al. 2016). Meteorites have been found in various geological environments and ecosystems (extreme, as well as non-extreme, see Figs. 11.2, 11.3, 11.4, 11.5 and 11.6) on Earth, though the number is highly variable between different countries even with comparable surface conditions. Only ten countries/areas feature more than 300 documented meteorites within their borders. Instead, the largest number of meteorites has been found in Antarctica (41,328), followed by the hot desert countries in Northern Africa, Arabia, the USA, South America, and Australia (Table 11.1).

The sources of meteorites available for research (and collectors) have changed over time. Rare observed falls and chance finds were the only available meteorites for a long time. The idea to actively search for meteorites was first applied by Harvey Nininger in the US Midwest, starting around the 1930s by informing farmers about

Table 11.1 Countries with either over 300 documented meteorites and falls and/or documented Lunar and Martian meteorites

Geographical location	Meteorites	Falls	Martian	Lunar
Algeria	753	7	9	4
Antarctica	41,328	0	30	35
Australia	749	18	0	2
Botswana	12	0	0	2
Brazil	82	28	1	0
Chile	776	1	0	0
China	232	64	0	0
Egypt	75	2	1	0
France	87	69	1	0
India	150	137	1	0
Libya	1500	1	8	7
Mali	19	3	5	1
Mauritania	48	6	1	5
Mexico	108	20	0	0
Morocco	1133	10	22	34
Nigeria	18	15	1	0
Northwest Africa	8784	0	106	135
Oman	4037	0	17	70
Russia	162	52	0	0
Sahara	479	0	0	0
Saudi Arabia	112	4	0	1
Sudan	9	8	0	1
Tunisia	70	5	1	0
USA	2157	161	2	0
Western Sahara	154	3	2	5

Data from 1 June 2016, http://www.lpi.usra.edu/meteor/metbull.php

meteorites (Nininger 1972). After first chance finds, systematic searches for meteorites were started in Antarctica in the early 1970s. Today, systematic meteorite collection expeditions organized by different nations, most systematically by the USA and Japan, continue to take place regularly (Cassidy 2003). Systematic searches for meteorites in hot deserts commenced shortly later in the Western USA, the Nullarbor Plain of Australia, the North African Sahara, Arabia, and the Atacama Desert. While search campaigns in Antarctica are carried out for purely scientific reasons, many hot desert searches are performed by collectors (including local inhabitants), at least partly motivated by the high prices obtained by some meteorites on the collectors market. Scientifically motivated searches in hot deserts make up a minor fraction of the finds in most areas. Since 2000, by far the largest amount of meteorites has been collected by locals in Northwest Africa, sold mainly through Morocco, and named "NWA" plus running number due to a lack of precise locality information. Many rare NWA meteorites have been recovered, but ordinary chondrites remain largely unclassified and essentially all information on pairing and strewn fields is lost. NWA

Fig. 11.2 Field images of meteorite finds in the hot desert in the Sultanate of Oman: (**a**) example of a find in the vast emptiness of the desert and (**b**) example of several meteorite finds in a rocky desert environment with many other dark rocks (scale bar 20 cm). Photos by Beda Hofmann and his team

Fig. 11.3 (**a**) Field image of the chondrite meteorite Al Huqf 070 (field No 0902–73) with lichen growth (*Ramalina maciformis*, *Aspicila* sp., *Diploica* sp.), (**b**) a closeup image of the meteorite Al Huqf 070, (**c**) field image of a limestone rock with *Ramalina maciformis* and other lichen species. Photos by Beda Hofmann and his team. Scale bar 5 cm

meteorites can therefore not be used for statistical purposes. The differently sourced groups of meteorites provide the following main advantages and disadvantages:

(i) Observed falls provide the freshest material, but are rare and large events are over represented.
(ii) Antarctic meteorites are available in large numbers, are relatively unweathered, but, due to ice transport information on pairing and strewn fields, are mostly lost.
(iii) Hot desert finds from scientific collection campaigns provide the best location-pairing information for large numbers of meteorites, but contain a high proportion of strongly weathered meteorites.

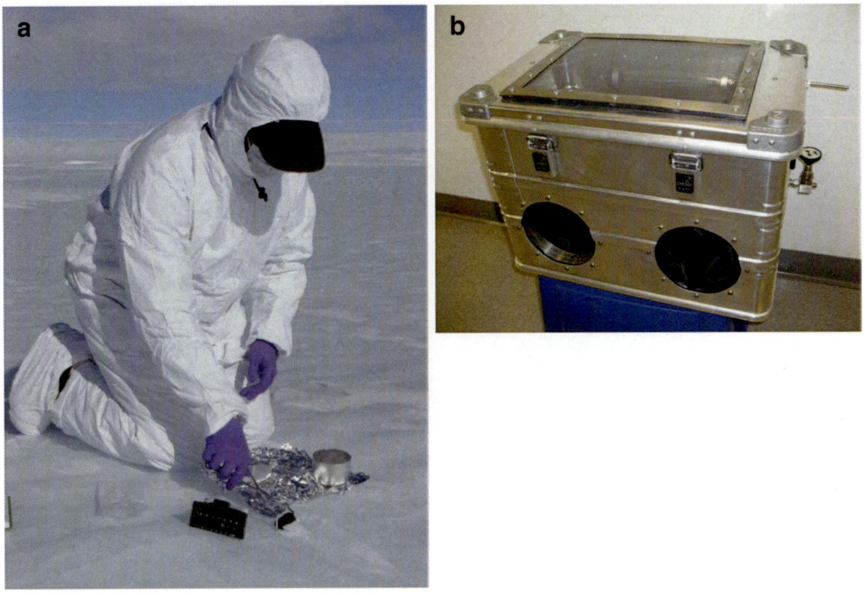

Fig. 11.4 (**a**) Meteorites in a cold desert/Antarctica, showing aseptic sampling of a meteorite in the field. The collector is wearing a full-body cleanroom garment, boot covers, sterile gloves, a facemask, sterile tongs, and sampling tray and the sampling is performed downwind from the meteorite. (**b**) A special prepared clean miniature glove box used for measurements on-site (e.g., of ATP as an indicator of microbial activity). The glove box uses filtered air and was cleaned in the field prior to use, and witness plates exposed in the box during the measuring protocol showed no organisms detectable by the same techniques used in the meteorite analysis. Both photos by the ANSMET team, USA

Fig. 11.5 Example of a meteorite found in a non-extreme environment. Strongly sculptured individual (size 6 cm, mass 144 g) of the Twannberg iron meteorite (mass TW453) found at a depth of 10 cm in forest soil using a metal detector. Mont Sujet, Jura Mts, Switzerland. Photo by Beda Hofmann

(iv) "Commercial" hot desert meteorites (mainly NWA) are available in the highest numbers providing the highest probability for the discovery of very rare types, but they mostly lack information on location and pairing.

Meteorites are classified based on their mineralogy; petrology; structure; elemental, chemical, and isotopic composition; origin in the Universe; size; and

Fig. 11.6 Example of a fossil meteorite, Österplana 065, from southern Sweden. The original meteorite is almost completely replaced by diagenetic minerals, primarily clay minerals, calcite, and barite. The only common relict mineral is chrome spinel. The Österplana 065 meteorite is the only fossil meteorite found in the Thorsberg quarry that is not an equilibrated ordinary chondrite. Probably the meteorite originates from the body that impacted into the L-chondrite parent body leading to its breakup. The coin in the image has a diameter of 2.5 cm. Photo by Birger Schmitz and his team

weathering state. Principally there are four main groups of meteorites (simplified summary from, e.g., Meibom and Clark 1999; Norton 2002):

(i) *Chondrites:* Primitive, non-melted rocky meteorites, which contain characteristic spherical chondrules representing former melt droplets with a typical size of 0.1–2 mm. Chondrites consist mainly of silicate minerals like olivine, pyroxene, and feldspar but can also contain a range of other compounds including metallic Fe-Ni, and iron sulfides. Depending on their composition, they can be classified broadly into ordinary chondrites (the most abundant type), carbonaceous chondrites (less abundant but more diverse), and a few rarer groups. Chondrites are the most common type of meteorites on Earth (>80 % of all finds so far). They are composed of different particles that were present during the early stage of the Solar System and are thus important for research on the origin and age of the Solar System, the synthesis of organic compounds, the origin of life, or the presence of water on Earth and the chemical and isotopic composition of Earth, including the presence of water.

(ii) *Achondrites and primitive achondrites:* Melted or partially melted rocky meteorites without chondrules, with distinct textures and traces indicative of igneous processes. They consist mainly of basalts or plutonic rocks. The achondrites are classified into seven groups, depending on their origin and history: primitive achondrites, aubrites, ureilites, angrites, HED, Martian, and Lunar. So far, 2329 achondrites have been found, which corresponds to ~3.6 % of all meteorites (data from 1 June 2016, http://www.lpi.usra.edu/meteor/metbull.php).

(iii) *Iron meteorites:* Meteorites with a high amount of meteoric iron in the form of Fe-Ni alloys kamacite and taenite. They also contain traces of many other elements, and based on this they can be classified into at least 13 groups and a number of grouplets and unique meteorites. Iron meteorites represent more than 100–130 different parent bodies. Their origin has mainly been linked to different types of larger, ancient asteroids. So far, 1132 iron meteorites have been found (~1.8 % of all meteorites, (data from 1 June 2016, http://www.lpi.usra.edu/meteor/metbull.php). However, despite this small amount, they represent the largest known meteorites (e.g., the Hoba meteorite in Namibia (weight 60 tons) and the Willamette meteorite in the USA (weight 14,500 kg) and account for almost 90 % of the mass of all so far known meteorites. The reason for this is that they are stronger than rocky meteorites and therefore are less subject to atmospheric fragmentation than chondrites and achondrites.

(iv) *Stony-iron meteorites:* These meteorites consist of both stony (silicates) and metallic (iron) components. They are further subdivided into mesosiderites and pallasites. Mesosiderites represent impact-generated mixtures of silicates and metal and are testimony of one (or several) giant collisions between asteroids. Pallasites consist of large grains of olivine embedded in an iron matrix. They have been interpreted as samples from the core–mantle boundary of asteroids. Based on new interpretations, they may also be impact-related (Yang et al. 2010).

While all meteorite groups may reveal different types of valuable information about our Solar System and weathering processes on Earth, two of these groups may be particularly useful for astrobiology: the carbonaceous chondrites for our understanding of prebiotic chemistry during the early stage of our Solar System and the achondrites for our understanding of Earth's early history, and for exploring whether past life forms have existed on other planetary bodies.

Lunar meteorites (achondrites) The Moon's history has been a matter of many disputes throughout time. The early researchers postulated that the Moon was chondritic and that it was thus initially cold, others claimed it was however initially hot (Baldwin 1963). Today, the most commonly accepted theory suggests that the Moon was formed not long after Earth, after a giant impact between Earth and a hypothetical planetary body called Theia (e.g., Canup and Asphaug 2001). Thus, rocky meteorites could bear records from this event and maybe even contain material remains testifying the early prebiotic or early life conditions on Earth. In the early 19th century, many scientists believed that the chondritic meteorites falling towards Earth were from the Moon. However, it was not until in 1981 that the first lunar meteorite (later called AH81005) was discovered in Antarctica by the ANSMET program from the USA (Marvin 1983, http://caslabs.case.edu/ansmet/). The AH81005 was first simply recognized as rather unusual, until comparisons were made with rock samples brought back from the Moon by the American Apollo and the Russian LUNA missions in the 1960s and 1970s, respectively. These rock samples (all together about 381 kg) provided compelling evidence on a common astrogeological history of Earth and our Moon. So far, 255 lunar meteorites have been found on Earth, mainly in Northern Africa, Antarctica, and the Sultanate of Oman (Table 11.1, data from 1 June 2016, http://www.lpi.usra.edu/meteor/metbull.php). All these samples produced several interesting novel insights, including the discovery of novel minerals, e.g., Armalcolite, named after the first three astronauts on the Moon, Armstrong, Aldrin, and Collins (Lind and Housley 1972). In contrast to the lunar mission samples that represent only a limited area of the lunar surface and geological history, the suite of lunar meteorites found on Earth provides a more representative, although still far from complete, composition of the global lunar surface.

Martian meteorites (achondrites) Martian meteorites are for the time being defined as rocks that are assumed to be from Mars, since their elemental and isotopic compositions are similar to rocks and atmospheric gases analyzed by spacecraft on Mars (Treiman et al. 2000; Schröder et al. 2008). So far, 171 hypothetical Martian meteorites have been found on Earth, mainly in the Northern Africa, Antarctica, and the Sultanate of Oman (Table 11.1, data from 1 June 2016, http://www.lpi.usra.edu/meteor/metbull.php). The two oldest Martian meteorites discovered so far, the heavily debated AHL 84001 found 1984 in Antarctica and the NWA 7034 (and pairings) found 2011 in the Sahara desert (Humayun et al. 2013; Cannon et al. 2015), are interesting for at least three reasons:

(i) While shergottites are relatively young volcanic to subvolcanic Martian rocks, the orthopyroxenite ALH84001 is of very high age and was formed 4.1 billion years ago. This ancient meteorite represents part of the early geological periods on Mars when there were higher amounts of water and possibly even life as we know it on Earth (e.g., Fairén et al. 2011).
(ii) They contain a large spectrum of compounds (including possibly also water) essential for terrestrial life or products of it (see further Sect. 11.4).
(iii) They contain an inventory of reduced carbon compounds (Steele et al. 2012).
(iv) The oldest of which, AHL840001, contains a series of signatures which have been postulated to be of relic biotic origin (McKay et al. 1996; see also further Sect. 11.5).

Thus, to shed further light on these highly disputed issues, it is pivotal to collect and explore more Martian meteorites.

Fossil meteorites on Earth Fossil meteorites are meteorites of ancient terrestrial age, since they fell on Earth a few million years ago. They have therefore been exposed to severe weathering (see Sect. 11.7) throughout time, buried, and thus incorporated into the geological record (Fig. 11.6). These fossil meteorites can provide valuable knowledge about the variations in the meteorite influx and major asteroid breakup events in the asteroid belt during the past ~3 Ga. One example is the Ordovician meteorite shower, a proposed shower of L-chondrite meteors that occurred roughly 470 million years ago (Schmitz 2013). It is likely that such meteorites can also be found in other parts of the Solar System like on Mars. So far, it is estimated that about a quarter of all meteorites falling on Earth originate from these events. Such meteorites have mainly been found in the Thorsberg quarry in Middle Ordovician marine limestone in southern Sweden at an unusual high frequency, which makes this one of the areas with the highest meteorite density on Earth. So far, about one hundred fossil meteorites from the few million years enduring Ordovician meteor event on Earth have been described (Schmitz 2013; Schmitz and Tassinari 2001; Schmitz et al. 2014). However, this amount represents only a minor fraction of all objects from this particular event present on Earth.

Meteorites from Earth So far, no meteorites from Earth have been found elsewhere in our Solar System. Some scientists speculate that it could be possible to find Earth meteorites on the surface of the Moon (Armstrong et al. 2002). If this is true, they could be extremely interesting, especially if they are older than 3.9 billion years, since all rocks on Earth older than that have been destroyed by different types of geological processes (Armstrong et al. 2002). As earlier mentioned in this section (Lunar meteorites), these could also contain remains of prebiotic or early life from Earth. However, other scientists argue that such speculations are rather unlikely because (i) meteorites impact the moon at high velocity since there is no atmosphere to slow their fall, thus only rare tiny fragments have a chance to survive, if at all, and (ii) even if such meteorites would survive, they would be exposed to intense irradiation and constant impact gardening on the Lunar surface, which would alter the rocks considerably. Since recently, caves and subsurface areas have also been discovered on the Lunar surface (Boyd et al. 2012); however so far, virtually nothing is known about whether more protected rock may be present there.

11.4 Chemical Composition of Meteorites and Their Role for Prebiotic Chemistry and Origin of Life on Earth

Mineralogical, petrological, isotopical, chemical, and cosmochemical analyses have played a key role in understanding the origin and evolution of terrestrial rocks on Earth, or meteorites from other parent bodies. Thanks to this, it has become possible to distinguish between terrestrial and extraterrestrial (e.g., meteorites) rocks and to differentiate these. Furthermore, it has also become possible to make attempts to correlate them to different cosmophysical, cosmochemical, and cosmogeological events throughout time, although they have been recovered in entirely different regions and in different states. Several novel strategies have been developed, such as remote sensing, space missions, and vehicles like Mars rovers, to enable further advanced chemical, analyses on Earth and on distant places within the Solar System, or even beyond. Both detailed analyses at the level of parts per billion of, e.g., single mineral grains, as well as of bulk major elemental analyses via remote sensing during space exploration are employed. Today, of meteorites or cosmic dust particles have been analyzed and extensive databases have been constructed (for a review, see, e.g., Nittler et al. 2004; Pizzarello and Shock 2010; for online databases, see, e.g., the NASA Ames website of the Astrophysics and Astrochemistry lab http://www.astrochemistry.org/db.php). Although these massive analyses represent only a minor fraction of all samples that still await to be discovered and analyzed, several striking observations have been made. These reveal interesting insights and hypotheses about the astrochemistry and extraterrestrial organic chemistry that occurred even before life emerged on Earth (see, e.g., Smith et al. 2013). As described in Sect. 11.3, the different categories of meteorites have different chemical compositions. Several of these elements, molecules and compounds, are essential for life as we know it on Earth – indirectly as well as directly, with respect to both inorganic as well as organic compounds, as exemplified below.

Inorganic compounds A large spectrum of inorganic elements and molecules are highly essential for life, since they may serve as building blocks for a variety of biomolecules, such as nucleic acids and proteins, or as electron acceptors or donators (for a brief overview, see the astrobiological periodic table by Charles S. Cockell, http://www2.ph.ed.ac.uk/~ccockell/astrobiologicalperiodictable.jpg). A crucial issue is the concentration, where either too low or too high concentrations of the elements or compounds may be either toxic or limit the chances for the origin, survival, and further development of life (see, e.g., Cockell 2015; Nixon and Cockell 2015). Two striking examples currently under intensive discussion are N and P, where some studies have postulated that neither N nor P were abundant on the early Earth and that meteorites must therefore have played an initial crucial role for the enrichment or the origin of biologically relevant N and P compounds on the early Earth (Gull 2014; Harries et al. 2015).

Organic compounds A surprisingly large number of molecules that are crucial for contemporary biochemistry on Earth are found in all kinds of different objects (including meteorites) in the Solar System, other stellar systems, or the Interstellar region of the Milky Way (Ehrenfreund and Cami 2010). Some of these are at least as far back as 27,000 light-years from Earth. These may have originated in gas, on interstellar dust or even on surfaces of ice grains (e.g., Belloche et al. 2014). Meteorites are thus precious objects for exploring the extraterrestrial organic chemistry that occurred in the Solar System or the interstellar medium, ahead of life's origin on the Earth. According to Pizzarello and Shock 2010, the organic content of meteorites is as diverse as kerogen-like macromolecules and simpler soluble compounds, such as amino acids and polyols. Many of the compounds in meteorites have in fact identical counterparts on the Earth's biosphere, e.g., methane, isopropyl cyanide, xanthine (the sugar glycolaldehyde needed to form ribonucleic acid RNA), RNA nucleobase (uracil), building blocks of DNA and proteins, such ethanamine and dipeptides, and amino acids like glycine, and polycyclic aromatic hydrocarbons

(PAHs) which may account for more than 20 % of the carbon in the universe (e.g., Blamey et al. 2015; Callahan et al. 2011; Cockell 2015; Jørgensen et al. 2012; Strasdeit 2005). Much of our understanding of organics in these meteorites has come from the analysis of the Murchison meteorite in the years since its fall in 1969. Indeed recent analyses show that this meteorite contains 10 of 100 of separate organic compounds (Schmitt-Kopplin et al. 2010).

Contribution to origin of life on Earth According to Pizzarello and Shock (2010), many of the biogenic elements and organics in meteorites have isotopic compositions that may have originated even in pre-solar environments. Just recently, a significant expansion of Miller's famous prebiotic experiment (Miller 1953) was made. Here, it could be demonstrated that not only amino acids but also complex DNA and RNA organic compounds of life, including uracil, cytosine, and thymine, can be synthesized in the laboratory under outer space conditions, using starting chemicals such as pyrimidine, which can be found in meteorites (e.g., Bera et al. 2016; Nuevo et al. 2008; Sandford et al. 2015; Saladino et al. 2015). Thus, the biogenic elements and organics in meteorites may indeed reveal a long cosmochemical process of changes throughout time. However, for the time being, it is still an open question whether these compounds, once they had been delivered to Earth, resumed their former cosmochemical process on Earth and thus contributed to the emergence of cell-based life forms, or whether they just increased the concentration of biogenic elements and organics already present on Earth but which had been reduced or even destroyed by the hot magma ocean on the early Earth. The fact that the unique L asymmetry of some meteoritic amino acids are also found in amino acids on Earth, do suggest their possible contribution to a terrestrial molecular evolution (Pizzarello 2016).

Meteorites today Earth is continuously bombarded with meteoroids and thus continues the import to Earth of different compounds and in some rare cases even water or ice. For example, such a transmission was observed during one of the intensive Leonid storms in 2002 (Pellinen-Wannberg et al. 2004). Leonid trail images from the same meteor on 19 November through a 423 (±14) nm calcium-iron and 589 (±10) nm sodium filter taken with the ALIS CCD imagers (Brändström 2003) are shown in Fig. 11.2. The Na component shows quite typical emission at the altitude range 95–110 km, while the Ca-Fe one was saturated with a very strong emission between 90 and 130 km reaching up to 145 km and down to 80 km, even though the ALIS imagers are much less sensitive for this wavelength. The altitude range is much higher than the one where ablation of metals occurs. The feature was traced back to suprathermal N_2^+ + H_2O emissions within the bandwidth. A meteoroid velocity of at least 50 km/s would be required for such strong emissions. A water molecule in a Leonid would collide with atmospheric nitrogen gas with the relative velocity of 72 km/s. The observed Leonid was ejected from the comet as late as the 1767 passage. Analyzing the possible sublimation of this meteoroid in terms of its closest distance to the Sun and its albedo, it is very probable that such a young meteoroid still includes ice (Fig. 11.7).

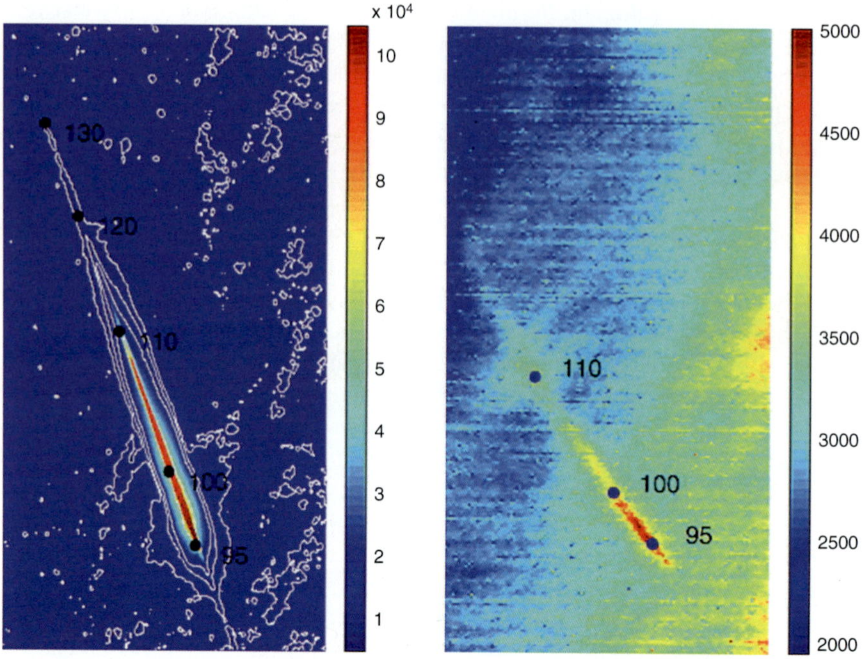

Fig. 11.7 Example of meteor trails recorded during the Leonid storm in 2002. The strong emission through the left-hand 409–437 nm filter is assumed to originate from the 433 hydrogen gamma emission line revealing presence of ice within the meteoroid (From Pellinen-Wannberg et al. (2004))

11.5 Extraterrestrial Life, Panspermia, and the Role of Meteorites

Extraterrestrial life in the scientific context can be defined as hypothetical life that may exist outside Earth. Although no evidence for extraterrestrial life has been obtained so far, the search for the origin of life in the Universe (be it on or beyond Earth), and for current or past signs of extraterrestrial life, is a very active and persistent objective of today's space research. One of the main rationales for this is based on at least two basic discoveries in astrochemistry and astrobiology: the hypothesis that prebiotic compounds were existent in the Universe already before even our Solar System, and the discovery of extremophiles on Earth. These studies have expanded our concepts of life and habitability on Earth and provided us with plausible hypotheses about life in other parts of our Solar System or even in the Universe. However, so far it has only been possible to test these hypotheses on Earth, in the lower orbit of Earth (LEO), and on the Moon in connection with the Apollo missions (e.g., de Vera et al. 2012; Moissl-Eichinger et al. 2016). We still lack knowledge about how life, as we know it today, be it on Earth or elsewhere, can spread and establish in other of parts of the Universe. For the time being, the most probable natural transport system (as opposed to designed space missions) is astronomical objects (e.g., dust, ice, rocks, comets, asteroids, and meteoroids), since this appears to be the most common mode to exchange material between different astronomical bodies, such as moons or planets. Thus, since the late 1950s, several scientists from different research fields have been seriously speculating about how high the probability could be that life may have emerged independently once or several times in the Universe and then spread by a hypothetical called panspermia (e.g., Cockell 2016; Moissl-Eichinger et al. 2016; Shapiro and Schulze-Makuch 2009).

The panspermia hypothesis is based on the speculation that life in the Universe may have a common origin and that it may exist elsewhere in the Universe and thus be distributed by different means – either interstellar (between star systems) or interplanetary (between planets in the same star system). Thus, life on Earth could have originated on other planets or star systems and *vice versa* (i.e., life on Earth could also spread to other parts of the Universe). The panspermia hypothesis was suggested already by the Greek philosopher Anaxagoras in the 5th century BC. The panspermia hypothesis has been periodically revived and expanded with additional concepts, such as cosmic pluralism, radiopanspermia, lithopanspermia, glaciopanspermia, accidental panspermia, molecular panspermia, and directed panspermia. These concepts of panspermia differ with respect to the proposed mechanism of transmission, and whether there is an intelligent purpose or not for the transmission. A wide range of natural scientists since the 19th century have contributed to these elaborations, such as Jöns Jacob Berzelius, William Herrschel, Lord Kelvin, Hermann von Helmholtz, and Svante Arrhenius; and during the last 50 years, Sir Fred Hoyle, Chandra Wickramasinghe, Carl Sagan, Francis Crick, and Stephen Hawkings, and a long list of contemporary exo- and astrobiologists (see, e.g. Horneck et al. 2008; Cockell et al. 2011; Nicholson 2009; Worth et al. 2013) (Fig. 11.8).

However, even though it is rather likely that astronomical objects such as meteorites delivered to the early Earth essential compounds produced elsewhere in the Universe (see Sect. 11.4), it has not yet been possible to provide unambiguous experimental evidence for the hypothesis on panspermia. The research efforts aiming to explore this issue have so far only been able to target local evolutionary processes, while the chances for tracing their origin are extremely low and prone to evoke misinterpretations and lively debates. For example, some researchers have postulated that detectable biosignatures, or even culturable species from ancient remains of amber, clay, sulfide deposits, sedimentary rocks, salt crystals, or even in the stratosphere as high as 41 km, may represent either extraterrestrial species or their descendants (Cano and Borucki 1995; Narlikar et al. 2003; Rasmussen 2000; Rosing 1999; Shen et al. 2001; Shivaji et al. 2009; DeLeon-Rodriguez et al. 2012; Vreeland et al. 2000; Wainwright 2003; Wainwright et al. 2006, 2009, 2010). Irrespective of if these observations are real, or rather the result of post-contamination in ancient or modern times despite careful efforts to avoid contamination, it is difficult to produce undisputable experimental evidence about their origin. For example, the issue about whether the stratosphere may allow the passage of extremophilic extraterrestrial species (e.g., UV radiation tolerant) has been highly debated and revisited several times along with new, often contradictory discoveries.

Furthermore, we must also consider that the transport via astronomical objects like meteoroids is full of risks, and the next plausible step is thus to speculate about whether and which type of organisms can actually survive an interplanetary or even

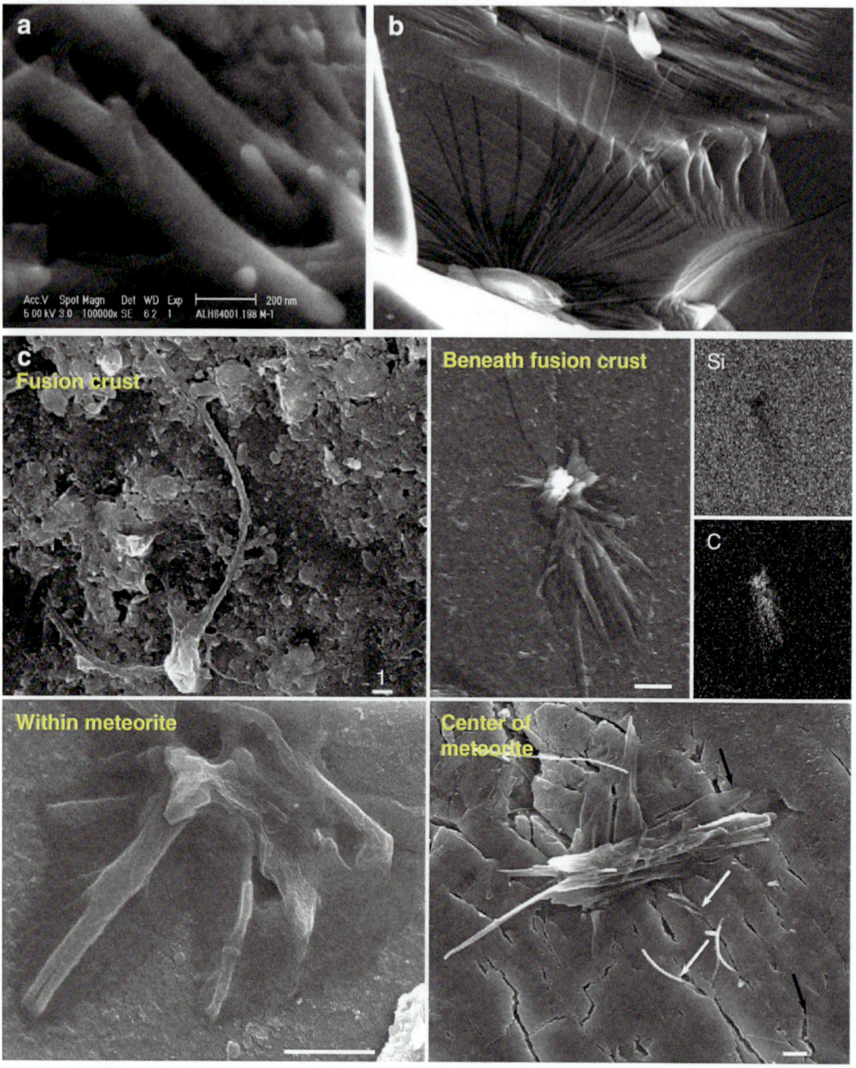

Fig. 11.8 Morphology alone is no indicator of extraterrestrial biogenicity, exemplified by scanning electron microscopy of meteorites discovered in Antarctica by the ANSMET team, USA. Studies made so far showed that ~14 meteorites, including Nakhla and ALH84001, have been contaminated by bacteria (Steele et al. 2001). (**a**) Hyphae-like structures; (**b**) terrestrial organisms on a depth profile of the meteorite Nakhla (From Toporski and Steele 2007); and (**c**) unknown potential contaminant on the meteorite Nakhla

interstellar transmission and by what means. This is probably one of the toughest challenges for cell-based life, be it terrestrial or even extraterrestrial. For example, the organisms or at least their seeds and spores on such material must survive the following four parameters (Fritz et al. 2005; Meyer et al. 2011):

(i) Ejection associated with extreme temperature swings, acceleration, jerk, and shock pressures from their original astronomical body
(ii) Interstellar or interplanetary travel within an undefined time (up to millions of years) through vacuum, deep-space radiation, and extreme cold
(iii) Infall onto another body possibly to include atmospheric entry with associated mass loss and aerodynamic heating
(iv) Adaptation to a new physiochemical environment in which they did not originally evolve to survive and produce offspring

These four steps demand that the organism(s) or their seeds and spores – on or rather within the meteorite – must tolerate multiple extreme parameters, such as desiccation, low temperatures, ionizing radiation levels, near vacuum pressures, extreme g-forces, starvation, and limited access to parameters essential for their metabolism, growth, proliferation, development, and evolution – under an unpredictable time period. Worse yet, the stressors they are subjected to during flight through space are inflicted, while the microbes are in a sporulated form otherwise they would be killed outright. While sporulated, these microbes are unable to perform damage mitigation functions such as radiation damage repair. Before the discovery of extremophiles, this would have been unimaginable for life on/from Earth to survive. But since the discovery of extremophiles and several recent studies using both space simulated facilities, rockets or space stations in the lower Earth orbit, it has become obvious that a large range of organisms, human-borne as well as extremophiles (heterotrophs, autotrophs), do exist that may survive at least some of these extreme conditions, especially if they are protected by soil, rock, or even real or artificial meteorite powder (Mastrapa et al. 2001; Brack et al. 2002; Fajardo-Cavos et al. 2005; Horneck et al. 2001, 2008; Meyer et al. 2011; Olsson-Francis and Cockell 2010, see also Sects. 11.6 and 11.7). Thus, the trajectory of research on panspermia is currently tending to shift from "is it even possible" to more detailed scenarios "who, how, when, from where and to where – and what are the risks, what is our responsibility?" (with respect to planetary protection issues, see, e.g., Rettberg et al. 2016).

The functional category and level of complexity of organisms surviving such polyextremophilic conditions associated with the entry to another astronomical body will most likely have a fundamental impact on the evolutionary course of the "inoculated planet." If the conditions of the "inoculated planet" are suitable to allow an evolutionary course of life as observed on our own planet Earth, then logically, useful traits would be autotrophy such as lithoautotrophic and photosynthetic organisms (anoxygenic and later oxygenic) as we know them on Earth. However, while several experiments have demonstrated that photosynthetic organisms can survive direct exposure to certain space conditions, they have so far not been shown to survive the entry to an Earth-like planet if they are located on the outside of e.g. meteoroids (Cockell et al. 2007, 2011; Cockell 2008). The highest chance for survival is within the meteoroids (depending on the composition and heat conductivity of the meteorite between a few mm down to 1 cm deep, e.g., Horneck et al. 2001). Heterotrophs and chemolithoautotrophs may have a higher survival rate than phototrophs, unless the phototrophs happen to be able to also survive under non-phototrophic conditions (e.g., certain mixotrophic cyanobacteria, Wan et al. 2015).

Another useful contribution to the evolutionary course of the "inoculated planet" is the level of cell complexity provided by the "visitor." Generally, prokaryotic species (e.g., *Bacillus* endospores, *Escherichia coli*, and *Paracoccus denitrificans*) can withstand hyperacceleration (up to 10,000 greater than Earth's gravity) better than eukaryotic species, which can only barely tolerate accelerated levels slightly greater than Earth's gravity. This is generally attributed to the fact that the gravitational potential is dependent on the size. As a consequence of this, both multicellular eukaryotes, single eukaryotes, such as yeast (*Saccharomyces cerevisiae*), as well as their larger eukaryotic organelles, such as the nucleus, are at a greater risk for mechanical deformation or collapse at accelerated levels than smaller singular prokaryotes and smaller organelles like ribosomes (Deguchi et al. 2011; Nicholson et al. 2000). Thus, at least certain species of prokaryotes are therefore usually regarded as more suitable model systems for research on extraterrestrial survival. However, recent observations by Vasanthan and coworkers suggest that at least one eukaryotic species, the Eutardigrade species *Hypsibius dujardini*, can tolerate short-term extreme accelerations up to 16,060 g (Vasanthan et al. 2016). Obviously, our current knowledge about the polyextremophilic tolerance spectra and conditions of different species on Earth is still rather scarce; thus, more knowledge about the mechanisms of hypergravity in different organisms is needed (Rea et al. 2016).

Large field missions and projects have now been established to screen for primitive or advanced signs or signals of present or past life, or even for postbiological phenomena – based on field expeditions to extreme ecosystems on Earth, radios, telescopes, or advanced space missions within or beyond our Solar System (e.g., Dick 2003; Horneck et al. 2016; Garber 1999, SETI webpage http://setiathome.ssl.berkeley.edu/). The goal is to continue collecting chemical–physical–geological–biological data for an expanded understanding of the potential habitability and evolutionary potential of extremophiles and how at least terrestrial life can travel and survive in the Universe. Chances for new, astonishing discoveries are most likely high, considering that we have still only mapped about 1 % of the whole prokaryotic diversity of today's Earth and even much less of the prokaryotes in earlier ages (e.g., Solden et al. 2016). Furthermore, recent indications in three different types of studies (Fernandes et al. 2013; Fritz et al. 2014; Bell et al. 2015) indicated that Earth could have been habitable already a few 100 million years earlier than formerly assumed. Although these findings are currently under debate, if this hypothesis is further proved, it would imply that life was emerged in fact 300 million years earlier than previously thought. Thus, current theories about the evolution of early Earth and how quickly life can actually arise – on Earth or elsewhere and whether life on a planet can develop quicker via the hypothetical panspermia – may possibly have to be revised. Thus, we anticipate that further research into extremophiles and ancient microfossils will reveal several novel insights, expand our current knowledge about the boundaries for the limits of terrestrial life, and enable us to postulate what life forms may exist in space as well as survive when traveling through space.

11.6 Terrestrial Life Associated with Meteorites: Biological Fossils

Terrestrial life refers here to organisms observable on our planet Earth, while fossils refer to their remains (e.g., whole bodies or part of them, or chemical biosignatures, see example in Figs. 11.9a–b) that have been preserved in rock or sediments for a long time period. Their origin may be truly terrestrial or even extraterrestrial (from outside Earth). Fossils are thus precious tools for tracing the evolution of life on Earth and, if even possible, their extraterrestrial origins. Although large amounts of fossils, mainly of macroscopic organisms, have been found, they still only represent

Fig. 11.9 Example of microbiologically induced textures in rock (biofabrics). (**a**) Field photograph showing mineralized microbial filaments encrusted with brown iron hydroxides and bluish chalcedony in a former cavity in dacite (volcanic rock), Cady Mountains, California. Scale bar 10 cm. (**b**) Photomicrograph of a thin section in transmitted light showing the transition between altered volcanic rock (*top*) and former void filled with mineralized microbial filaments and chalcedony (*bottom*). Same locality as a). Photographs by Beda Hofmann

a minor part of all life that has existed on Earth (see, e.g., Mc Kinney 1997). It is even more problematic with microbes since only very few of them can produce fossils, such as certain species among cyanobacteria, endolithic bacteria, and metal transforming bacteria such as iron-oxidizing bacteria like *Leptothrix* and *Gallionella*, or magnetotactic bacteria, preferably in microbial mats like stromatolites (e.g., Riding 1999). Thus, exploring microbial fossils on Earth as well as in the Solar System is a most challenging scientific field, demanding rigorous strict systematic methodology and critical evaluation (e.g., Brasier et al. 2015; Edwards et al 2007; Hofmann et al. 2008; Lindsay et al. 2003; McLoughlin et al. 2007; McLoughlin and Grosch 2016; Wacey et al. 2011, 2013, 2014).

According to Briggs (1963), the first claims that meteorites contain fossilized bacteria were made already in the late 19th century (Hahn 1880; Weinland 1882), though these claims were later turned down. However, since then, several studies have continued to report on detecting either microscopic structures resembling fossilized microbes in meteorites or even isolating bacteria from meteorites. Roy (1935) isolated *Bacillus subtilis* and *Staphylococcus albus* from the achondrite Johnston meteorite. Oró and Tornabene (1965) investigated three different types of chondrite meteorites (the Orgueil, the Murray, and the Mokoia meteorite) and found that bacteria (*Bacillus cereus*, *Bacillus badius*, and *Staphylococcus epidermis*) could be easily isolated from two of these meteorites (the Murray and the Mokoia). They explained this by different degrees of contamination levels and suggested already then that it is pivotal to explore microbial contamination of meteorites in order to achieve a better understanding of the origin and transformation processes of organic compounds in meteorites. A few studies have postulated that the observable microbes on meteorites could have an extraterrestrial origin. The most famous modern case which has been seriously investigated by a large group of scientists is the 1.93 kg Mars meteorite ALH84001, which was found on the Allan Hills in Antarctica 1984. For the time being, it is the oldest known igneous rock from Mars, with an estimated crystallization age of 4.1 billion years. Current estimations suggest that they are from one of the geological periods when there was still liquid water and therefore possibly also even at least primitive cellular species on Mars (Lapen et al. 2010). Based on ^{14}C measurements, the terrestrial age of AHL84001 is estimated to be ~13,000 years (Jull 2001). The hypothesis on its Martian origin was made in 1994 by Mittlefehldt (1994). Shortly after, David McKay and his team observed that the AHL84001 contained carbonate globules, and they therefore postulated that the AHL84001 contained fossils of prokaryotes, proclaiming this as the first evidence of extraterrestrial life on Earth (McKay et al. 1996). Later investigations suggested, however, that those structures were more likely abiotically formed from organic molecules (see Fig. 11.8a–c), or were signs of contamination by earthly biofilms (e.g., Jull et al. 1998). Steele et al. (2000) suggested that the meteorites must have been colonized by microbes in its surrounding, since the structures were similar to *Actinomycetales*-like species found in Antarctic cryptoendolithic communities. However, the debate about the origin of these microbe-like structures on ALH84001 is far from over. A reinterpretation of these findings was made when more advanced reexaminations showed that the meteorite contained magnetite crystals, which resemble those found in terrestrial bacteria that may be either of biotic or abiotic origin (Bada et al. 1998; Barber and Scott 2002; Fairén et al. 2011; Golden 2001; García-Ruiz et al. 2002, 2003; Thomas-Keptra et al. 2002, 2009).

Despite the lively debates around ALH84001, several other spectacular claims on discoveries of traces of life in meteorites have been made by other researchers – terrestrial or even extraterrestrial:

(i) Following the first presentations of the results on ALH84001, a highly controversial discussion about the presence of biomorphic forms on lunar meteorites, rocks, and other items sampled during the Apollo missions blossomed up (Zhmur et al. 1997; Zhmur and Gerasimenko 1999; Joseph 2016). These observations have, however, not been taken seriously by the majority of the established research community.

(ii) Gillet et al. (2000) declared that they had observed bacteria via electron microscopy, as well as cultivated bacteria from the Tatahouine meteorite. They described two types of bacteria-like forms on the mineral surfaces: (a) rod-shaped forms (RSF) about 70–80 nm wide and ranging from 100 nm to 600 nm in length and (b) ovoid forms (OVF) with diameters between 70 and 300 nm. Although the samples contained no magnetite, iron oxides, silicates, or carbonates, they were similar to the bacteria-like forms described on the Martian meteorite ALH84001. The origin of these bacteria is unknown, but it is postulated that their origin is terrestrial.

(iii) In 2001, Italian researchers reported to have found novel, live, extraterrestrial heat- and pressure-tolerant bacteria related to species of today like *Bacillus* and *Staphylococcus*. The authors claim that despite the fact that the bacteria had been kept inside for millions of years in different ancient rocks and even in two meteorites (chondrites) from Transylvania and Somalia, they would still be alive and capable of enduring extreme temperatures up to 550 ° C (Abbott 2001; Geraci et al. 2001). However, these results have neither been published in peer-reviewed journals nor reproduced by other studies, these are results are therefore not taken seriously by the majority of the established research community.

(iv) The most recent claim was made in 2013 on a new kind of carbonaceous meteorite called Polonnaruwa, which was found in Sri Lanka in 2012. The authors claimed that fossil diatom frustules had been found and that this would thus indicate that a population of extraterrestrial cyanobacteria must have been present in the initial sample (Wallis et al. 2013; Wickramasinghe et al. 2013a, b). However, since this meteorite has still not been included in the meteorite database (http://www.lpi.usra.edu/meteor/), these results are therefore not taken seriously by the majority of the established research community.

(v) Claims of life have also been made within the Tissint meteorite on the basis of both microfossils and carbon isotopes (Steele et al. 2016).

Additional examples can be found in a short review by Mignan (2011). Even though all these authors claim to have avoided contamination, a large part of the research community does not take any of these claims seriously (see, e.g., Abbot 2001 and the blogs by Plait in 2013), simply because several precautions for undisputable examinations and interpretations of the biological observations from the meteorites were not met, or at least followed up as seriously as the studies on AHL84001. Since accurate identification of microbial fossils on Earth or elsewhere in the Universe has such high importance, it is pivotal to improve current methodologies and expand existing databases as outlined in the astrobiology roadmaps both in Europe and in the USA (e.g., Horneck et al. 2015; NASA roadmaps 2015, http://www.nasa.gov/offices/oct/home/roadmaps/index.html; Pullan et al. 2008).

11.7 Terrestrial Life Associated with Meteorites: Today

This section will deal exclusively with terrestrial life associated with meteorites today – not whether they may have an extraterrestrial origin or not, as elaborated in Sect. 11.6. As mentioned in Sect. 11.3, meteorites can be classified into several different groups, depending on their origin in the Universe, mineralogical composition, and weathering state. Most of these analyses have focused on mineralogical, geochemical, and geophysical characteristics, while only a handful of these have been described from a biological point of view. However, several interesting research objectives can be suggested, such as:

(i) What kind of terrestrial life can colonize meteorites on Earth?
(ii) How will they influence the meteorites over time und different conditions? Does colonization by terrestrial life forms influence the weathering process?
(iii) Can this knowledge be useful for novel astro- or geobiotechnological processes, such as using meteorite powder as a protective agent for terrestrial life exposed to the conditions in the Universe (see, e.g., Horneck et al. 2001; Slobodkin et al. 2015) or for asteroid mining (e.g., Cockell 2010; Cousins and Cockell 2015)? What possibilities are there to develop asteroid farming, and how can meteoritic material or other regolith types like on the surface of the Moon be transformed into soil suitable for terrestrial plants?

Fate of meteorites on Earth – weathering: Once a meteorite has fallen on Earth, it may be present in the terrestrial environment as identifiable meteorite for long time periods – depending on their mineralogical composition and location, typically for 1000 to 100s of 1000s of years, with few exceptions older than a million years (Welten et al. 1997). Meteorites may become near completely unrecognizable due to alteration or mechanically destroyed, e.g., by wind ablation. Most of the mineral phases in meteorites are unstable at the Earth's surface, the stability of the dominant phases roughly being iron metal < iron sulfides < olivine < pyroxene < feldspar < chromite. Meteorites will successively alter into other phases that are more stable in their terrestrial surroundings. This alteration process is called weathering, and depending on the original composition, time, and location on Earth, it may be caused by different chemical, physical, geological, and biological parameters, the most important being oxidation, hydrolysis, and mechanical disintegration, caused by oxygen, water, geological processes, biological colonization, and so forth (Stelzner et al. 1999; Lee and Bland 2004; Al-Kathiri et al. 2005; Bland et al. 2006; Zurfluh et al. 2013, 2016). There are several benefits in gaining more knowledge about the weathering processes of different types of meteorites collected from different parts of the world:

(i) To expand our general fundamental understanding of terrestrial rock weathering
(ii) To use them as a global "chronometer" of environmental (including climatic) conditions on Earth throughout time
(iii) To interpret their pre-terrestrial history
(iv) To explore specific processes in the Universe such as the alteration of asteroids (for a review, see, e.g., Al-Kathiri et al. 2005; Bland et al. 2006)

Depending on the original mineralogical composition of the meteorite, the weathering parameters may cause a wide range of changes, such as oxidation of iron (rust), hydrolysis of silicates, hydration, carbonation, elemental contamination (e.g., by iodine, carbon, sulfur, strontium, noble gases), or incorporation of terrestrial oxygen into the minerals. The time of arrival of the meteorite on Earth (terrestrial age) can be determined by measuring the abundance of cosmogenic radionuclides with different half-life times (Jull 2006). Because the meteorite will be shielded from cosmic rays since its fall on Earth, its concentration of different cosmogenic radionuclides will decline in a predictable way. For an adequate investigation of meteorites, it is therefore crucial to examine their terrestrial residence age based on the decline of their nuclides and their weathering state. For a systematic evaluation, different weathering indices or scales, depending on their location (e.g., hot *versus* cold regions) have been applied to indicate extent of alteration. Weathering of meteorites is assessed by microscopic examination of thin sections. Stages W1-6 are assigned based on Wlotzka (1993). A refined weathering scale has been introduced by Zurfluh et al. (2016). A special evaluation approach was developed for fossil meteorites, since the terrestrial alteration of meteorites that fell into the Ordovician sea in Sweden caused the destruction of all primary minerals with the exception of chromite. The involved redox reactions led to an accumulation of a number of elements (U, V, Ba), while Fe was mobilized from the sediment near the meteorites. In a geochemical study of the Brunflo meteorite (the first fossil meteorite found in Sweden), the involvement of microorganisms in these alteration processes was therefore postulated based on the need for a powerful redox catalyst (Hofmann et al. 2000). However, further studies are needed to confirm such hypotheses.

Impact craters Impact craters are produced by high velocity impacts of large meteorites. The instantaneous release of an enormous amount of kinetic energy generates an explosion resulting in a crater formation. During this process, only a tiny amount of the meteorite survives and is thus seldomly ever found (Maier et al. 2006). The formation of a large crater is usually accompanied by profound changes of the site, morphologically, geochemically, as well as biologically (e.g., Veski et al. 2004; Cockell et al. 2002). Cockell et al. (2012) studied a 1.76 km drill core obtained from the deep subsurface of the ~35 million-year-old Chesapeake Bay impact structure in the USA and reported that the effects of the impact could still be observed through different disturbance patterns of the local microbiota. However, exploring impact craters is a complex field, because of the manifold parameters caused by the surrounding macro- and the micro-ecosystems throughout time, especially if the crater is large and of high age. So far, only 190 impact craters on Earth have been officially confirmed, but they are all distributed in different geographical regions, with different climates (from http://www.passc.net/EarthImpactDatabase/Worldmap.html). Estimated ages are from a few 100 years up to 2400 million years, and their diameter range from 0.0135 to 160 km (data from May 2016: http://www.passc.net/EarthImpactDatabase/index.html). Most of the research on impact craters has been of a geological nature, but some biological studies have also been performed. One example is the saline and alkaline Lonar Lake in India, which is characterized by an interesting basaltic structure similar to the Mars or the Moon, where several novel interesting methylotrophs and other taxa were observed (Antony et al. 2010; Paul et al. 2016). However, due to lack of sufficient amount of knowledge about the former natural history of this lake as well as of the microbial biodiversity on Earth in general (e.g., Locey and Lennon 2016), it is difficult to speculate about the origin and history of this unique microbial community. The same applies to most other biological studies of other impact craters (Pache et al. 2001). However, despite the apparent difficulties, it is an extremely interesting research field. Cockell et al. (2012) suggested that research on such impact fields are promising targets to gather experience for research on survival of life after meteorite impacts, which may also within the longer context, have implications for our theories about the state of the past and present habitability of Mars.

Biological experiments with meteorites So far, four different classical biological approaches have been employed to explore the colonization and impact of terrestrial life on meteorites: (i) different types of microscopy, (ii) isolation of microbes from meteorites, (iii) extraction of DNA from meteorites for amplification and Sanger sequencing of molecular markers like the 16S rRNA gene and functional genes via standard clone libraries, and iv) in situ experiments on meteorites with a selection of prokaryotic species. According to Briggs (1963), even Louis Pasteur conducted experiments on meteorites as part of the discussions on the concept of spontaneous generation on Earth *versus* in the Universe (trying to retrieve bacterial spores from meteorites, but failed). It was not until in the 1930s that the first reports of successful isolation of prokaryotic species like Bacillus subtilis, Staphylococcus albus, or other unidentifiable species from meteorites were published (Roy 1935; Briggs 1963). These studies were heavily debated during the following decades with regard to the validity of the observations and methods used, or whether the isolated species were from Earth, or not (see also Sect. 11.6). We anticipate that these discussions will continue for a long period. It is therefore pivotal that we continue to screen and analyze different categories of meteorites and gain a better understanding of the contamination sources of the natural background of the locations of the meteorites, as well as of the natural biodiversity and evolution of life on our own planet. To accomplish this, it is also pivotal to develop better sampling techniques and devices (see, e.g., Fig. 11.5a–b), methods, controls, and construction of curated reference databases. This will allow a much more critical and systematic experimental approach and evaluation of results than is possible even today.

To complement studies aiming at exploring the extraterrestrial biology or terrestrial contamination of meteorites (see above and Sect. 11.6), some few studies have been conducted to simply explore the post-colonization (post-contamination) of the meteorite in their natural surroundings, in order to explore their role in weathering. These studies have been conducted in either hot or cold deserts. Hot desert meteorites are both a challenging substrate for life and a niche in an otherwise extremely inhospitable environment (Figs. 11.3 and 11.4). The specific challenges for life in hot desert meteorites include high-temperature swings (typically reaching 65 °C in summer during daytime and close to freezing during night based on measurements in, e.g., the Sultanate of Oman) due to the typically dark surface of meteorites, high salt contents in pore waters, including Mg-Cl brines (Zurfluh et al. 2013) and a high availability of some otherwise rare heavy metals (Ni, Co). On the other hand, meteorites may represent local niches for life due to the availability of chemical energy sources for non-photosynthetic life, such as iron metal (and resulting H_2), ferrous iron, and iron sulfide. It is, however, currently not known whether microbial activity plays a long-term role in meteorite weathering.

One of the first studies indicating that microbes can be found on meteorites in the hot desert were published by Benzera et al. (2006), who explored the colonization of a Tatahouine meteorite fragment collected in the western border of the Sahara desert in 2000, after its estimated fall in 1931. With scanning electron microscopy, they could observe calcite precipitation, pyroxene weathering, and numerous microbial-like forms on the meteorite fragments. The DNA concentration of the meteorite fragment was about 0.5 μg/g of meteorite, which corresponds to a rough estimate of 2×10^8 prokaryotic cells per g of meteorite fragment. The microbial diversity was explored with standard clone libraries of partial 16S rRNA genes (450 nucleotides), consisting of 89 clones affiliated to the *Bacteria* and 30 clones to the *Archaea* – retrieved both from the meteorite fragments as well as from surrounding sand. Both the meteorite fragments and the soil were colonized by a highly diverse microbial community, where both similarities as well as differences could be observed. Eleven different bacterial divisions (*Cytophaga-Flexibacter-Bacteroides (CFB), Cyanobacteria* and *Acidobacteria,* α-, β-, and δ-*Proteobacteria, Actinobacteria, Firmicutes, Gemmatimonadetes,* OP10, and *Planctomycetes*, as well as one non-culturable lineage of nonthermophilic *Crenarchaeota*) were obtained. The majority of these were affiliated to unculturable species, but interestingly, many showed affiliations to sequences retrieved from unculturable novel species in other deserts or arid soil in other countries. Although only a low amount of the clones were sequenced, some first general hypotheses could be suggested with regard to the role of these various microbial taxa. For example, it was postulated that the *Cyanobacteria* fix N and produce organic polymers via photosynthesis, which can then be consumed and recycled by a range of versatile heterotrophic bacteria, such as the *Proteobacteria* and the *Actinobacteria*. Furthermore, the *Actinobacteria* were postulated to contribute to the weathering of silicates and the *Crenarchaeota* to oxidize ammonium. However, the authors also stated that several microbial groups that would be expected to be present, such as heavy metal (e.g., Fe)-transforming bacteria, were not discovered. They postulated that this could either be a result of a methodological bias or that the colonization time of 70 years was too short for this. A similar study was performed on a selection of meteorites in the hot deserts of the Sultanate of Oman by Lee et al. (2012, and manuscript in preparation). These meteorites were found in the desert on different locations, at different weathering conditions, where some of them were even partly covered by different lichen species (Fig. 11.4a–c). Standard clone library studies retrieved different taxa affiliated to, e.g., *Actinobacteria*, *Cyanobacteria*, *Firmicutes*, *Gemmatimonadetes*, *Proteobacteria*, *Synergistetes*, with a 16S rRNA gene similarity range from 85 to 98 % to so far sequenced microorganisms. The majority of the closest affiliations belonged to other unculturable taxa, many of them to UV-tolerant and radio-resistant extremophiles. From the two abovementioned studies on meteorites in hot deserts, it seems to be evident that meteorites can get colonized by different types of microbes – the future challenge will however be to determine their origin and their long-term impact on the weathering of meteorites.

Cold desert meteorites represent also a challenging substrate for life, such as extremely low temperatures and long winters (e.g., in the Arctic, the Antarctic, or on Greenland), but in contrast to hot deserts they have generally higher precipitation levels. However, the vegetation and fauna are considerably poorer, and thus the contamination level and weathering state may be lower in cold deserts than in hot deserts (Fig. 11.4a), but they may nevertheless also serve as a niche for certain extreme life forms. An advanced attempt to characterize the post-colonization of meteorites in the cold deserts was performed by Fries et al. (2010, 2012) where the aim was to explore the background level of microbial contamination in meteorites from Antarctica during different time periods and states (as-found condition versus post-curation and long-term storage). To achieve this, a range of advanced sampling devices, field-portable instruments, and molecular methods, based on DNA extraction, PCR, and sequencing, was employed. Preliminary results suggest that the meteorites in Antarctica on-site are apparently not contaminated, at least not by those Gram-negative bacteria that respond positively to the methods employed. Furthermore, the impact of other microbial species like Gram-positive species and eukaryotes, such as fungi, could so far not yet be tested in greater detail. Although this study suggests that certain species could not be found, it does unfortunately not reveal the presence of other species that are not detectable by the methods used. This again suggests the importance of expanding our knowledge and methodology to explore meteorites as well as life on in general on Earth, in order to enable a better interpretation of retrieved results with regard to the pattern and origin of the microbial diversity observed on meteorites.

Meteorites as substrate for microorganisms As mentioned in Sect. 11.3, meteorites contain a variety of compounds that are essential for life as we know it on Earth, including a number of compounds that are suitable as energy sources for life (metallic Fe, iron sulfide, Fe^{2+}-bearing minerals). Mautner and his colleagues demonstrated that extraterrestrial organic compounds in meteorites can survive extreme conditions, such as high temperature and pressure as experienced when entering the Earth atmosphere, and then still serve as substrates for microorganisms of today and possibly also for microorganisms on the early Earth (Mautner et al. 1995, 1997; Mautner 2002a, b). This was demonstrated by different experiments performed on different prokaryotic, plant, and fungal species like *Flavobacterium oryzihabitans*, *Pseudomonas maltophilia*, the genetically modified *Pseudomonas fluorescens*, *Nocardia asteroides*, *Arthrobacter pascens*, *Asparagus officinalis*, and *Solanum tuberosum* (potato) (Mautner et al. 1995, 1997; Mautner 2002a, b). Furthermore, two other studies demonstrated that iron-oxidizing bacteria (*Leptospirillum ferrooxidans*, *Acidithiobacillus ferrooxidans*) can grow by using iron meteorites as their energy source (González-Toril et al. 2005; Gronstal et al. 2009; Kutlucinar et al. 2015; Reed 2015). Both these studies demonstrate that interesting microcosmos studies can be performed with meteorites to explore metabolism of inorganic as well as organic compounds by terrestrial prokaryotic and eukaryotic species. In the longer context, this would suggest that experimental astroecological experiments on, e.g., carbonaceous asteroids and Martian basalts, can be useful to rate planetary materials as targets for astrobiology exploration and as potential space bioresources.

Eukaryotes The amount of eukaryotic species, which can tolerate at least cosmic conditions in the lower Earth orbit or in space simulated chambers, is lower compared to the amount of so far tested prokaryotes, which can tolerate such conditions quite impressively. However, recent studies have shown that there are nevertheless several polyextremophilic eukaryotes, such as tardigrades, nematodes, insect larvae, lichens, and fungi, which can also tolerate such conditions (de Vera et al. 2003; Sancho et al. 2008; Cockell 2015). Despite this, only few studies have been performed to explore what kind of eukaryotes can be observed in association with meteorites – be it of terrestrial or even extraterrestrial origin. Some few spectacular reports claim to have found eggs of tardigrades on recently discovered Martian meteorites (http://universe2go.com/en/egg-of-a-waterbear-found-in-a-mars-meteorite/); however, these results have received large critics, and thus such observations must therefore be followed up by more rigorous, systematic studies. For the present, the most commonly reported eukaryote-colonizing meteorites on Earth are lichens. So far, lichens growing on meteorites have been reported from several countries, such as Greenland (Hansen and Graff-Petersen 1986), Antarctica (Edwards et al. 2005), the UK (Pillinger et al. 2011), the Sultanate Oman (Lee et al. 2012, and unpublished results), and from some different sources with less precise descriptions (http://astrobob.areavoices.com/2011/01/07/lichens-call-this-space-rock-home/). These studies have altogether described more than 20 different lichen species. The number of so far recognized lichen species is however probably too low for further advanced interpretations on correlative parameters between colonization of a specific lichen species and meteorite type. But in the longer context, lichens can possibly become useful for conclusions about, e.g., post-colonization, exposure time on Earth, and location and climatic conditions endured.

Viruses Viruses are for the time being not defined as terrestrial life, although they are the most abundant biological entity on Earth and have a great influence on all cell-based terrestrial species. So far, viruses have also not been discovered elsewhere in the Universe. Thus, our planet Earth contains two unique phenomena in the Universe not yet found elsewhere so far: cells and viruses. Science fiction authors like Fred Hoyle and his colleagues have made several speculative suggestions about the origin and role of viruses in the Universe (Hoyle et al. 1986; Wickramasinghe et al. 2013c). However, these works have rather been regarded with some skepticism and have thus not been followed up by the established research community. Since recently, viruses are however obtaining a renewed interest from another perspective, due to some recent unexpected discoveries (Griffin 2013; Mahy and van Regenmortel 2009; Roossinck 2011; Steele 2014). Apart from groundbreaking hypotheses, such as that viruses may have invented DNA (Forterre et al. 2004), there are other observations that are equally groundbreaking for our further understanding of the biology of extremophiles, for survival of life in general, and for a renewed evaluation of geological records of fossils and rocks (Jalasvuori and Bamford 2009; Pacton et al. 2014). For example, some viruses have shown an amazing ability to interact with their hosts and help in adaptation to extreme environments. Such is the case of the *Curvularia* thermal tolerance virus (CThTV), which confers tolerance to high temperatures to its fungal host *Curvularia protuberate*, an endophyte of the grass *Dichanthelium lanuginosum*. This tritrophic interaction enables a thermal tolerance to the grass to more than 50 °C in geothermal soils (Yellowstone National Park, USA; Roossinck 2011). Another striking example is how pathogenic viruses can induce tolerance to other abiotic conditions such as drought to their hosts (Xu et al. 2008). Although the extent of viruses in other extremophiles is not that well explored, there is growing evidence that at least some of the so far investigated extremophiles have unique viruses/bacteriophages (Prangishvili 2013, 2015). After all, viruses with very interesting characteristics can be found in a wide range of extreme environments in both archaeal and bacterial hosts (Le Romancer et al. 2007). This suggests that these viruses are involved in the adaptation of their hosts and that a virus that has found a balance with its hosts will interact rather synergistically with it and avoid causing disease. Thus, "viruses are the ways by which genetic information has adapted to survive in this biosphere" (Jalasvuori 2012). So far, no explorations of viruses have been made on meteorites – it is therefore not yet possible to explore the hypothetical role of viruses for the fitness of cells associated with meteorites. However, based on the evidence we have from the viruses explored so far on Earth, this does suggest that it could be at least interesting to explore whether viruses are also associated with the life forms found on meteorites – in the past as well as in the present – and if present, what their role would be.

11.8 Summary and Outlook

It is evident that research on meteorites may offer a whole range of interesting interdisciplinary insights for astronomy, astrophysics, astrochemistry, astrobiology, and the extreme biology and geology on Earth – along with many other scientific and engineering fields, including space technology and astrobiotechnology. The origin of life – be it on Earth or elsewhere and whether terrestrial or extraterrestrial life can be transported throughout the Universe – belongs to some of the most intriguing scientific questions of today. However, in order to approach this research topic in an adequate way to produce unambiguous results, it is crucial that further fundamental research and improvements are undertaken for several issues:

(i) More efficient search tools for missing as well as approaching meteorites – on Earth and beyond Earth (e.g., missions to the Moon or Mars), in order to expand our understanding of the evolution of our Solar System and the Universe in general
(ii) Rigorous sampling protocols for an adequate variety of analyses (chemical, geological, physical, biological), which include not only appropriate handling of the meteorites during sampling but also sampling of the surroundings of the meteorite, in order to gain a better understanding of possible contamination sources, age of meteorite, and so forth
(iii) More advanced investigations of so far found meteorites, for example, with regard to the biology to explore the detected organism's ecology, physiology, adaptation, biochemistry, molecular biology, genetics, and epigenetics
(iv) Expanded understanding of the evolution and diversity of cells and viruses on Earth, throughout time, including fossils, a variety of biosignatures, and molecular biological methods, in order to address common evolutionary or unique (terrestrial *versus* extraterrestrial) traits in representative model species
(v) Adequate reporting of results and construction of databases

References

Abbott A (2001) Resuscitated 'alien' microbes stir up an Italian storm. Nature 411:229

Agee CB, Wilson NV, McCubbin FM, Ziegler K, Polyak VJ, Sharp ZD, Asmerom Y, Nunn MH, Shaheen R, Thiemens MH, Steele A, Fogel ML, Bowden R, Glamoclija M, Zhang Z, Elardo SM (2013) Unique meteorite from early Amazonian Mars: water-rich basaltic breccia Northwest Africa 7034. Science 339:780–785

Alexander CM, Bowden R, Fogel ML, Howard KT, Herd CD, Nittler LR (2012) The provenances of asteroids, and their contributions to the volatile inventories of the terrestrial planets. Science 337:721–723

Al-Kathiri A, Hofmann BA, Jull AJT, Gnos E (2005) Weathering of meteorites from Oman: correlation of chemical and mineralogical weathering proxies with ^{14}C terrestrial ages and the influence of soil chemistry. Meteorit Planet Sci 40:1215–1239

Antony CP, Deepak Kumaresan D, Lucia Ferrando L, Boden R, Moussard H, Scavino AF, Shouche YS, Colin Murrell JC (2010) Active methylotrophs in the sediments of Lonar Lake, a saline and alkaline ecosystem formed by meteor impact. ISME J 4:1470–1480

Armstrong JC, Wells LE, Gonzalez G (2002) Rummaging through Earth's attic for remains of ancient life. Icarus 160:183–196

Bada JL, Glavin DP, McDonald GD, Becker L (1998) A Search for endogenous amino acids in Martian meteorite ALH84001. Science 279:362–365

Baldwin RB (1963) The measure of the Moon. University of Chicago Press, Chicago, p. 488

Barber DJ, Scott ERD (2002) Origin of supposedly biogenic magnetite in the Martian meteorite Allan Hills ALH84001. Proc Natl Acad Sci U S A 99:6556–6561

Bell EA, Boehnke P, Harrison TM, Mao WL (2015) Potentially biogenic carbon preserved in a 4.1 billion-year-old zircon. Proc Natl Acad Sci U S A 112:14518–14521

Belloche A, Garrod RT, Müller HS, Menten KM (2014) Detection of a branched alkyl molecule in the interstellar medium: iso-propyl cyanide. Science 345:1584–1587

Benzerara K, Chapon V, Moreira D, Lopez-Garcia P, Guyot F, Heulin T (2006) Microbial diversity on the Tatahouine meteorite. Meteorit Planet Sci 41:1249–1265

Bera PP, Nuevo M, Materese CK, Sandford SA, Lee TJ (2016) Mechanisms for the formation of thymine under astrophysical conditions and implications for the origin of life. J Chem Phys 144:144308. doi:10.1063/1.4945745

Blamey NJF, Parnell J, McMahon S, Mark DF, Tomkinson T, Lee TM, Shivak J, Izawa MRM, Banerjee NR, Flemming RL (2015) Evidence for methane in Martian meteorites. Nat Commun 6:7399. doi:10.1038/ncomms8399

Bland PA, Smith TB, Jull AJT, Berry FJ, Bevan AWR, Cloudt S, Pillinger CT (1996) The flux of meteorites to the Earth over the last 50 000 years. Mon Not R Astron Soc 283:551–565

Bland PA (2001) In: Peucker-Ehrenbrink B, Schmitz B (eds) Accretion of extraterrestrial matter throughout Earth's history. Springer Publisher, Chapter 15, pp. 267–303

Bland PA, Zolensky ME, Benedix GK, Sephton MA (2006) Weathering of chondritic meteorites. In: Lauretta DS, McSween JHY (eds) Meteorites and the early solar system II. University of Arizona Press, Tucson, pp. 853–867

Bouvier A, Wadhwa M (2010) The age of the Solar System redefined by the oldest Pb–Pb age of a meteoritic inclusion. Nat Geosci 3:637–641

Boyd AK, Hiesinger H, Robinson MS, Tran T, van der Bogert CH, Wagner RV and the LROC Science Team (2012) Lunar pits: sublunarean voids and the nature of Mare Emplacement. 42nd LPSC. http://www.lpi.usra.edu/meetings/lpsc2011/pdf/2771.pdf

Brack A, Baglioni P, Borruat G, Brandstätter F, Demets R, Edwards HGM, Genge M, Kurat G, Miller MF, Newton EM, Pillinger CT, Roten C-A, Wäsch E (2002) Do meteoroids of sedimentary origin survive terrestrial atmospheric entry? The ESA artificial meteorite experiment STONE. Planet Space Sci 50:763–772

Brändström U (2003) The auroral large imaging system—design, operation, and scientific results. PhD Thesis, IRF Sci. Rep. 279, Swedish Institute of Space Physics, Kiruna

Brasier MD, Antcliffe J, Saunders M, Wacey D (2015) Changing the picture of Earth's earliest fossils (3.5-1.9 Ga) with new approaches and new discoveries. Proc Natl Acad Sci U S A 112:4859–4864

Briggs MH (1963) Biological problems of meteorites. Tuatara J Biol Soc (Wellington, New Zeeland) 11: 2–16. http://nzetc.victoria.ac.nz/tm/scholarly/tei-Bio11Tuat01-t1-body-d1.html

Buchner E, Schmieder M, Gero K, Brandstätter F, Kramar U, Ntaflos T, Kröchert J (2012) Buddha from space – an ancient object of art made of a Chinga iron meteorite fragment. Meteorit Planet Sci 47:1491–1501

Caleb I, Fassett CI, Minton DA (2013) Impact bombardment of the terrestrial planets and the early history of the Solar System. Nat Geosci 6:520–524

Callahan MP, Smith KE, Cleaves HJ, Ruzica J, Stern JC, Glavi DP, House CH, Dworkin JP (2011) Carbonaceous meteorites contain a wide range of extraterrestrial nucleobases. Proc Natl Acad Sci U S A 108:13995–13998

Cannon KM, Mustard JF, Agee CB (2015) Evidence for a widespread basaltic breccia component in the Martian low-albedo regions from the reflectance spectrum of northwest Africa 7034. Icarus 252:150–153

Cano RJ, Borucki MK (1995) Revival and identification of bacterial spores in 25- to 40-million-year-old Dominican amber. Science 268:1060–1064
Canup R, Asphaug E (2001) Origin of the Moon in a giant impact near the end of Earth's formation. Nature 412:708–712
Cassidy WA (2003) Meteorites, ice and Antarctica. Cambridge University Press, Cambridge, p. 364
Ceplecha Z, Borovička J, Graham EW, Revelle DO, Hawkes RL, Porubčan V, Šimek M (1998) Meteor phenomena and bodies. Space Sci Rev 84:327–471
Chladni EFF (1794) Über den Ursprung der von Pallas gefundenen und anderer ihr ähnlicher Eisenmassen und über einige damit in Verbindung stehende Naturerscheinungen [On the origin of the iron masses found by Pallas and others similar to it, and on some natural phenomena associated with them]. Johann Friedrich Hartknoch, Riga
Chung L (1976) Studies on the iron blade of a Shang dynasty bronze axe unearthed at Kao-ch'ng. Acta Archaeologia Sinica 2:17–34
Cockell CS, Lee P (2002) The biology of impact craters – a review. Biol Rev Camb Philos Soc 77:279–310
Cockell CS (2008) The interplanetary exchange of photosynthesis. Orig Life Evol Biosph 38:87–104
Cockell CS (2010) Geomicrobiology beyond Earth: microbe-mineral interactions in space exploration and settlement. Trends Microbiol 18:308–314
Cockell CS (2015) Astrobiology – understanding life in the universe. John Wiley & Sons Ltd UK, Chichester
Cockell CS (2016) The similarity of life across the universe. Mol Biol Cell 15:1553–1555
Cockell CS, Lee P, Osinski G, Horneck G, Broady P (2002) Impact-induced microbial endolithic habitats. Meteorit Planet Sci 37:1287–1298
Cockell CS, Brack A, Wynn-Williams DD, Baglioni P, Brandstätter F, Demets R, Edwards HG, Gronstal AL, Kurat G, Lee P, Osinski GR, Pearce DA, Pillinger JM, Roten CA, Sancisi-Frey S (2007) Interplanetary transfer of photosynthesis: an experimental demonstration of a selective dispersal filter in planetary island biogeography. Astrobiology 7:1–9
Cockell CS, Rettberg P, Rabbow E, Olsson-Francis K (2011) Exposure of phototrophs to 548 days in low Earth orbit: microbial selection pressures in outer space and on early earth. ISME J 5:1671–1682
Cockell CS, Voytek MA, Gronstal AL, Finster K, Kirshtein JD, Howard K, Reitner J, Gohn GS, Sanford WE, Horton JW Jr, Kallmeyer J, Kelly L, Powars DS (2012) Impact disruption and recovery of the deep subsurface biosphere. Astrobiology 12:231–246
Comelli D, D'orazio M, Folco ML, El-Halwagy M, Frizzi T, Alberti R, Capogrosso V, Elnaggar A, Hassan H, Nevin A, Porcelli F, Rashed MG, Valentini G (2016) The meteoritic origin of Tutankhamun's iron dagger blade. Meteorit Planet Sci 51:1301–1309
Cousins CR, Cockell CS (2015) The geobiology in space exploration. Topical team. An ESA roadmap for geobiology in space exploration. Acta Astronaut 118:286–295
Davis AM (2011) Stardust in meteorites. Proc Natl Acad Sci U S A 108:19142–19146
Deguchi S, Shimoshige H, Tsudome M, Mukai S, Corkery RW, Ito S, Horikoshi K (2011) Microbial growth at hyperaccelerations up to 403,627× g. Proc Natl Acad Sci U S A 108:7997–8002
DeLeon-Rodriguez N, Lathem TL, Rodriguez-R LM, Barazesh JM, Anderson BE, Beyersdorf AJ, Ziemba LD, Bergin M, Nenes A, Konstantinidis KT (2012) Microbiome of the upper troposphere: Species composition and prevalence, effects of tropical storms, and atmospheric implications. Proc Natl Acad Sci U S A 110:2575–2580
De Vera JP, Horneck G, Rettberg P, Ott S (2003) The potential of the lichen symbiosis to cope with extreme conditions of outer space – I. Influence of UV radiation and space vacuum on the vitality of lichen symbiosis germination capacity. Int J Astrobiol 1:285–293
De Vera JP et al (2012) Supporting Mars exploration: BIOMEX in low Earth orbit and further astrobiological studies on the Moon using Raman and PanCam technology. Planet Space Sci 74:103–110
Dick SJ (2003) Cultural evolution, the postbiological universe and SETI. Int J Astrobiol 2:65–74

Edwards HGM, Villar SEJ, Pullan D, Hargreaves MD, Hofmann BA, Westall F (2007) Morphological biosignatures from relict fossilised sedimentary geological specimens: a Raman spectroscopic study. J Raman Spectroscopy 38:1352–1361

Edwards HGM, Moody CD, Jorge-Villar SE, Wynn-Williams DD (2005) Raman spectroscopic detection of key biomarkers of cyanobacteria and lichen symbiosis in extreme Antarctic habitats: evaluation for Mars Lander missions. Icarus 174:560–571

Ehrenfreund P, Cami J (2010) Cosmic carbon chemistry: from the interstellar medium to the early Earth. In: Deamer D, Szostak JW (eds) The origins of life. Cold Spring Harbor Press. pp. 318. http://cshperspectives.cshlp.org/content/2/3/a002105.short

Evatt GW, Coughlan MJ, Joy KH, Smedley AR, Connolly PJ, Abrahams ID (2016) A potential hidden layer of meteorites below the ice surface of Antarctica. Nat Commun 16:10679

Fairén AG, Dohm JM, Baker VR, Thompson SD, Mahaney WC, Herkenhoff KE, Rodríguez JAP, Davila AF, Schulze-Makuch D, El Maarry MR, Uceda ER, Amils R, Miyamoto H, Kim KJ, Anderson RC, McKay CP (2011) Meteorites at Meridiani Planum provide evidence for significant amounts of surface and near-surface water on early Mars. Meteorit Planet Sci 46:1832–1841

Fajardo-Cavazos P, Lindsey L, Melosh HJ, Nicholson WL (2005) *Bacillus subtilis* spores on artificial meteorites survive hypervelocity atmospheric entry: implications for lithopanspermia. Astrobiology 5:726–736

Fassett CI, Minton DA (2013) Impact bombardment of the terrestrial planets and the early history of the solar system. Nat Geosci 6:520–524

Fernandes VA, Fritz J, Weiss B, Garrick-Bethel I, Shuster D (2013) The bombardment history of the Moon as recorded by 40Ar–39Ar chronology. Meteorit Planet Sci 48:241–269

Forterre P, Filée J, Myllykallio H (2004) Origin and evolution of DNA and DNA replication machineries. In: Ribas de Pouplana L (ed) The genetic code and the origin of life. pp. 145–168. http://www.ncbi.nlm.nih.gov/books/NBK6360/

Forterre P, Gribaldo S (2007) The origin of modern terrestrial life. HFSP J 1:156–168

Fries M, Harvey R, Wainwright N, Jull AJT, Steele A (2010) Microbial contamination study of Antarctic meteorites: as-found, post-curation, and long-term storage. 73rd Annual Meteoritical Society Meeting. Meteorit Planet Sci 73:5363. http://www.lpi.usra.edu/meetings/metsoc2010/pdf/5363.pdf

Fries M, Harvey R, Jull AJT, Steele A, Wainright N, and the ANSMET 0708 Team (2012) The microbial contamination state of as found on Antarctic meteorites. Conference proceeding. http://www.lpi.usra.edu/meetings/lifedetection2012/pdf/6036.pdf

Fries M, Le Corre L, Hankey M, Fries J, Matson R, Schaefer R, Reddy V (2014) Detection and rapid recovery of the Sutter's Mill meteorite fall as a model for future recoveries worldwide. Meteorit Planet Sci 49:1989–1996

Fritz J, Artemieva NA, Greshake A (2005) Ejection of Martian meteorites. Meteorit Planet Sci 40:1393–1411

Fritz J, Bitsch B, Kührt E, Morbidelli A, Tornow C, Wünnemann K, Fernandes VA, Grenfell JL, Rauer H, Wagner R, Werner SC (2014) Earth-like habitats in planetary systems. Planet Space Sci 98:254–267

Garber SJ (1999) Searching for good science: the cancellation of NASA's SETI program. J Br Interplanet Soc 52:3–12

García-Ruiz JM, Carnerup A, Christy AG, Nicholas JW, Hyde ST (2002) Morphology: an ambiguous indicator of biogenicity. Astrobiology 2:353–369

García-Ruiz JM, Hyde ST, Carnerup AM, Christy AG, Van Kranendonk MJ, Welham NJ (2003) Self-assembled silica-carbonate structures and detection of ancient microfossils. Science 302:1194–1197

Geraci G, del Gaudio R, D'Argenio B (2001) Microbes in rocks and meteorites: a new form of life unaffected by time, temperature, pressure. Rendiconti Lincei 12:51–68

Gillet P, Barrat JA, Heulin T, Achouak W, Lesourd M, Guyot F, Benzerara K (2000) Bacteria in the Tatahouine meteorite: nanometric-scale life in rocks. Earth Planet Sci Lett 175:161–167

Golden DC (2001) A simple inorganic process for formation of carbonates, magnetite, and sulfides in Martian meteorite ALH84001. Am Min 86:370–375

González-Toril E, Martínez-Frías J, Gómez Gómez JM, Rull F, Amils R (2005) Iron meteorites can support the growth of acidophilic chemolithoautotrophic microorganism. Astrobiology 5:406–414

Gronstal A, Pearson V, Kappler A, Dooris C, Anand M, Poitrasson F, Kee TP, Cockell CS (2009) Laboratory experiments on the weathering of iron meteorites and carbonaceous chondrites by iron-oxidizing bacteria. Meteorit Planet Sci 44:233–247

Griffin DW (2013) The quest for extraterrestrial life: what about the viruses? Astrobiology 13:774–783

Gull M (2014) Prebiotic phosphorylation reactions on the early Earth. Challenges 5:193–212

Hahn O (1880) Die Meteorite (Chondrite) und ihre Organismen. Verlag der H. Laupp'schen Buchhandlung, Tübingen

Halliday I (2001) The present-day flux of meteorites to the Earth. In: Peucker-Ehrenbrink B, Schmitz B (eds) Accretion of extraterrestrial matter throughout Earth's history. Springer Publisher, Boston, pp. 305–318

Hansen ES, Graff-Petersen P (1986) Lichens growing on the Ella Island meteorite, Central East Greenland. Lichenologist 18:71–78

Harries P, Hoppe P, Langenhorst F (2015) Reactive ammonia in the solar protoplanetary disk and the origin of Earth's nitrogen. Nat Geosci Lett 8:97–101

Hofmann BA, Nyström JO, Krähenbühl U (2000) The Ordovician chondrite from Brunflo, central Sweden III. Geochemistry of terrestrial alteration. Lithos 50:305–324

Hofmann BA, Gnos E, Al-Kathiri A (2004) Harvesting meteorites in the Omani desert: Implications for Astrobiology. Proc. III Eur. Workshop on Exo-Astrobiology, Madrid, 18–20 Nov 2003. ESA SP-545, March 2004, pp. 73–76

Hofmann BA, Farmer JD, von Blanckenburg F, Fallick AE (2008) Subsurface filamentous fabrics: an evaluation of origins based on morphological and geochemical criteria, with implications for exopaleontology. Astrobiology 8:87–117

Hofmann BA (2010) Meteorites: messengers from the early solar system. Chimia (Aarau) 64:736–740

Horneck G, Rettberg P, Reitz G, Wehner J, Eschweiler U, Strauch K, Panitz C, Starke V, Baumstark-Khan C (2001) Protection of bacterial spores in space, a contribution to the discussion on panspermia. Orig Life Evol Biosph 31:527–547

Horneck G, Stöffler D, Ott S, Hornemann U, Cockell CS, Moeller R, Meyer C, de Vera JP, Fritz J, Schade S, Artemieva NA (2008) Microbial rock inhabitants survive hypervelocity impacts on Mars-like host planets: first phase of lithopanspermia experimentally tested. Astrobiology 8:17–44

Horneck G, Walter N, Westall F, Grenfell JL, Martin WF, Gomez F, Leuko S, Lee N, Onofri S, Tsiganis K, Saladino R, Pilat-Lohinger E, Palomba E, Harrison J, Rull F, Muller C, Strazzulla G, Brucato JR, Rettberg P, Capria MT (2016) AstRoMap European Astrobiology Roadmap. Astrobiology 16:201–243

Hoyle F, Wickramasinghe C, Watkins J (1986) Viruses from space. Univ Coll Cardiff Press, Cardiff, p. 128

Humayun M, Nemchin A, Zanda B, Hewins RH, Grange M, Kennedy A, Lorand JP, Göpel C, Fieni C, Pont S, Deldicque D (2013) Origin and age of the earliest Martian crust from meteorite NWA 7533. Nature 503:513–516

Jalasvuori M, Bamford JK (2009) Did the ancient crenarchaeal viruses from the dawn of life survive exceptionally well the eons of meteorite bombardment? Astrobiology 9:131–137

Jalasvuori M (2012) Revolutionary struggle for existence: Introduction to four intriguing puzzles in virus research. In: Witzany G (ed) Viruses: essential agents of life. Springer Publisher, Dordrecht, pp. 1–19. doi:10.1007/978-94-007-4899-6

Jenniskens P, Fries MD, Yin QZ, Zolensky M, Krot AN, Sandford SA, Sears D, Beauford R, Ebel DS, Friedrich JM, Nagashima K, Wimpenny J, Yamakawa A, Nishiizumi K, Hamajima Y, Caffee MW, Welten KC, Laubenstein M, Davis AM, Simon SB, Heck PR, Young ED, Kohl IE, Thiemens MH, Nunn MH, Mikouchi T, Hagiya K, Ohsumi K, Cahill TA, Lawton JA, Barnes D, Steele A, Rochette P, Verosub KL, Gattacceca J, Cooper G, Glavin DP, Burton AS, Dworkin JP, Elsila JE, Pizzarello S, Ogliore R, Schmitt-Kopplin P, Harir M, Hertkorn N, Verchovsky A,

Grady M, Nagao K, Okazaki R, Takechi H, Hiroi T, Smith K, Silber EA, Brown PG, Albers J, Klotz D, Hankey M, Matson R, Fries JA, Walker RJ, Puchtel I, Lee CT, Erdman ME, Eppich GR, Roeske S, Gabelica Z, Lerche M, Nuevo M, Girten B, Worden SP; Sutter's Mill Meteorite Consortium (2012) Radar-enabled recovery of the Sutter's Mill meteorite, a carbonaceous chondrite regolith breccia. Science 338:1583–1587

Johnson D, Tyldesley J, Lowe T, Withers PJ, Grady MM (2013) Analysis of a prehistoric Egyptian iron bead with implications for the use and perception of meteorite iron in ancient Egypt. Meteorit Planet Sci 48:997–1006

Joseph RG (2016) Was life discovered on the moon? The truth about the moon microbes. http://cosmology.com/MoonLife1.html

Jørgensen JK, Favre C, Bisschop S, Bourke T, Dishoeck E, Schmalzl M (2012) Detection of the simplest sugar, glycolaldehyde, in a solar-type protostar with ALMA. Astrophys J Lett 757:1–6

Jull AJT, Courtney C, Jeffrey DA, Beck JW (1998) Isotopic evidence for a terrestrial source of organic compounds found in the martian meteorites ALH 84001 and EETA79001. Science 279:365–369

Jull AJT (2001) Terrestrial age of meteorites. In: Peucker-Ehrenbrink B, Schmitz B (eds) Accretion of extraterrestrial matter throughout Earth's history. Springer Publisher, Boston, pp. 241–266

Jull AJT (2006) Terrestrial ages of meteorites. In: Lauretta DS, McSween HY (eds) Meteorites and the early solar system II. University of Arizona Press, Tucson, pp. 889–905. http://www.lpi.usra.edu/books/MESSII/9011.pdf

Kleine T (2011) Geoscience: earth's patchy late veneer. Nature 477:168–169

Krot AN, Yurimoto H, Hutcheon ID, MacPherson GJ (2005) Chronology of the early Solar System from chondrule-bearing calcium-aluminium-rich inclusions. Nature 434:998–1001

Kutlucinar K, Kleiner F, Zebec Z, Rittmann S, Schleper C, Kutlucinar K, Ferrière L, .Milojevic T (2015) Acquiring the microbial-mineral interface of *Metallosphaera sedula* grown on stony meteorite. 15th EANA (European Astrobiology Network Association) Workshop, 6–9 Oct 2015, ESTEC Noordwijk, Netherlands

Lapen TJ, Righter M, Brandon AD, Debaille V, Beard BL, Shafer JT, Peslier AH (2010) A younger age for ALH84001 and its geochemical link to shergottite sources in Mars. Science 328:347–351

Le Romancer M, Gaillard M, Geslin C, Prieur D (2007) Viruses in extreme environments. Rev Environ Sci Biotechnol 6:17–31

Lee MR, Bland PA (2004) Mechanisms of weathering of meteorites recovered from hot and cold deserts and the formation of phyllosilicates. Geochim Cosmochim Acta 68:893–916

Lee NM, Ganzenko O, Beck A, Gnosq E, Zurfluh FJ, Hofmann BA (2012) Meteorites as habitats for lichens and microorganisms in hot deserts. Abstr P5.01, 12th EANA (European Astrobiology Network Association) Workshop, 15–17 Oct 2012, Stockholm

Lind MD, Housley RM (1972) Crystallization studies of lunar igneous rocks: crystal structure of synthetic armalcolite. Science 175:521–523

Lindsay JF, McKay DS, Allen CC (2003) Earth's earliest biosphere— a proposal to develop a collection of curated Archean geologic reference materials. Astrobiology 3:739–758

Locey KJ, Lennon JT (2016) Scaling laws predict global microbial diversity. Proc Natl Acad Sci U S A 113:5970–5975

Mahy B, van Regenmortel M (2009) Desk encyclopedia of general virology. Academic Press, Oxford, p. 672

Maier WD, Andreoli MAG, McDonald I, Higgins MD, Boyce AJ, Shukolyukov A, Lugmair GW, Ashwal LD, Graser P, Ripley EM, Hart RJ (2006) Discovery of a 25-cm asteroid clast in the giant Morokweng impact crater, South Africa. Nature 441:203–206

Martin D, Arcones A, Nazarewicz W, Olsen E (2016) Impact of nuclear mass uncertainties on the r process. Phys Rev Lett 116:121101

Marvin UB (1983) The discovery and initial characterization of Allan Hills 81005: the first lunar meteorite. Geophys Res Lett 10:775–778

Mastrapa R, Glanzberg H, Head J, Melosh H, Nicholson W (2001) Survival of bacteria exposed to extreme acceleration: implications for panspermia. Earth Planet Sci Lett 189:1–8

Mautner MN, Leonard RL, Deamer DW (1995) Meteorite organics in planetary environments: hydrothermal release, surface activity, and microbial utilization. Planet Space Sci 43:139–147
Mautner MN, Conner AJ, Killham K, Deamer DW (1997) Biological potential of extraterrestrial materials. 2. Microbial and plant responses to nutrients in the Murchison carbonaceous meteorite. Icarus 129:245–253
Mautner MN (2002a) Planetary resources and astroecology. Planetary microcosm models of asteroid and meteorite interiors: electrolyte solutions and microbial growth – implications for space populations and panspermia. Astrobiology 2:59–76
Mautner MN (2002b) Planetary bioresources and astroecology: 1. Planetary microcosm bioassays of Martian and carbonaceous chondrite materials: nutrients, electrolyte solutions, and algal and plant responses. Icarus 158:72–86
McCall GJH, Bowden AJ, Howarth RJ (eds) (2006) The history of meteoritics and key meteorite collections: fireballs, falls and finds, Geological Society Special Publication no, vol 256. Geological Society, London
McKay DS, Gibson EK, Thomas-Keprta KL, Vali H, Romanek CS, Clemett SJ, Chillier XDF, Maechling CR, Zare R (1996) Search for past life on Mars: possible relic biogenic activity in Martian meteorite ALH84001. Science 273:924–930
Mc Kinney ML (1997) How do rare species avoid extinction? A paleontological view. In: Kunin WE, Gaston KJ (eds) The biology of rarity: causes and consequences of rare – common differences. Springer Science & Business Media B.V. Chapman & Hall, London, pp. 110–129
McLoughlin N, Brasier MD, Wacey D, Green OR, Perry RS (2007) On biogenicity criteria for endolithic microborings on early Earth and beyond. Astrobiology 7:10–26
McLoughlin N, Grosch EG (2016) A hierarchical system for evaluating the biogenicity of metavolcanic- and ultramafic-hosted microalteration textures in the search for extraterrestrial life. Astrobiology 15:901–921
Meibom A, Clark BE (1999) Invited review: Evidence for the insignificance of ordinary chondritic material in the asteroid belt. Meteorit Planet Sci 34:7–24
Meyer C, Fritz J, Misgaiski M, Stöffler D, Artemieva NA, Hornemann U, Moeller R, De Vera J-P, Cockell C, Horneck G, Ott S, Rabbow E (2011) Shock experiments in support of the Lithopanspermia theory: the influence of host rock composition, temperature, and shock pressure on the survival rate of endolithic and epilithic microorganisms. Meteorit Planet Sci 46:701–718
Mignan A (2011) Claims of indigenous life forms in meteorites: a short review. Aust Met Mag 17:34–38
Miller SL (1953) Production of amino acids under possible primitive earth conditions. Science 117:528–529
Mittlefehldt DW (1994) ALH84001, a cumulate orthopyroxenite member of the SNC meteorite group. Meteoritics 29:214–221
Moissl-Eichinger C, Cockell C, Rettberg P (2016) Venturing into new realms? Microorganisms in space. FEMS Microbiol Rev 40:722–737 pii: fuw015
Narlikar JV, Lloyd D, Wickramasinghe NC, Harris MJ, Turner MP, Al-Mufti S, Wallis MK, Wainwright M, Rajaratnam P, Shivaji S, Reddy GSN, Ramadurai S, Hoyle F (2003) A balloon experiment to detect microorganisms in the outer space. Astrophys Space Sci 285:555–562
Nicholson WL, Munakata N, Horneck G, Melosh HJ, Setlow P (2000) Resistance of *Bacillus* endospores to extreme terrestrial and extraterrestrial environments. Microbiol Mol Biol Rev 64:548–572
Nicholson WL (2009) Ancient micronauts: interplanetary transport of microbes by cosmic impacts. Trends Microbiol 17:243–250
Nininger HH (1972) Find a falling star. Paul S Eriksson Inc, New York, 254 pp
Nittler LR, McCoy TJ, Clark PE, Murphy ME, Trombka JI, Jarosewich E (2004) Bulk element compositions of meteorites: a guide for interpreting remote-sensing geochemical measurements of planets and asteroids. Antarct Meteorite Res 17:233–253
Nixon SL, Cockell CS (2015) Nonproteinogenic D-amino acids at millimolar concentrations are a toxin for anaerobic microorganisms relevant to early Earth and other anoxic planets. Astrobiology 15:238–246

Norton OR (2002) The Cambridge encyclopedia of meteorites. Cambridge University Press, Cambridge, p. 354

Nuevo M, Auger G, Blanot D, d'Hendecourt L (2008) A detailed study of the amino acids produced from the vacuum UV irradiation of interstellar ice analogs. Orig Life Evol Biosph 38:37–56

Olsson-Francis K, Cockell CS (2010) Experimental methods for studying microbial survival in extraterrestrial environments. J Microbiol Methods 80:1–13

Oró J, Tornabene T (1965) Bacterial contamination of some carbonaceous meteorites. Science 150:1046–1048

Pache M, Reitner J, Arp G (2001) Geochemical evidence for the formation of a large Miocene "travertine" mound at a sublacustrine spring in a soda lake; Wallerstein castle rock, Nördlinger Ries, Germany. Facies 45:211–230

Pacton M, Wacey D, Corinaldesi C, Tangherlini M, Kilburn MR, Gorin GE, Danovaro R, Vasconcelos C (2014) Viruses as new agents of organomineralization in the geological record. Nat Commun 5:4298. doi:10.1038/ncomms5298

Pasek M, Lauretta D (2008) Extraterrestrial flux of potentially prebiotic C, N, and P to the early Earth. Orig Life Evol Biosph 38:5–21

Paul D, Kumbhare SV, Mhatre SS, Chowdhury SP, Shetty SA, Marathe NP, Bhute S, Shouche YS (2016) Exploration of microbial diversity and community structure of Lonar Lake: the only hypersaline meteorite crater lake within basalt rock. Front Microbiol 6:1553. doi:10.3389/fmicb.2015.01553

Pellinen-Wannberg A, Murad E, Gustavsson B, Brändström U, Enell CF, Roth C, Williams IP, Steen Å (2004) Optical observations of water in Leonid meteor trails. Geophys Res Lett 31:L03812. doi: 10.1029/2003/GL018785,2004

Pellinen-Wannberg A, Kero J, Häggström J, Mann I, Tjulin A (2016) The forthcoming EISCAT_3D as an extra-terrestrial matter monitor. Planet Space Sci 123:33–40. doi:10.1016/j.pss.2015.10.009,2016

Peucker-Ehrenbrink B, Ravizza G, Winckler G (2016) Geochemical tracers of extraterrestrial matter in sediments. Elements 12:191–196

Pillinger JM, Pillinger CT, Powell M, Butler M (2011) Application of lichenometry to meteorite studies. 74th Annual Meeting of the Meteoritical Society, Abstract 5277, London

Pizzarello S, Shock E (2010) The organic composition of carbonaceous meteorites: The evolutionary story ahead of biochemistry. In: Deamer D, Szostak JW (eds). The origins of life. Cold Spring Harbor Press, pp. 318. http://cshperspectives.cshlp.org/content/2/3/a002105.short

Pizzarello S (2016) Molecular asymmetry in prebiotic chemistry: An account from meteorites. Life 6:E18. doi:10.3390/life6020018

Plait P 2013 Blogs from January 2013. http://www.slate.com/blogs/bad_astronomy/2013/03/11/meteorite_life_claims_of_fossils_in_a_meteorite_are_still_wrong.html, http://www.slate.com/blogs/bad_astronomy/2013/01/15/life_in_a_meteorite_claims_by_n_c_wickramasinghe_of_diatoms_in_a_meteorite.html

Prangishvili D (2013) The wonderful world of archaeal viruses. Annu Rev Microbiol 67:565–585

Prangishvili D (2015) Archaeal viruses: living fossils of the ancient virosphere? Ann N Y Acad Sci 1341:35–40

Pullan D, Westall F, Hofmann BA, Parnell J, Cockell CS, Edwards HG, Villar SE, Schröder C, Cressey G, Marinangeli L, Richter L, Klingelhöfer G (2008) Identification of morphological biosignatures in Martian analogue field specimens using in situ planetary instrumentation. Astrobiology 8:119–156

Rasmussen R (2000) Filamentous microfossils in a 3,235-million-year-old volcanogenic massive sulphide deposit. Nature 405:676–679

Rea G, Cristofaro F, Pani G, Pascucci B, Ghuge SA, Corsetto PA, Imbriani M, Visai L, Rizzo AM (2016) Microgravity-driven remodeling of the proteome reveals insights into molecular mechanisms and signal networks involved in response to the space flight environment. J Proteomics 137:3–18

Reed C (2015) Earth microbe prefers living on meteorites. Science. doi:10.1126/science.aab2499
Rehren T, Belgya T, Jambonc A, György K, Zsolt K, Kis Z, Kovácse I, Boglárka M, Martinón-Torres M, Miniacif G, Pigott VC, Radivojević M, Rosta L, Szentmiklósi L, Szőkefalvi-Nagye Z (2013) 5,000 years old Egyptian iron beads made from hammered meteoritic iron. J Archaeol Sci 40:4785–4792
Rettberg P, Anesio AM, Baker VR, Baross JA, Cady SL, Detsis E, Foreman CM, Hauber E, Ori GG, Pearce DA, Renno NO, Ruvkun G, Sattler B, Saunders MP, Smith DH, Wagner D, Westall F (2016) Planetary protection and Mars special regions – a suggestion for updating the definition. Astrobiology 16:119–125
Riding R (1999) The term stromatolite: towards an essential definition. Lethaia 32:321–330
Roossinck M (2011) The good viruses: viral mutualistic symbioses. Nat Rev Microbiol 9:99–108
Rosing MT (1999) ^{13}C-depleted carbon in >3700 Ma seafloor sedimentary rocks from West Greenland. Science 283:674–676
Roy SK (1935) The question of living bacteria in stony meteorites. Geol Ser Field Museum Nat Hist 6:180–198
Rubin AE, Grossman JN (2010) Meteorite and meteoroid: new comprehensive definitions. Meteorit Planet Sci 45:114–122
Saladino R, Carota E, Botta G, Kapralov M, Timoshenko GN, Rozanov AY, Krasavin E, Di Mauro E (2015) Meteorite-catalyzed syntheses of nucleosides and of other prebiotic compounds from formamide under proton irradiation. Proc Natl Acad Sci U S A 112:E2746–E2755
Sancho LG, de la Torre R, Pintado A (2008) Lichens, new and promising material from experiments in astrobiology. Fungal Biol Rev 22:103–109
Sandford SA, Bera PP, Lee TJ, Materese CK, Nuevo M (2015) Photosynthesis and photo-stability of nucleic acids in prebiotic extraterrestrial environments. Top Curr Chem 356:123–164
Schmitt-Kopplin P, Gabelica Z, Gougeon RD, Fekete A, Kanawati B, Harir M, Gebefuegi I, Eckel G, Hertkorn N (2010) High molecular diversity of extraterrestrial organic matter in Murchison meteorite revealed 40 years after its fall. Proc Natl Acad Sci U S A 107:2763–2768
Schmitz B, Tassinari M (2001) Fossil Meteorites. In: Peucker-Ehrenbrink B, Schmitz B (eds) Accretion of extraterrestrial matter throughout Earth's history. Springer Publisher, Boston, pp. 319–331
Schmitz B (2013) Extraterrestrial spinels and the astronomical perspective on Earth's geological record and evolution of life. Chemie der Erde 73:117–145
Schmitz B, Huss GR, Meier MMM, Peucker-Ehrenbrink B, Church RP, Cronholm A, Davies MB, Heck PR, Johansen A, Keil K, Kristiansson P, Ravizza G, Tassinari M, Terfelt F (2014) A fossil winonaite-like meteorite in Ordovician limestone: a piece of the impactor that broke up the L-chondrite parent body? Earth Planet Sci Lett 400:145–152
Schröder C, Rodionov DS, McCoy TJ, Jolliff BL, Gellert R, Nittler LR, Farrand WH, Johnson JR, Ruff SW, Ashley JW, Mittlefehldt DW, Herkenhoff KE, Fleischer I, Haldemann AFC, Klingelhöfer G, Ming DW, Morris RV, de Souza PA, Squyres SW, Weitz C, Yen AS, Zipfel J, Economou T (2008) Meteorites on Mars observed with the Mars exploration rovers. J Geophys Res Planets 113(E6):E06S22 2156–2202
Shapiro R, Schulze-Makuch D (2009) The search for alien life in our solar system: strategies and priorities. Astrobiology 9:335–343
Shen Y, Buick R, Canfield DE (2001) Isotopic evidence for microbial sulphate reduction in the early Archean era. Nature 410:77–81
Shivaji S, Chaturvedi P, Begum Z, Pindi PK, Manorama R, Padmanaban DA, Shouche YS, Pawar S, Vaishampayan P, Dutt CB, Datta GN, Manchanda RK, Rao UR, Bhargava PM, Narlikar JV (2009) *Janibacter hoylei* sp. nov., *Bacillus isronensis* sp. nov. and *Bacillus aryabhattai* sp. nov., isolated from cryotubes used for collecting air from the upper atmosphere. Int J Syst Evol Microbiol 59:2977–2986
Slobodkin A, Gavrilov S, Ionov V, Iliyin V (2015) Spore-forming thermophilic bacterium within artificial meteorite survives entry into the Earth's atmosphere on FOTON-M4 satellite landing module. PLoS One 10:e0132611

Smith DJ (2013) Microbes in the upper atmosphere and unique opportunities for astrobiology research. Astrobiology 13:981–990

Smith IWM, Cockell CS, Leach S (eds) (2013) Astrochemistry and astrobiology. Springer Verlag, Berlin/Heidelberg, p. 349

Solden L, Lloyd K, Wrighton K (2016) The bright side of microbial dark matter: lessons learned from the uncultivated majority. Curr Opin Microbiol 31:217–226

Steele A, Goddard DT, Stapleton D, Toporski JK, Peters V, Bassinger V, Sharples G, Wynn-Williams DD, McKay DS (2000) Investigations into an unknown organism on the martian meteorite Allan Hills 84001. Meteorit Planet Sci 35:237–241

Steele A, Toporski JKW, McKay DS (2001) The terrestrial contamination of meteorites. An update. Abstr. 64th Annual Meteoritical Society Meeting, 10–14 Sept 2001, Gregorian Pontifical University, Vatican City

Steele A, McCubbin FM, Fries M, Kater L, Boctor NZ, Fogel ML, Conrad PG, Glamoclija M, Spencer M, Morrow AL, Hammond MR, Zare RN, Vicenzi EP, Siljeström S, Bowden R, Herd CD, Mysen BO, Shirey SB, Amundsen HE, Treiman AH, Bullock ES, Jull AJ (2012) A reduced organic carbon component in martian basalts. Science 337:212–215. doi:10.1126/science.1220715 Epub 2012 May 24

Steele A (2014) What do viruses, viroids and prions reveal about the formation of life on Earth. In: Dating the origin of life: present-day molecules and first fossil record. GAIA inform 7. http://www.geobiologie.uni-goettingen.de/images/pdf/origin%20of%20life_1.pdf

Steele A, McCubbin FM, Fries MD (2016) The provenance, formation and implications of reduced carbon phases in Martian meteorites. Meteorit Planet Sci 51:2203–2225 doi: 10.11111/maps.12670

Stelzner T, Heide K, Bischoff A, Weber D, Scherer P, Schultz L, Happel M, Schrön W, Neupert U, Michel R, Clayton RN, Mayeda TK, Bonani G, Haidas I, Ivy-Ochs S, Suter M (1999) An interdisciplinary study of weathering effects in ordinary chondrites from the Acfer region, Algeria. Meteorit Planet Sci 34:787–794

Strasdeit H (2005) New studies on the Murchison meteorite shed light on the pre-RNA world. Chembiochem 6:801–803

Thomas-Keprta KL, Clemett SJ, Bazylinski DA, Kirschvink JL, McKay DS, Wentworth SJ, Vali H, Gibson Jr EK Jr, Romanek CS (2002) Magnetofossils from ancient Mars: a robust biosignature in the martian meteorite ALH84001. Appl Environ Microbiol 68:3663–3672

Thomas-Keprta KL, Clemett SJ, McKay DS, Gibson EK, Wentworth SJ (2009) Origins of magnetite nanocrystals in Martian meteorite ALH84001. Geochim Cosmochim Acta 73:6631–6677

Toporski J, Steele A (2007) Observations from a 4-year contamination study of a sample depth profile through Martian meteorite Nakhla. Astrobiology 7:389–401

Treiman AH, Gleason JD, Bogard DD (2000) The SNC meteorites are from Mars. Planet Space Sci 48:1213–1230

Vasanthan T, Alejaldre L, Hider J, Patel S, Husain N, Umapathisivam B, Ston J (2016) G-equivalent acceleration tolerance in the eutardigrade species *Hypsibius dujardini*. Astrobiology (in review/in press)

Veski S, Heinsalu A, Lang V, Kestlane I, Possnert G (2004) The age of the Kaali meteorite craters and the effect of the impact on the environment and man: evidence from inside the Kaali craters, island of Saaremaa, Estonia. Veg Hist Archaeobotany 13:197–206

Vreeland RH, Rosenzweig WD, Powers DW (2000) Isolation of a 250 million-year-old halotolerant bacterium from a primary salt crystal. Nature 407:896–900

Wacey D, Kilburn MR, Saunders M, Cliff J, Brasier MD (2011) Microfossils of sulphur-metabolizing cells in 3.4-billion-year-old rocks of Western Australia. Nat Geosci 4:698–702

Wacey D, McLoughlin N, Kilburn MR, Saunders M, Cliff JB, Kong C, Barley ME, Brasier MD (2013) Nanoscale analysis of pyritized microfossils reveals differential heterotrophic consumption in the ~1.9-Ga Gunflint chert. Proc Natl Acad Sci U S A 110:8020–8024

Wacey D, Saunders M, Roberts M, Menon S, Green L, Kong C, Culwick T, Strother P, Brasier MD (2014) Enhanced cellular preservation by clay minerals in 1 billion-year-old lakes. Sci Rep 5841: doi:10.1038/srep05841

Wainwright M (2003) A microbiologist looks at panspermia. Astrophys Space Sci 285:563–570

Wainwright M, Alharbi S, Wickramasinghe NC (2006) How do microorganisms reach the stratosphere? Int J Astrobiol 5:13–15

Wainwright M, Laswd A, Alshammari F (2009) Bacteria in amber coal and clay in relation to lithopanspermia. Int J Astrobiol 8:141–143

Wainwright M, Fawaz Alshammari F, Alabri K (2010) Are microbes currently arriving to Earth from space? J Cosmol 7:1692–1702

Wallis J, Miyake N, Hoover RB, Oldroy A, Wallis DH, Samaranayake A, Wickramarathne K, Wallis MK, Gibson CH, Wickramasinghe NC (2013) The Polonnaruwa meteorite: oxygen isotope, crystalline and biological composition. J Cosmol 22:10004–10011

Wan N, Abernathy M, Tang JKH, Tang YJ, You L (2015) Cyanobacterial photo-driven mixotrophic metabolism and its advantages for biosynthesis. Front Chem Sci Eng 9:308–316

Wang W, Lang MG, Diehl R, Halloin H, Jean P, Knödlseder J, Kretschmer K, Martin P, Roques JP, Strong AW, Winkler C, Zhang XL (2009) Spectral and intensity variations of Galactic 26Al emission. Astron Astrophys 496:713–724

Weinland DF (1882) Über die in Meteoriten entdeckten Thierreste. Esslingen, Germany

Welten KC, Alderliesten C, van der Borg K, Lindner L, Loeken T, Schultz L (1997) Lewis Cliff 86360: an Antarctic L-chondrite with a terrestrial age of 2.35 million years. Meteorit Planet Sci 32:775–780

Wickramasinghe C, Wallis J, Wallis DH, Samaranayake A (2013a) Fossil diatoms in a new carbonaceous meteorite. J Cosmol 21:9560–9571

Wickramasinghe C, Wallis J, Miyake N, Oldroyd A, Wallis DH, Samaranayake A, Wickramarathne K, Hoover RB, Wallis MK (2013b) Authenticity of the life-bearing Polonnaruwa meteorite. J Cosmol 21:9772–9777

Wickramasinghe C, Vaidya PG, Wainwright M, McLean R, Netea M, Joseph G (2013c) Diseases from space: astrobiology, viruses, microbiology, meteors, comets, evolution. Cosmology Science Publishers. pp. 198. ISBN/10:1938024184

Wlotzka F (1993) A weathering scale for the ordinary chondrites. Meteoritics 28:460

Worth RJ, Sigurdsson S, Hous CH (2013) Seeding life on the moons of the outer planets via lithopanspermia. Astrobiology 13:1155–1165

Xu P, Chen F, Mannas J, Feldman T, Sumner L, Roossinck M (2008) Virus infection improves drought tolerance. New Phytol 180:911–921

Yang J, Goldstein JI, Scott ERD (2010) Main-group pallasites: thermal history, relationship to IIIAB irons, and origin. Geochim Cosmochim Acta 74:4471–4492

Zhmur SI, Gerasimenko LM (1999) Biomorphic forms in carbonaceous meteorite Allende and possible ecological system – producer of organic matter chondrites. In: Hoover RB (ed) Instruments, methods and missions for astrobiology II, Proc. SPIE 3755, pp 48–58

Zhmur SI, Rozanov AY, Gorlenko VM (1997) Lithified remnants of microorganisms in carbonaceous chondrites. Geochem Int 35:58–60

Zurfluh FJ, Hofmann BA, Gnos E, Eggenberger U (2013) "Sweating meteorites" – water-soluble salts and temperature variation in ordinary chondrites and soil from the hot desert of Oman. Meteorit Planet Sci 48:1958–1980

Zurfluh FJ, Hofmann BA, Gnos E, Eggenberger U, Jull AJT (2016) Weathering of ordinary chondrites from Oman: correlation of weathering parameters with ^{14}C terrestrial ages and a refined weathering scale. Meteorit Planet Sci 51:1685–1700. doi:10.1111/maps.12690

Organism Index

A

H. Stan-Lotter, S. Fendrihan (eds.), *Adaption of Microbial Life to Environmental Extremes*, DOI 10.1007/978-3-319-48327-6

R

S

T

Subject Index

H. Stan-Lotter, S. Fendrihan (eds.), *Adaption of Microbial Life to Environmental Extremes*, DOI 10.1007/978-3-319-48327-6

Zeitfracht Medien GmbH
Ferdinand-Jühlke-Straße 7
99095 Erfurt, Deutschland
produktsicherheit@kolibri360.de